Recent Advances in Cryobiology

Recent Advances in Cryobiology

Edited by **Ray Phillips**

New York

Published by Callisto Reference,
106 Park Avenue, Suite 200,
New York, NY 10016, USA
www.callistoreference.com

Recent Advances in Cryobiology
Edited by Ray Phillips

International Standard Book Number: 978-1-63239-529-0 (Hardback)

Contents

Preface

It is often said that books are a boon to humankind. They document every progress and pass on the knowledge from one generation to the other. They play a crucial role in our lives. Thus I was both excited and nervous while editing this book. I was pleased by the thought of being able to make a mark but I was also nervous to do it right because the future of students depends upon it. Hence, I took a few months to research further into the discipline, revise my knowledge and also explore some more aspects. Post this process, I begun with the editing of this book.

This book on cryobiology focuses on the recent developments and expansion in the field. Lately, there have been certain severe tectonic shifts in cryobiology though not visible on the surface but will have significant impact on both the advancement of novel cryopreservation techniques and the future of cryobiology. This comprehensive book discusses the existing applications and emerging practical protocols for the purpose of cryopreservation along with the description of the novel cryobiological ideas. The topics discussed in this book are as follows: Cryopreservation of Wildlife Genome (Terrestrial Animals), Cryopreservation of Aquatic Species, Cryopreservation of Plants, and Equipment and Assays. It consists of contributions made by researchers as well as scientists from across the planet and will be an important source for those interested in cryobiology.

I thank my publisher with all my heart for considering me worthy of this unparalleled opportunity and for showing unwavering faith in my skills. I would also like to thank the editorial team who worked closely with me at every step and contributed immensely towards the successful completion of this book. Last but not the least, I wish to thank my friends and colleagues for their support.

Editor

Part 1

Cryopreservation of Wildlife Genome (Terrestrial Animals)

Wildlife Cats Reproductive Biotechnology

Regina Celia Rodrigues da Paz
Departamento de Ciências Básicas e Produção Animal,
Faculdade de Agronomia, Medicina Veterinária e Zootecnia,
Universidade Federal de Mato Grosso – DCBPA/FAMEV/UFMT
Brazil

1. Introduction

The techniques used in assisted reproduction for wildlife or domestic animals are similar, and consist in the collection, evaluation, and cryopreservation of semen; artificial insemination (AI), *in vitro* fertilization (IVF), and embryo transfer (ET).

Considering that the maintenance of genetic diversity is reproduction dependent, reproduction technologies are important tools for the preservation of endangered species, throught the development of methods to increase the fertility in these animals (Howard, 1993).

Several studies show that hormone therapies may induce ovulation and/or superovulation in wild animals (Pope, 2000). The artificial induction of ovulation and superovulation are important components in assisted reproduction techniques, especially in cats, because their ovulation is induced by external stimulations (Hamner et al., 1970).

According to The Word Zoo Conservation Strategy ([IUDZG/CBSG], 1993), some of the many benefits from using assisted reproduction techniques in conservation management programs are:

1. To enable the exchange of genetic material between two or more zoos collaborating in the same program, including animals in captivity and in the wild. Semen transportation is economical and reduces the risks involved in animal transfers and consequently diseases transmission;
2. To enable reproduction in animals with physical and reproductive behavioral disabilities. This is important for animals that are representative of genetic lines that cannot be lost;
3. To enable rapid population growth when only a small founder population is available;
4. To assist in maintaining the ratio between males and females by selectively transplanting embryos of one sex;
5. To determine the number of offspring per individual;
6. To promote the formation of databases for gametes and embryos from species of interest.

The reproductive techniques used for domestic animals are gradually being used for zoo animals (Comizzoli et al., 2000; Dresser et al., 1986).

Artificial insemination has been conducted in different species of carnivores such as cougar (*Felis concolor*), leopard (*Panthera pardus saxicolor*), cheetah (*Acinonyx jubatus*), tiger (*Panthera tigris altaica*), ocelot (*Leopardus pardalis*) tigrina (*Leopardus tigrinus*), and jaguar (*Panthera onca*) (Dresser et al., 1982; Donoghue et al.; 1993, Howard et al., 1992a; Jimenez et al., 1999; Moore et al., 1981; Moraes et al., 1997; Silva et al., 2000; Swanson et al., 1996a).

In vitro fertilization (IVF) has also been performed in captive wild cats such as tiger (*Panthera tigris altaica*), jaguar (*Panthera onca*), ocelot (*Leopardus pardalis*) and tigrina (*Leopardus tigrinus*) (Donoghue et al., 1990; Morato et al., 2000; Swanson & Brown, 2004).

The application of artificial reproductive methods in wild animals has not being successful, showing low reproductive rates. Some of the several reasons for these low reproductive rates are lack of knowledgement on the species' physiology, poor sperm or oocytes quality, and difficulties in adapting from the methodologies used in experimental models.

Considering the zoos' limitation to maintain genetically viable populations of threatened species, the establishment of genetic banks containing semen, oocytes, embryos and cells emerge as a strategy to ensure the genetic diversity of populations (Lasley et al., 1994). The potential of assisted reproduction for endangered species should be emphasized by the possibility of semen and embryos cryopreservation, which are genetically valuable for the future populations.

Although there are some records related to the female reproductive physiology (Table 1) and semen characteristics for various Neotropical felids species (Table 2), there is a little knowledge about the fertilization ability using artificial methods (Table 3) in these animals. Studies about the application of assisted reproduction techniques in Neotropical felids showed limitations because of the lack of basic knowledge on the physiology of the species.

2. General reproductive characteristics in wildlife cats

The large number of felines and their wide geographical distribution determine many of the species' particularities, mainly in the reproductive aspects. The reproductive seasonality is one of the most variable aspect between species.

In domestic cats (*Felis catus*), the reproductive seasonality is related to photoperiod (Johnston et al., 1996; Shille et al., 1979; Tsuitsui & Stabenfeldt, 1996), whereas in wild animals, it is also related to high or low food supply during the seasons (Ewer, 1975).

The ovulation mechanism is another variable aspect among cat species. Studies evaluating serum hormone levels confirmed that some wild cats like tiger (*Panthera tigris*) (Seal et al., 1985), snow leopard (*Panthera uncia*) (Schmidt et al., 1993), jaguars (*Panthera onca*) (Wildt et al., 1979) and cougars (*Puma concolor*) (Bonney et al., 1981), have induced reflex ovulation, similar to the domestic cat (Johnston et al., 1996; Shille et al., 1979; Tsuitsui & Stabenfeldt, 1996).

However, female leopards (*Panthera pardus*) presented two ovulation mechanisms in two different situations. When kept isolated, they showed typical hormonal profile for ovulation reflex mechanism; but, when housed in pairs with another female, the ovulation was probably stimulated by physical contact (Schmidt et al., 1988).

Lionesses (*Panthera leo*), when isolated from the males, showed an ovulation pattern distinct from other cats. Schmidt et al. (1979), using serum hormone levels and corpus luteum visualization, demonstrated that this species presents spontaneous ovulation in a higher frequency compared to what is described for other felines.

The genus Leopardus is polyestral and can cycle all year round (Morais et al., 1996; Moreira et al., 2001); the margay (*Leopardus wiedii*) is the only species in this genus presenting spontaneous ovulation (Moreira et al., 2001).

According to Tebet (1999), the estrous cycle in ocelots is characterized by the presence of serum estradiol peaks associated with relatively low levels of serum progesterone (<2.61 ng/mL). This demonstrates the polyestral characteristic of this species, similar to previously observed for other species such as cats (Shille al., 1979; Tsuitsui & Stabenfeldt, 1996; Verhage et al., 1976), tigers (Seal et al., 1985), and cheetahs (Brown et al., 1996).

Ocelot females that ovulated and were not fertilized showed a period of increased progesterone serum concentration and estrus inhibition (Tebet, 1999), called pseudopregancy or diestrus, similar to domestic cats (Feldman & Nelson, 1996) and leopards (Schmidt et al., 1988).

Under an evolutionary analysis and among other factors, the process of spontaneous or induced ovulation may be related to the sociability of the species. Thus, solitary cats would require a longer estrus period, extended viability of oocytes, and extended time for the ovulation to occur, after the couple meet in the wild (Ewer, 1975).

The detection of fecal estrogens and progestins, throught the analyses of fecal metabolites in domestic and wild cats such as the leopard cat (*Felis bengalensis*), cheetah (*Acinonyx jubatus*), clouded leopard (*Neofelis nebulosa*), and snow leopard (*Panthera uncia*) was successfully performed by Brown et al. (1994). Likewise, this methodology has been widely used to monitor ovarian function in Neotropical felines such as the ocelot and margay (Morais et al, 1996; Moreira et al., 2001).

In this context, noninvasive methods such as the quantification of fecal hormonal metabolites are increasingly being used in wildlife animals.

The estrus cycle, gestational time, and number of pups observed in different species of Neotropical wild cats are presented in Table 1.

Common Name	Scientific Name	Estrus (days)	Estrous Cycle (days)	Gestation (days)	Pups (n°)
Ocelot	*Leopardus pardalis*	4.63±0.63[1]	16.5 ± 1.5[2] 18.4±1.6[4]	70-85[6]	1-4[6]
Tigrinus	*Leopardus tigrinus*	3.0-9.0[5]	15.8 ± 1.5[2] 16.7±1.3[4]	73-78[6]	1-4[6]
Margay	*Leopardus wiedii*	4.0-10.0[5]	19.5 ± 2.1[2] 17.6±1.5[4]	81-84[6]	1[6]
Geofroy's cat	*Oncifelis geoffroyi*	2.5±0.5[1]	20.0[5]	72-76[6]	1-3[6]

Common Name	Scientific Name	Estrus (days)	Estrous Cycle (days)	Gestation (days)	Pups (n°)
Pallas cat	*Oncifelis colocolo*	-	-	80-85[6]	1-3[6]
Jaguarondi	*Herpaelurus yaguarundi*	3.17±0.75[1]	53.63±2.41[1]	72-75[6]	1-4[6]
Puma	*Puma concolor*	8.0[5]	23.0[5]	84-98[6]	1-6[6]
Jaguar	*Panthera onca*	12.0±1.0[3]	47.2±5.4[3]	90-111[6]	1-2[6]

Mellen 1989[1]; Morais et al., 1996[2]; Morato & Paz, 2001[3;] Moreira et al., 2001[4;] Oliveira, 1994[5]; Oliveira & Cassaro, 1997[6].

Table 1. Neotropical wildlife cats reproductive characteristics.

3. Sperm collection

Two methods are used for the collection of semen from wild animals: the first is the epididymis' semen collection from dead animals or after castration or vasectomy. The second is through an artificial vagina, electroejaculation, or digital manipulation.

3.1 Post mortem sperm collection

In the post mortem sperm collection, immediately after death or castration, the testes should be kept cold at 5°C (Howard, 1993). Generally, postmortem epididymidal spermatozoa remain viable for several days after the animal's death. However, this period depends on the species, the storage conditions of the testes, and methods used for sperm collection. Several techniques could be used including: totally cutting (homogenization) the cauda epididymis; washing out or aspiration (flushing) through sperm duct; making cuts (mincing) in the cauda epididymis and squeezing of content; or just squeezing (pressing). The first two techniques are the most frequently used (Maksudov et al., 2008).

For cats, the epididymis is washed and homogenized in HEPES medium plus Ham's F10, and the sperm is recovered after centrifugation. Live sperm can be recovered within the first twelve hours of the epididymis isolation using this technique (Howard, 1993). In vasectomized animals, the semen is collected by aspiration from the epididymis' tail using syringe and needle. The syringe should contain HEPES medium plus HAM'S F10 and the sperm is recovered after centrifugation.

Sperm recovered from the epididymis of cats are known to be motile, viable, and capable of penetrating oocytes (Goodrowe & Hay, 1993; Hay & Goodrowe, 1993). However, epididymal sperm have naturally more cytoplasmic droplets, which normally are lost during transport through the duct (Briz et al., 1995).

According to Tebet et al. (2006) there were no significant differences between the fresh or frozen-thawed domestic cat spermatozoa for the variables: sperm motility, plasma membrane integrity and morphology of electroejaculated and epididymal spermatozoa analyzed immediately after collection, and after freezing and thawing. Jewgenow (Jewgenow et al., 1997) reported motility of frozen-thawed epididymal sperm of lions (*Panthera leo*), tigers (*Panthera tigris*), leopards (*Panthera pardus*), pumas (*Felis concolor*) and jaguar (*Panthera onca*).

Epididymal spermatozoa were collected from immobilized adult male lion by caudal epididymectomy, cryopreserved, and used for IVF of *in vitro* matured lionesses' ova. The post-thawed motility ranged from 55-65%, and the percentages of fertilized ova were 12.7% and 11.5% for 30 and 36 hours of *in vitro* maturation, respectively (Bartels et al., 2000).

It has been shown, for several species, that the methods developed for ejaculated sperm are effective for freezing epididymidal spermatozoa. However, the physiology of ejaculated and epididymidal spermatozoa are different thus, it can be assumed that optimum methods of freezing and thawing may be different.

Postmortem material that can be retrieved in zoos usually belonged to the aged animals with different diseases or that died because of accidents (stress, fights). Moreover, some species display seasonal breeding. All these factors can influence on the spermatozoa's presence and quality in epidydimis, at the collection time (Maksudov et al., 2008).

Analysis of different factors showed that the concurrency of death with breeding season has the strongest affect on the spermatozoa content in testicles of dead animals. The spermatozoa content and quality were almost equal in males that died of sickness (cancer, chronic cardiovascular or excretory system disorders) and in males that died in accidents (stress, fights). However, the quality of postmortem semen of animals that died of natural death is worse than that of animals that died accidentally (Maksudov et al., 2008).

The energy invested by a male with copulation is minimal compared to that of the female, for whom the costs continue throughout pregnancy and lactation. Therefore, it is a significant genetic advantage for cats to be reproductively active throughout the year, or at least to remain active well outside of the usual female breeding season (Spindler & Wildt, 1999).

This strategy is apparent in wild felids. The female Siberian tiger exhibits more estrual activity in January-June than at any other time of the year (Seal et al., 1985), however, the sperm quality is fairly consistent throughout the year (Byers et al, 1990). The quantity and quality of the sperm from some felids can vary when a distinctive female reproductive seasonality is know, as in the snow leopard (Johnston et al., 1994) and Pallas' cat (Swanson et al., 1996), the males remains reproductively active for longer periods than the females of the same species.

The collection of the postmortem sperm of recently dead animals belonging to endangered species can be of substantial importance, and therefore, the method of choice for the preservation of reproductive cells, in the wild, at zoos, and in national parks. The preservation and utilization of postmortem represent the last chance to obtain offspring from the dead males in cases of unexpected loss of valuable animals (Maksudov et al., 2008).

3.2 In vivo sperm collection

The collection of semen, through digital manipulation or using an artificial vagina are indicated because they promote a natural and normal ejaculation, however, these methods require intensive animal training. These techniques are effective for wild dogs, but no routinely used in wild cats.

The electroejaculation is the most used method in wild cats because it can be performed in anesthetized animals. However, one of the disadvantages of using this method in cats is the urine contamination of the semen. The urine contamination occurs when the voltage exceeds the minimum necessary level for ejaculation or when the electrode is positioned cranially. One alternative to minimize this problem would be the catheterization or cystocentesis before start procedure.

The Tiletamine-Zolazepam combination has been the most used anesthetic protocol for this procedure because it produces insignificant changes in the ejaculate. The ketamine hydrochloride and xylazine association in the same syringe has been used for semen collection in small cats, but according to Dooley et al. (1991), these drugs, in combination, seems to be related to retrograde ejaculation.

The electroejaculator should indicate both the voltage and amperage. The voltage should reach up to 12V, controlled by the command button, which should provide smooth control and gradual increase in power output.

The rectal bipolar electrode used for electroejaculation should have three longitudinal strips of copper. The copper strips must have a 0.4 cm apart and protruding approximately 0.2 cm. The electrode, previously lubricated with mineral oil, should be introduced into the rectum with the longitudinal strips ventrally positioned, applying light pressure to increase contact with the pelvic plexus region. The diameter of the electrode should be specific to each species (Table 2).

The electrical series follows a specific protocol with 80 stimuli divided into three series: 30 (series 1: 10 stimuli at 2, 3 and 4V), 30 (Series 2: 10 stimulations in 3, 4 and 5V) and 20 (series 3: 10 stimulations in 4 and 5V) (Howard 1993). In jaguars, the last series of stimuli can reach up to 6V to achieve ejaculation (Paz et al., 2000).

The stimulation cycle starts at 1 second from 0 voltage to the desired voltage, 2 to 3 seconds at the desired voltage, and 3 seconds returning to 0 voltage. An interval of 10 minutes should be used for resting between sets.

Before the start of the series, the penis must be exposed, examined, and washed with saline solution and gauze. The semen collection should be performed in plastic tubes that are maintained warm at 37° C water bath. For each series, the tubes should be replaced in order to avoid urine contamination. All ejaculates should be used; the total volume of semen is the sum of each ejaculate's volume (Figure 1).

4. Sperm evaluation

The appearance is the first evaluation of the semen: changes in color may be associated with diseases in the accessory organs and testes. The ejaculate's volume is the second aspect to be evaluated and must be determined immediately after collection. The volume provides information about the semen production in different species. The pH determination is important because it may indicate urine (acidic pH) or bacteria contamination (basic pH).

For sperm evaluation, an aliquot of the ejaculate is placed on a microscope slide warmed at 37°C, covered with a warmed glass coverslip, and examined at 400 X magnification. The semen should be evaluated for motility and progressive sperm motility. Motility is expressed in percentages, with 0% being the value for immobile spermatozoa and 100% for maximum spermatozoa performance. The sperm type of movement is evaluated by the progressive sperm motility in scale from 0 to 5 (0 - no motility, 1 - poor lateral movement with some progression, 2 - moderate lateral movement with occasional progression, 3 - slow progression, 4 - progression, 5 - rapid progression) (Howard, 1993).

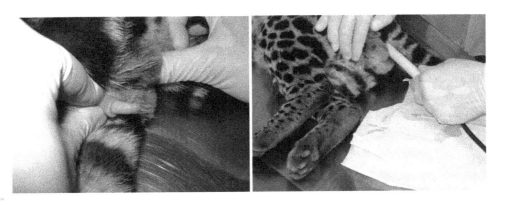

Fig. 1. Ocelot's penis exposed before semen collection; the penis' spines can be observed (left). Ocelot's electroejaculation procedure (right). Pictures: Regina Paz.

The sperm morphology and concentration can be evaluated by fixing an aliquot of semen (1:3 dilution) in a 10% formaldehyde saline solution or in a 2.5% glutaraldehyde solution after the preparation of samples in a humidified chamber.

For the determination of the sperm morphology, 200 cells per slide were counted at 1000 X magnification under light microscopy; the abnormalities were classified as primary or secondary defects expressed as percentages. According to the primary defects presented in the sample, the sperm can be classified as macrocephalic, microcephalic, bicephalic, pyriform head, rounded head, abnormal acrosome, abnormal midpiece, no midpiece, tightly coiled tail and biflagelat. According to the secondary defects presented in the sample, the sperm can be classified as bent midpiece with or without droplet, bent tail with or without droplet, and proximal or distal droplet.

The concentration can be evaluated using a Neubauer chamber at 400 X magnification under ligth microscopy. The volume, concentration, motility, vigor and abnormal sperm data in different species are presented in Table 2.

Species	Probe (cm)	N° Ejaculates	Volume (mL)	Concentr. (x10⁶/mL)	Motility (%)	Vigour (0-5)	Normals (%)
Ocelot (L. pardalis)	1.0[1]	5[1]	0.3±0.1[1]	28.0±17.0[1]	72.0±12.5[1]	4.0±0.5[1]	80.8±0.9[1]
		38[2]	0.62±0.08[2]	53.8±17.8[2]	70.4±2.3[2]	-	58.4±5.8[2]
Tigrinus (L. tigrinus)	1.0[1]	18[2]	0.11±0.02[2]	78.5±33.8[2]	62.1±5.7[2]	-	35.6±6.0[2]
Margay (L. wiedii)	1.0[1]	11[1]	0.2±0.1[1]	79.9±28.1[1]	86.0±3.3[1]	4.6±0.2[1]	48.5±6.1[1]
		27[2]	0.31±0.05[2]	14.2±5.3[2]	62.8±5.3[2]	-	39.5±7.7[2]
Jaguarondi (H. yagouarundi)	1.0[1]	3[1]	0.1±0.1[1]	12.5±9.4[1]	50.0±9.9[1]	3.5±0.4[1]	35.4±14.3[1]
		21[2]	0.08±0.02[2]	7.2±4.0[2]	57.8±2.5[2]	-	25.7±4.6[2]
Pampas cat (O. colocolo)	1.0[1]	5[1]	0.3±0.1[1]	10.8±5.7[1]	36.7±6.6[1]	2.8±0.2[1]	65.9±23.8[1]
		2[2]	0.08±0.01[2]	364.0±326.0[2]	81.3±6.3[2]	-	56.5±0.5[2]
Geofroy's cat (O. geofroy)	1.0[1]	8[1]	0.2±0.1[1]	300.0±233.2[1]	73.0±4.4[1]	4.0±0.3[1]	29.0±11.5[1]
		24[2]	0.21±0.03[2]	66.5±24.4[2]	64.0±4.7[2]	-	46.9±5.0[2]
Puma (P. concolor)	1.6[1]	12[1]	2.8±0.5[1]	20.2±4.7[1]	52.0±8.0[1]	3.5±0.2[1]	23.4±3.7[1]
Jaguar (P. onca)	3.0[1]	5[1]	2.7±0.6[1]	12.0±1.9[1]	82.0±5.8[1]	4.1±0.3[1]	58.2±11.1[1]
	2.3[3]	38[3]	5.7±1.71[3]	13.16±10.76[3]	56.9±9.35[3]	3.02±0.77[3]	65.73±6.7[3]

Howard, 1993[1]; Morais, 2001[2]; Paz et al., 2000[3];

Table 2. Neotropical wildlife cats seminal characteristics.

The poor semen quality in carnivores may be related to the nutritional status of the animals. Rodrigues da Paz et al. (2006), studying the reproduction of jaguars, observed a positive correlation between the improvement in the semen quality and the decrease of primary defects, after diet supplementation with vitamins and minerals. Ocelots, tigrinus and margays showed an increase in the number of ejaculates and 20-30% improvement related to sperm defects after receiving vitamin and mineral supplementation (Morais, 2001).

The seminal plasma constituents compromise the sperm viability in some species. Thus, washing the ejaculates in culture media (HEPES, HAM'SF-10) by centrifugation is efficient in removing the seminal plasma, which could contain bacteria and other undesirable microorganisms, especially when the semen will be used for intrauterine artificial insemination.

5. Sperm cryopreservation

The semen cryopreservation procedures should be initiated only after the centrifugation at 300g for 10 minutes in culture medium or HEPES HAM'SF-10. This procedure is essential for eliminating microorganisms and seminal plasma remove.

The "Double Step" cryopreservation method, using glycerol as cryoprotectant, is in general, used for the semen of most carnivores. This method use two fractions: fraction A containing nutritional constituents and antibiotics; and fraction B containing nutritional constituents, antibiotics, and the cryoprotectant. The PDV medium is used for the semen cryopreservation of the majority of wildlife cats. The fraction A contains 20% egg yolk, 11% lactose, 1000 IU penicillin/mL, 1000 mg streptomycin/mL; the fraction B contains 20% egg yolk, 11% lactose, 8% glycerol, 1000 IU penicillin/mL and 1000 mg streptomycin/mL.

After being removed from liquid nitrogen, the semen straws should be immediately thawed for 1 min in waterbath at 37°C, evaluated for total motility (%) and progressive sperm motility (scale, 0-5) before use.

The first step in the process of freezing semen is the removal of the supernatant after centrifugation of the semen collected in HAM'S F-10 or HEPES culture media, and the subsequent ressuspension of the pellet in PDV fraction A at 37 °C. This mixture should be kept in the refrigerator for 2 hours followed by a subsequent slow addition of the PDV fraction B. The material is them transferred to cooled 0.25 mL straws and kept in the refrigerator for 30 minutes. Afterwards, each straw is placed in liquid nitrogen vapor for 20 min, immersed in liquid nitrogen, transferred to the racks, and loaded into the canisters for long-term liquid nitrogen storage at -196°C.

The straws, racks and canister identification are extremely important, being the determining factor for the germplasm bank establishment and successful operation. The material collected might be extremely valuable for populations in the future and the safety use of this material depends to the correct identification.

The straws identification must contain the animal species (scientific name), tattoo or microchip number, the institution to which the animal belongs, and the date. For free-living animals, the straws must contain the species, the location where the animal was captured, and the date. The racks may be identified by numbers or if applicable, by species. A registry, which can be computerized, is essential to record all of the straws, racks, and canisters, thereby facilitating the location of the material.

The reasons for the poor quality of wildlife semen after thawing are still unknown and involve a range of information and specific characteristics for each species, which are also not yet clearly understood. New tests with different protocols and different cryoprotectors for each species of interest are required in order to maximize the spermatozoa viability after cryopreservation procedures.

6. Ovarian activity induction and superovulation

The currently used ovarian stimulation and superovulation protocols require injections of exogenous gonadotropins, which consist of large complexes of glycoproteins. Equine Chorionic Gonadotropin (eCG) and Human Chorionic Gonadotrophin (hCG) are frequently used due to their long half-life in circulation (24-48h) and good ovarian response with a single application. Other hormones used are the porcine Follicle Stimulating Hormone (pFSH) and porcine Luteinizing Hormone (pLH), these hormones are characterized as short half-life (~ 2h) gonadotropins, therefore, they present the disadvantage of requiring multiple applications to produce a good ovarian response (Crichton et al., 2003; Dresser et al., 1988; Pope, 2000; Wildt et al., 1981).

Studies on wild cats report the use of eCG/hCG in combination, mainly to avoid the stress associated with multiple injections of FSH (Roth et al., 1997). However, the use of porcine FSH/LH determined equivalent number of oocytes compared to the established protocol for eCG/hCG used in tigers (*Panthera tigris*) (Crichton et al., 2000), ocelot (*Leoparuds pardalis*) and tigrinas (*Leopardus tigrinus*) (Paz et al., 2005, 2006), demonstrating that the stress caused by daily injections did not influence the ovarian response.

The eCG/hCG combination has been used successfully in tigers (*Panthera tigris*), cheetahs (*Acinonyx jubatus*), clouded lepards (*Neofelis nebulosa*), pumas (*Puma concolor*), ocelots (*Leopardus pardalis*) and tigrinas (*Leopardus tigrinus*) (Barone et al., 1994; Donoghue, 1993; Donoghue et al., 1990; Howard, 1992b; Moore et al., 1981; Moraes et al., 1997; Morato et al., 2000; Swanson, 1996a). The FSH and hCG combination was used successfully in the Indian desert leopard (*Felis sylvestris ornata*) (Pope et al., 1989), and the pFSH and pLH combination in tigers (Crichton et al., 2000, 2003) jaguars (*Panthera onca*) (Morato et al., 2000), ocelots (*Leopardus pardalis*) and tigrinas (*Leopardus tigrinus*) (Paz et al., 2005, 2006).

Swanson et al. (1995, 1996a) suggest that the repeated administration of exogenous gonadotropins, within short time intervals is a problem because it causes a reduction in the ovarian stimulation which is immunologically mediated. The repeated administration of exogenous gonadotropins has been associated with the production of neutralizing immunoglobulin, which prevents the ovarian response to superovulation protocols.

Alternating gonadotropins regimens in sequential treatments are indicated because of variable immunoglobulin affinities to different exogenous gonadotropins (Maurer et al., 1968; Swanson et al., 1995).

Ocelots and tigrinas treated four to six times, at 4-month intervals, with alternating exogenous gonadotropin regimens (eCG/hCG and pFSH/pLH) did not show a reduction in ovarian response (total follicles and Corpora Lutea), oocyte maturation or exogenous gonadotropins antibodies production over time (Paz et al., 2005, 2006). The findings suggest that, these endangered cat species may be managed intensively with the use of alternating exogenous gonadotropin regimens in assisted reproduction procedures without compromising ovarian responsiveness to these hormones.

Specie	Procedure	Treatment 1		Treatment 2	
		ECG (UI)	hCG (UI)	pFSH (UI)	pLH (UI)
Ocelot (*Leopardus pardalis*)	AI[6]	400[6]	200[6]	-	-
	IVF[4,5]	500[4,5]	225[4,5]	50[4,5]	20[4,5]
Tigrinus (*Leopardus tigrinus*)	AI[4,5]	75[4,5]	100[4,5]	30[4,5]	10[4,5]
	IVF[6]	200[6]	150[6]	-	-
Gato mourisco (*H. yagouarundi*)	AI[7]	100[7]	75[7]	-	-
	IVF[7]	200[7]	150[7]	-	-
Puma (*Puma concolor*)	AI[1]	200[1]	100[1]	-	-
	IVF	-	-	-	-
Jaguar (*Panthera onca*)	AI[2]	200[2]	150[2]	-	-
	IVF[3]	-	-	50[3]	25[3]

Barone et al., 1994[1]; Jimenez et al., 1999[2]; Morato et al., 2000[3]; Paz et al., 2005[4]; Paz et al., 2006[5]; Swanson et al., 1996a[6]; Swanson (Personal Communication)[7].

Table 3. Ovarian stimulation with exogenous gonadotropins used in wildlife cats reproduction (AI= Artificial Insemination and IVF= *In vitro* fertilization).

The exogenous gonadotropins dosage used for ovarian stimulation is another important factor to the fertilization rate and subsequent embryonic development (Donoghue et al., 1993). Species with similar size and weight may require varying dosages, possibly because they have different sensitivities to exogenous gonadotropins (Roth et al., 1997; Swanson et al., 1996b). The nutritional status of the animal also influences the fertilization success rate. Swanson et al. (2002a), studying ocelots and tigrinas in Brazil, observed better quality of oocytes and increase fertilization rates after supplementing the diet with vitamins and minerals.

7. Artificial insemination and oocytes collection

The rate of artificial insemination success in carnivores is influenced by the localization of the semen deposition. Non-surgical methods of semen deposition in the vagina has shown inferior results compared with the surgical method, with semen deposition directly into the uterus. This can be explained by the chemical restraint need in wild animals, with the anesthesia compromising the sperm transportation in non-surgically insemination (Howard, 1993).

The artificial insemination success with semen deposition in the uterine horn is described in several species of wild cats such as puma (*Puma concolor*) (Moore et al., 1981); leopard (*Panthera pardus saxicolor*) (Dresser et al., 1982); cheetah (*Acinonyx jubatus*) (Howard, 1992); tiger (*Panthera tigris altaica*) (Donoghue et al., 1993); ocelot (*Leopardus pardalis*) (Swanson, 1996a) and tigrina (*Leopardus tigrinus)* (Moraes et al., 1997*)*.

Artificial insemination using video-laparoscopy technique has been developed for the semen deposition directly into the uterine horn, close to the oviduct where fertilization occurs, in addition to being a less invasive method. In this procedure, the ovaries and uterine horns can be accessed and evaluated for thickness, consistency and color in all species. In cats, the ovaries are easily observed, facilitating the counting and characterization of pre-ovulatory follicles (brighter small elevated areas) and post-ovulatory corpus luteum (yellow-red area).

According to ovarian stimulation protocols, the animals should be inseminated within 24 to 48 hours after hCG or pLH administration, or after the ovulation process. Inhalatory anesthesia is necessary to perform this procedure (Figure 2).

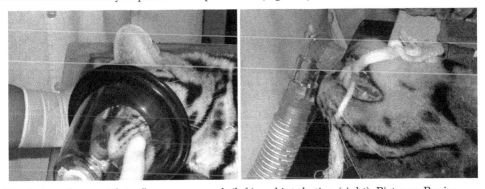

Fig. 2. Anesthesia with isoflurane gas mask (left) and intubation (right). Pictures: Regina Paz.

The anesthetized cats should be secured in dorsal recumbency with the use of leg ties on a tilting surgical table, and the abdominal region of each female should be clipped and prepped with alternating applications of Betadine scrub and alcohol. A pneumoperitoneum should be created by means of CO_2 gas introduced through a *Verres* needle inserted transcutaneously into the central abdominal cavity. A 7-mm-diameter laparoscope should be inserted through a 1cm skin incision slightly cranial to the umbilicus. The ovaries could be manipulated with the *Verres* needle probe, and each ovary should be closely examined to determine the number of mature follicles ($\geq$ 2 mm diameter), recently-formed corpora lutea, and corpora albicans.

To stabilize the uterine horn, where the cannula will be introduced to deposit the semen, grasping forceps is inserted laterally, 4 to 5 cm of the umbilicus. This procedure maintains the uterus close to the abdominal wall. The horn to be inseminated is the ovary that shows the corpus luteum after ovulation. The procedure should be performed in both uterine horns if both ovaries present the corpus luteum.

For the semen deposition, the uterine horn is cannulated using a 20G sterile needle catheter inserted through the abdominal cavity, near the uterine lumen. As soon as the needle pierces the uterine horn, it is removed, keeping the catheter in place. Inside the catheter, a sterile polypropylene tube must be inserted, which will be connected to a syringe containing the semen. (Figure 3).

The intrauterine artificial insemination by laparoscopy is less invasive because the semen deposition ocurrs directly in the uterine horn without laparotomy. This methodology resulted in a 46.2% increase in cheetah's pregnancy rates (Howard, 1992).

Similarly, Donoghue et al. (1993) reported the first birth of a tiger cub (*Panthera tigris altaica*) in Siberian Tiger Species Survival Plan (SSP Program) after intrauterine insemination by video-laparoscopy in females stimulated with eCG and hCG. This result demonstrates the importance in using assisted reproduction methods in the production of genetically viable population with recommended breeding by a management program.

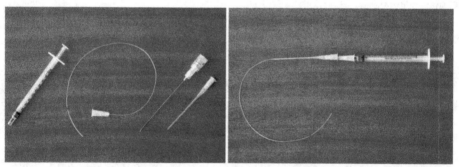

Fig. 3. Intrauterine insemination set: 1mL syringe, polypropylene tube and 22G needle (left). Conected set (right). Pictures: Regina Paz.

The procedures for oocyte retrieval in *in vitro* fertilization are performed using the laparoscopy technique. The ovaries' visualization and the oocyte retrieval in carnivores are species-specific. Cats' ovaries are easily accessed and the follicles easily aspirated.

General anesthesia is needed to perform this procedure; isoflurane inhalation anesthesia is generally used. After anesthetized, the animals are placed in the supine position (45 degrees) and pneumoperitoneum is created with the *Verres* needle, which can be coupled to a CO_2 gas automatic insulflator or a manual pump.

The endoscope is placed near the umbilicus for the evaluation of ovaries and observation of the follicles. Only mature follicles, larger than 2 mm, should be aspirated. The *Verres* needle is used for the follicles' size determination and for the maintenance of the ovary in an adequate position, near to the abdominal wall.

The mature follicles are aspirated with a 22G needle attached to polypropylene tubing connected to a sterile collection tube (15mL) containing M199 culture medium and heparin, which is attached to a vacuum aspiration pump (Figure 4). After the collection, the oocytes are placed in petri dishes with culture medium and observed by stereomicroscopy at 400 X magnification for their classification.

The oocytes from carnivores are dark and contain lipidic drops. The maturation status is characterized in I, II and III according morphological aspects. Oocytes I are of excellent quality, characterized by a uniformly dark cytoplasm, nucleus with a distinct corona radiate, and an expansive cumulus cell mass. Oocytes II are of regular quality, characterized by a non-uniformly cytoplasm, nucleus with an indistinct corona radiate, and a non-expansive cumulus cell mass. Oocytes III are degenerated and characterized by an abnormal cytoplasm, nucleus without corona radiata or cumulus cell mass (Goodrowe et al., 1988; Johnston et al., 1989).

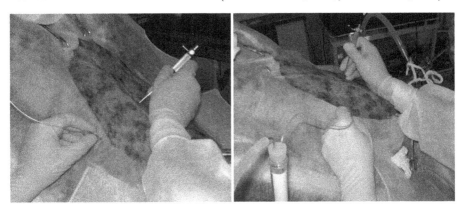

Fig. 4. Follicular aspiration using video-laparoscopy and *Verres* needle in ocelot (left). Aspiration follicular system (right). Pictures: Regina Paz.

8. *In vitro* fertilização and embryo transfer

The *in vitro* fertilization technique has been applied in wild animals after follicular aspiration using laparoscopy tecnhique, oocyte retrieval post-mortem, or after ovariohysterectomy. According Swanson (1998), oocytes collected from ovaries can be refrigerated at 5°C for 24 hours without maturation and changes in the fertilization potential.

Oocytes recovery from refrigerated ovaries can be achieved using the follicular aspiration technique with a syringe and needle, or through ovary laceration and oocyte harvest using stereomicroscopy. The second technique is used in small animals, which present ovaries with small diameters because the follicle aspiration would be difficult. The M199 culture medium is used for follicular aspiration and ovaries laceration.

However, laparoscopic follicular aspiration is the most used oocyte retrieval technique for IVF, which should be preceded by hormonal treatment. According Howard (1999), the treatment with exogenous gonadotropins and laparoscopy are the basic requirements for ovarian stimulation and oocyte retrieval for IVF procedures.

The immature oocytes collected ($\cong$ 60%) in cats become mature in 24-32 hours in culture media. These about 70% are fertilized, however, only a small percentage, from 20 to 30% develops in blastocysts (Johnston et al., 1989).

Donoghue et al. (1990) reported the birth of the first wild cat from IVF after an embryo transfer. Tiger cubs (*Panthera tigris*) were produced *in vitro* using excellent quality embryos, containing two to four cells, and surgically transferred to the oviducts of two females. Pregancy was successful in one of these females and three kittens were born after 107 days.

The first embryos produced by IVF in Brazil were jaguar (*Panthera onca*) embryos. The ovarian stimulation with pFSH/LH produced $\cong$ 25 follicles/female (> 80%), however, despite the recovery of high quality oocytes, the fertilization percentages were low (> 25%) (Morato et al., 2000).

In a Project involving the São Paulo University/Brazil (USP), the Mata Ciliar Association/Brazil (AMC) and the Cincinnati Zoo/USA (CREW), 128 ocelot (*Leopardus pardalis*) and tigrina (*Leopardus tigrinus*) embryos were produced. The animals were treated with eCG/hCG and produced $\cong$ 10 follicles/female. Follicles were aspirated by laparoscopy and 7-9 of excellent quality oocytes/female were recovered. Of these, 60% were fertilized *in vitro*, resulting in 76 ocelot and 52 tigrina embryos (Swanson & Brown, 2004). Two ocelots became pregnant with normal pregnancy development and birth, one from the Cincinnati Zoo/USA and another from the Sao Bernardo do Campo Zoo/Brazil. However, only the offspring born to the American female survived after birth (Swanson, 2002b).

Among all the Neotropical cats, the ocelot is the only species that produced offspring after the transfer of frozen embryos (Swanson, 2001, 2002b). These births were the result of a cooperative effort between Brazil and the USA for the development of adequate management programs for this species in captivity, and for the establishment of genetically viable populations between the Brazilian and American populations.

Based on the percentage of successful embryonic cleavage after thawing domestic cat embryos, which is less than 70% (Pope, 2000), the difficulties to develop feline embryos used post-freezing is recognized. The difficulties may be related to inappropriate timing for the embryo implantation or fetal survivel (Swanson & Brown, 2004) and not only related to the quality of the embryos. Thus, it is necessary to detect the female natural receptor estrus in order to perform the transfer of thawed embryos. This determination is achieved through measurement of fecal steroid levels according to the enzyme-immunoassay (EIA) technique for the fecal estrogens metabolites.

After the determination of the natural estrus, the ovulation should be induced with GnRH (Gonadotropin Releasing Hormone) and the embryos should be transferred by video-laparoscopy directly in the ostium of the oviduct.

9. Transmission of reproductive diseases

The methods used for assisted reproduction in wild animals should be free of contamination or diseases, therefore, some measures must be taken to avoid compromising the procedures. The semen centrifugation and seminal plasma removal are essential for artificial insemination with fresh or frozen semen and *in vitro* fertilization procedures.

The methods for processing semen without centrifugation and removal of the seminal plasma were responsible for the development of pyometra in 40% of inseminated domestic cats, regardless of the semen being diluted in culture medium containing antibiotics penicillin and streptomycin (Howard, 1993). It is believed that the donors might carry bacteria in their normal flora, eg *E. coli*, which would cause infection in the females.

In a study conducted by Paz et al. (1999), aiming at determining preputial microbiota in nine adult male jaguars (*Panthera onca*), the most frequently observed microorganism were *Staphylococcus sp* (40%), followed by *Streptococcus sp* (30%), *Escherichia coli* (20%) and *Corynebacterium sp* (10%).

The preputial microbiota in the genus Leopardus was assessed by Guido et al. (2000), and the results were *Escherichia coli* (40%), *Proteus rettgeri* (40%) and *Yersinia psedotuberculosis* (20%) in tigrina (*Leopardus tigrinus*) (n = 5), *Staphylococcus sp* (42.9%), *Escherichia coli* (28.5%), *Streptococcus sp* (14.3%), *Staphylococcus sp* + *Streptococcus sp* (14.3%) in ocelots (*Leopardus pardalis*) (n = 6) and only *Staphylococcus sp* in margay (*Leopardus wieddi*) (n = 1).

The feline immunodeficiency virus is present in the semen of domestic cats and can be transmitted to females by AI (Jordan et al., 1995, 1996). This aspect should be taken into consideration during assisted reproduction procedures performed in wild cats.

In addition, a reproductive evaluation and clinical examination should be performed in wild cats before the animal inclusion in the management programs using assisted reproduction.

10. References

Bartels, P.; Lubbe, K.; Killian, L.; Friedmann, Y.; Van Dyk, G. & Mortimer, D. (2000). In vitro maturation and fertilization of lion (*Panthera onca*) oocytes using frozen-thawed epididymal spermatozoa recovered by cauda epididymectomy of an immobilized lion. *Theriogenology*, Vol. 53, pp. 325.

Barone, M. A.; Wildt, D. E.; Byers, A. P.; Roelke, M. E.; Glass, C. M. & Howard, J. G. (1994). Gonadotrophin dose and timing of anaesthesia for laparoscopic artificial insemination in the puma (*Felis concolor*). *Journal of Reproduction and Fertility*, Vol.101, pp. 103-108.

Bonney, R. C.; Moore, H. D. M. & Jones, D. M. (1981). Plasma concentrations of oestradiol-17β and progesterone, and laparoscopic observations of the ovary in the puma (*Felis concolor*) during oestrus, pseudopregnancy and pregnancy. *Journal of Reproduction and Fertility*, Vol.63, pp. 523-532.

Briz, M. D.; Bonet, S.; Pinart, B.; Egozcue, J. & Campos, R. (1995). Comparative study of boar sperm coming from the caput, corpus, and cauda regions of the epididymis. *Journal of Andrology*, Vol. 16, pp. 175-188.

Brown, J. L.; Wasser, S. K.; Wildt, D.E. & Graham, L. H. (1994). Comparative aspects of steroid hormone metabolism and ovarian activity in felids, measured noninvasively in feces. *Biology of Reproduction*, Vol.51, pp. 766-86.

Brown, J. L.; Terio, K. A. & Graham, L. H. (1996). Fecal androgen metabolite analysis for non invasive monitoring of testicular steroidogenic activity in felids. *Zoo Biology*, Vol.15, pp. 425-434.

Byers, A. P.; Hunter, A. G.; Seal, E. S.; Graham, E.F. & Tilson, R.L. (1990). Effect of season on seminal traits and serum hormone concentration in male Siberian tiger (*Panthera tigris*). *Journal of Reproduction and Fertility*, Vol. 90, pp. 119-125.

Comizzoli, P.; Mermillod, P. & Mauget, R. (2000). Reproductive biotechnologies for endangered mammalian species. *Reproduction Nutrition Development*, Vol.40, pp. 493-504.

Crichton, E. G.; Armstrong, D. L.; Vajta, G.; Pope, C. E. & Loskutoff, N. M. (2000). Development competence in vitro of embryo produce from siberian tigers (*Panthera tigris altaica*) cryopreserved by controlled rate freezing versus vitrification. *Theriogenology*, Vol.53, No.1, pp.328. Abstract.

Crichton, E. G.; Bedows, E.; Miller-Lindholm, A. K.; Baldwin, D. M.; Armstrong, D. L.; Graham, L. H.; Ford, J. J.; Gjorret, J. O.; Hyttel, P.; Pope, C. E.; Vajta, G. & Loskutoff, N. M. (2003). Efficacy of porcine gonadotropins for repeated stimulation of ovarian activity for oocyte retrievel and in vitro embryo production and cryopreservation in siberian tigers (Panthera tigris altaica). *Biology of Reproduction*, Vol.68, pp. 105-113.

Donoghue, A. M.; Johnston, L. A.; Seal, U. S.; Armstrong, D. L.; Tilson, R. L.; Wolf, P.; Petrini, L. G.; Simmons, T.; Groos, T. & Wildt, D. E. (1990). In vitro fertilization and embryo development in vitro and in vivo in the tiger (*Panthera tigris*). *Biology of Reproduction*, Vol.43, pp. 733-44.

Donoghue, A. M.; Johnston, L. A.; Armstrong, D. L.; Simmons, L. G. & Wildt, D. E. (1993). Birth of a siberian tiger cub (*Panthera tigris altaica*) following laparoscopic intrauterine artificial insemination. *Journal of Zoo and Wildlife Medicine*, Vol.24, No.2, pp. 185-89.

Dooley, M. P.; Pineda, M. H; Hopper, J. G. & Hsu, W. H. (1991). Retrograde flow of spermatozoa into the urinary bladder of cats during eletroejaculation, collection of semen with an artificial vagina, and mating. *American Journal Veterinary Research,* Vol.52, No.5, pp. 687-91.

Dresser, B. L.; Kramer, L.; Reece, B. & Russel, P. T. (1982). Induction of ovulation and successful artificial insemination in a Persian Leopard (*Panthera pardus saxicolor*). *Zoo Biology*, Vol.1, pp. 55-57.

Dresser, B. L.; Sehlhorst, C. S.; Keller, G.; Kramer, L. W. & Reece, R. W. (1986). Artificial insemination and embryo transfer in the felidae. *Proceedings of International Symposium for World Conservation Strategies for Tigers*, pp. 4-18, Cincinnati, Ohio, USA.

Dresser, B. L.; Gelwicks, E. J.; Wachs, K. B. & Keller, G. L. (1988). First successful transfer of cryopreserved feline (*Felis catus*) embryos resulting in live offspring. *The Journal of Experimental Zoology*, Vol.246, pp. 180-186.

Ewer, R. F. (1975). The evolution of mating system in the felidae. In: *The world's cats: biology, behavior and management of reproduction*. Eaton, R. L. pp. 11-20, London: Unimark Publishing.

Feldman, E. C. & Nelson, R. W. (1996).*Canine and Feline Endocrinology and Reproduction*. pp. 529-46, 2 ed, Philadelphia: W.B. Saunders.

Goodrowe, K. L.; Hay, M. (1993).Characteristics and zona binding ability of fresh and cooled domestic cat epididymal spermatozoa. *Theriogenology*, Vol. 40, pp. 967-75.

Goodrowe, K. L.; Wall, R. J.; O'Brien, S. J.; Schmidt, P. M. & Wildt, D. E. (1988). Developmental competence of domestic cat follicular oocytes after fertilization in vitro. *Biology of Reproduction*, Vol.39, pp. 355-372.

Guido, M. C.; Paz, R. C. R.; Costa, E. O.; Zuge, R. M.; Benites, N. R.; Barnabe, V. H. & Barnabe, R. C. , (2000). Microbiota prepucial e vaginal de felinos neotropicais mantidos em cativeiro. *Proceedings of Combravet XVII Congresso Brasileiro de Medicina Veterinária*, pp.03, Águas de Lindóia, São Paulo, Brasil, 2000.

Hamner, C. E.; Jennings, L. L. & Sojka, N. J. (1970). Cat (*Felis catus L.*) spermatozoa require capacitation. *Journal Reproduction and Fertility*, Vol.23, pp. 477-480.

Hay, M. A.; Goodrowe, K. L. (1993).Comparative cryopreservation and capacitation of spermatozoa from epididymides and vasa deferentia of the domestic cat. *Journal Reproduction and Fertility Supplement*, Vol. 47, p. 297-30.

Howard, J.; Donoghue, A. M.; Barone, M. A.; Goodrowe, K. L.; Blumer, E. S.; Snodgrass, B. S.; Sarnes, D.; Michael, T.; Bush, M. & Wildt, D. E. (1992). Successful induction of ovarian activity and laparoscopic intrauterine artificial insemination in the cheetah (*Acinonyx jabatus*). *Journal of Zoo and Wildlife Medicine*, Vol.23, No.3, pp. 288-300a.

Howard, J. (1992). Embryogenesis in conservation biology or how to make an endangered species embryo. *Theriogenology*, Vol.37, No.1 b.

Howard, J. G. (1993). Semen collection and analysis in carnivores. In: *Zoo & Wild Animal Medicine Current Therapy*, Fowler, M. E, pp.390-399, 3 ed. Philadelphia: W.B. Saunders.

Howard J. G. (1999). Assisted reproductive techniques in nondomestic carnivores. In: *Zoo & Wild Animal Medicine*, Fowler, M.E. & Miller, R.E.. pp.449-457, 4 ed., Philadelphia: W.B. Saunders.

IUDZG/CBSG. (1993). *The word zoo conservation strategy: the role of the zoos and aquaria of the world in global conservation.* Chicago: Zoological Society, 76 p.

Jewgenow, K.; Blottner, S.; Lengwinat, T. & Meyer, H.H.D. (1997). New methods for gametes rescue from gonads of nondomestic felids. *Journal of Reproduction and Fertility Supplement*, Vol. 51, pp. 33-39.

Jimenez, T. G.; Zuge, R.; Paz, R. C. R.; López, J. E. & Crudeli, G. A. (1999). Sincronización de celo e inseminación artificial por video laparoscopia en yaguareté (*Panthera onca*) en cautiverio. *Comunicaciones Científicas y Tecnológicas*, Vol.4, pp. 67-70.

Johnston, L.A.; O'Brien, S. J. & Wildt, D. E. (1989). In vitro maturation and fertilization of domestic cat follicular oocytes. *Gamete Research*, Vol. 24, No.3, pp. 343-56.

Johnston, L. A.; Armstrong, D. L. & Brown, J. L. (1994). Seasonal effects on seminal and endocrine traits in the captive snow leopard (*Panthera uncia*). *Journal of Reproduction and Fertility*, Vol. 102, pp. 229-236.

Johnston, S. D.; Root, M. V. & Olson, P. N. S. (1996). Ovarian and testicular functions in the domestic cat: clinical management of spontaneous reproductive disease. *Animal Reproduction Science*, Vol.42, pp. 261-274.

Jordan, H. L.; Howard, J. G. & Tompkins, W. A. (1995). Detection of feline immunodeficiency virus in semen from seropositive domestic cats (*Felis catus*). *Journal Virology*, Vol.69, pp. 7328-7333.

Jordan, H. L.; Howard, J. G & Sellon, R. K. (1996). Transmission of feline immunodeficiency virus in domestic cats via artificial insemination. *Journal Virology*, Vol.70, pp. 8224-8228.

Lasley, B. L.; Loskutoff, N. M. & Anderson, G. B. (1994). The limitation of conventional breeding programs and the need and promise of assisted reproduction in non-domestic species. *Theriogenology*, Vol.41, pp. 119-132.

Maksudov GYu, Shishova NV, & Katkov II. 2009. In the Cycle of Life: Cryopreservation of Post-Mortem Sperm as a Valuable Source in Restoration of Rare and Endangered Species. In: Endangered Species: New Research. Eds.: A.M. Columbus and L. Kuznetsov. NOVA Science Publishers, Hauppauge NY, Ch. 8, pp 189-240.

Maurer, R. R.; Hunt, W. L. & Foote, R. H. (1968). Repeated superovulation following administration of exogenous gonadotrophins in dutch-belted rabbits. *Journal Reproduction and Fertility*, Vol.15, pp.93-102.

Mellen, J.D. 1989. *Reproductive behavior of small captive exotic cats (Felis sp).* (Doctoral thesis) University of California. Davis/CA.

Moore, H. D. M.; Bonney, R. C. & Jones, D. M. (1981). Successful induced ovulation and artificial insemination in the puma (*Felis concolor*). *Veterinary Record*, Vol.108, pp.282-83.

Moraes, W.; Morais, R.; Moreira, N.; Lacerda, O.; Gomes, M. L. F.; Mucciolo, R. G. & Swanson, W. F. (1997). Successful artificial insemination after exogenous gonadotropin treatment in the ocelot (*Leopardus pardalis*) and tigrina (*Leopardus tigrina*). Proceedings of *American Associatian of Zoo Veterinarians Congress*, p. 334-335.

Morais, R. N.; Moreira, W.; Mucciolo, R. G.; Lacerda, O.; Gomes, M. L. F.; Swanson, W. F.; Graham, L. H. & Brown, J. L. (1996). Testicular and ovarian function in south american small felids assessed by fecal steroids. Proceedings of *American Associatian of Zoo Veterinarians Congress*, pp. 561-565.

Morais, R. N. (2001). Reproduction in small felid males. In: *Biology, Medicine and Surgery of South American Wild Animals*, Fowler, M. E. & Cubas, Z. S. (ed). 1 ed. Ames: Iowa State University Press. pp.312-316.

Morato, R. G.; Crichton, E. G.; Paz, R. C. R.; Zugue, R. M.; Moura, C. A.; Nunes, A. V. L.; Teixeira, R. H.; Porto-Filho, L.; Guimarães, M. A. B. V.; Correa, S. H. R.; Barnabe, R. C.; Armstrong, D. L. & Loskutoff, N. M. (2000). Ovarian stimulation and sucessful in vitro fertilization in the jaguar (*Panthera onca*). *Theriogenology*, Vol.53, No.1, pp.339. Abstract.

Morato, R. G. & Paz, R. C. R. (2001). *Reproduction in jaguars.* In: *Biology, Medicine and Surgery of South American Wild Animals*, Fowler, M. E. & Cubas, Z. S. (ed). 1 ed. Ames: Iowa State University Press. pp.308-312.

Moreira, N.; Monteiro-Filho, E. L. A.; Moraes, W.; Swanson, W. F.; Graham, L. H.; Pasqual, O. L.; Gomes, K. L. F.; Morais, R. N.; Wildt, D. E. & Brown, J. L. (2001). Reproductive steroid hormones and ovarian activity in felids of the leopardus genus. *Zoo Biology*, Vol.20, pp.103-116.

Oliveira, T. G. (1994). *Neotropical Cats Ecology and Conservation.* São Luis: EDUFMA, 220 p.

Oliveira, T. G. & CASSARO, K. , (1997). *Guia de Identificação dos Felinos Brasileiros.* São Paulo: Fundação Parque Zoológico de São Paulo, 60p.

Paz, R. C. R.; Guido, M. C.; Costa, E. O.; Züge, R. M.; Morato, R. G.; Guimarães, M. A. B. V.; Nunes, A. L. V.; Felippe, P. A. N.; Jimenez, G. T.; Crudeli, G. A.; Barnabe, V. H. & Barnabe, R. C. (1999). Microbiota Prepucial de onças pintadas (*Panthera onca*) mantidas em cativeiro. Proceedings of *Congresso Brasileiro para Conservação de Felinos Neotropicais*, Jundiaí, São Paulo, Brasil.

Paz, R. C. R.; Züge, R. M.; Morato, R. G.; Barnabe, V. H.; Barnabe, R. C. & Felippe, P. A. N. (2000). Capacidade de Penetração de sêmen congelado de onça pintada (*Panthera onca*) em oócitos heterólogos. *Brazilian Journal of Veterinary Research and Animal Science*, Vol.37, No.6, pp. 462-466.

Paz, R. C. R.; Swanson, W. F.; Dias, E. A.; Adania, C. H.; Barnabe, V. H. & Barnabe, R. C. (2005). Ovarian and immunological responses to alternating exogenous gonadotropin regimens in the ocelots (*Leopardus pardalis*) and tigrina (*Leopardus tigrinus*). Zoo Biology, Vol.24, pp.247-260.

Paz, R. C. R.; Dias, E. A.; Adania, C. H.; Barnabe, V. H. & Barnabe, R. C. (2006). Ovarian response to repeated administration of alternating exogenous gonadotropin regimens in the ocelot (*Leopardus pardalis*) and tigrinus (*Leopardus tigrinus*). *Theriogenology*, Vol. 66, pp. 1787-1789.

Pope, C. E.; Gelwicks, E. J.; Wachs, K. B.; Keller, G. L.; Maruska, E. J. & Dresser, P. L. (1989). Successful interspecies transfer of embryos from the Indian desert cat (*Felis silvestris ornata*) to the domestic cat (*Felis catus*) following in vitro fertilization. *Biology of Reproduction*, Vol.40, pp. 61. Supplementum 40.

Pope, C. E. (2000). Embryo technology in conservation efforts for endangered felids. *Theriogenology*, Vol.53, pp. 163-174.

Rodrigues da Paz, R. C.; Morato, R .G.; Carciofi, A. C.; Guimarães, M.A.B.V.; Pessuti, C.; Ferraz, E. S.; Ferreira, F. & Barnabe, R.C. (2006). Influence of nutrition on the quality of sêmen in Jaguars (*Panthera onca*) in Brazilian zoos. *Internationa Zoo Yearbook*, Vol.40, p 351-359.

Roth, T. L.; Wolfe, B. A.; Long, J. A.; Howard, J. & Wildt, D. E. (1997). Effects of equine chorionic gonadotropin, human chorionic gonadotropin, and laparoscopic artificial insemination on embryo, endocrine, and luteal characteristics in the domestic cat. *Biology of Reproduction*, Vol.57, pp.165-71.

Seal, U. S.; Plotka, E. D.; Smith, J. D.; Wright, F H.; Reindl, N. J.; Taylor, R. S. & Seal, M F. (1985). Immunoreactive luteinizing hormone estradiol, progesterone, testosterone, and androstenedione levels during the breeding season and anestrus in siberian tigers. *Biology of Reproduction*, Vol. 32, pp. 361-368.

Schmidt, A. M.; Nadal, L. A.; Schmith, M. J. & Beamer, N. B. (1979). Serum concentrations of oestradiol and progesterone during the normal oestrus cycle and early pregancy in the lion (*Panthera leo*). *Journal of Reproduction and Fertility*, Vol. 57, pp. 267-272.

Schmidt, A. M.; Hess, D. L.; Schmith, M. J.; Smith, R. C. & Lewis, C. R. (1988). Serum concentration of oestradiol and progesterone, and sexual behavior during the normal oestrus cycle in the leopard (*Panthera pardus*). *Journal Reproduction and Fertility*, Vol. 82, pp. 43-49.

Schmidt, A. M.; Hess, D. L.; Schmith, M. J. & Lewis, C. R. (1993). Serum concentration of oestradiol and progesterone and frequency os sexual behavior during the normal oestrous cycle in the snow leopard (*Panthera uncia*). *Journal Reproduction and Fertility*, Vol. 98, pp. 91-95.

Seal, U. S.; Plotka, E. D.; Smith, J. D.; Wright, F. H.; Reindl, N. J.; Taylor, R. S. & Seal, M. F. (1985). Immunoreactive luteinizing hormone estradiol, progesterone, testosterone, and androstenedione levels during the breeding season and anestrus in siberian tigers. *Biology of Reproduction*, Vol.32, pp.361-368.

Shille, V. M.; Lundström, K. E. & Stabenfeldt, G. H. (1979). Follicular function in the domestic cat as determined by estradiol-17β concentration in plasma: relation to estrous behavior and cornification of exfoliated vaginal epithelium. *Biology of Reproduction*, Vol.21, pp. 953-963.

Silva, J. C.; Leitão, R. M.; Lapão, N. E.; Cunha, M. B.; Cunha, T. P.; Silva, J. P.& Paisana, F. C. (2000). Birth of siberian tiger (*Panthera tigris altaica*) cubs after transvaginal artificial insemination. *Journal of Zoo and Wildlife Medicine*, Vol.34, pp. 566-569.

Spindler, R. E.; Wildt, D. E. (1999). Circannual variations in intraovarian oocyte but not epididymal sperm quality in the domestic cat. *Biology of Reproduction*, Vol. 61, pp. 188-194.

Swanson, W. F.; Brown, J. L.; Wildt, D. E. (1996). Influence of season on reproductive traits of the male Pallas'cat (*Felis Manul*) and implications for captive management. *Journal of Zoo and Wildlife Medicine*, Vol. 27, pp. 234-240.

Swanson, W. F.; Horohov, D. W. & Godke, R. A. (1995). Production of exogenous gonadotrophin-neutralizing immunoglobulins in cats after repeated eCG-hCG treatment and relevance for assisted reproducton in felids. *Journal of Reproduction and Fertility*, Vol.105, pp. 35-41.

Swanson, W. F.; Howard, J. G.; Roth, T. L.; Brown, J. L.; Alvarado, T.; Burton, M.; Starnes, D. & Wildt, D. E. (1996). Responsiveness of ovaries to exogenous gonadotrophins and laparoscopic artificial insemination with frozen-thawed spermatozoa in ocelots (*Felis pardalis*). *Journal of Reproduction and Fertility*, Vol.106, pp. 87-94 a.

Swanson, W. F.; Roth, T. L.; Graham, K.; Horohov, D. W. & Godke, R. A. (1996). Kinetics of the humoral immune response to multiple treatments with exogenous gonadotropins and relation to ovarian responsiveness in domestic cats. *American Journal Veterinary Research*, Vol.57, No.3, pp. 302-307 b.

Swanson, W. F. (1998). Reproduction in domestic cats: model for endangered nondomestic cat especies. In: *Curso de extensão - Felinos selvagens: biotécnicas reprodutivas e conservação*, Morais, R. N. & Gomes, M. L. F. Curitiba: Departamento de Fisiologia da Universidade Federal do Paraná, 100 p.

Swanson, W. F. (2001). Reproductive biotechnology and conservation of the forgotten felids – the small cats. *Proocedings of 1st International Symposium on Assisted Reproductive Technologies: Conservation & Genetic Management Wildlife*, Omaha, Nebraska, USA, pp. 100-120.

Swanson, W. F.; Paz, R. C. R.; Morais, R. N.; Gomes, M. L. G.; Moraes, W. & Adania, C. H. (2002). Influence of species and diet on efficiency of in vitro fertilization in two endangered brazilian felids – The ocelot (*Leopardus pardalis*) and tigrina (*Leopardus tigrinus*). *Theriogenology*, Vol.57, No.1, pp. 593. Abstract.a

Swanson, W. F. (2002). The role of science and reproductive biotechnology in establishing and managing the Brazilian ocelot population in U.S. and Brazilian Zoos. *Proocedings of Annual Conference of the American Zoo & Aquarium Association*, Fort Worth, Texas, USA. pp.75-78.b

Swanson, W. F. & Brown, J. L. (2004). International training programs in reproductive sciences for conservation of Latin American felids. *Animal Reproduction Science*, Vol.82-83, pp. 21-34.

Tebet, J. M. (1999). *Aspectos clínicos e fisiológicos do ciclo estral da jaguatirica (Leopardus pardalis, L.1758)*. (Dissertação Mestrado em Reprodução Animal) - Faculdade de Medicina Veterinária e Zootecnia, Universidade Estadual Paulista "Júlio de Mesquita Filho", Botucatu, São Paulo. pp. 102.

Tebet, J. M.; Martins, M. I. M; Chirinea, V. H.; Souza, F. F.; Campagnol, D.; Lopes, M. D. (2006). Cryopreservation effects on domestic cat epididymal versus electroejaculated spermatozoa. *Theriogenology*, Vol. 66, pp. 1629-1632.

Tsuitsui, T. & Stabenfeldt, G. H. (1996). Biology of ovarian cycles, pregnancy and pseudopregnancy in the domestic cat. *Journal of Reproduction and Fertility*, pp. 29-35. Supplementum 47.

Verhage, H. G.; Beamer, N. B. & Brenner, R. M. (1976). Plasma levels of estradiol and progesterone in the cat during polyestrus, pregnancy and pseudopregnancy. *Biology of Reproduction*, Vol.14, pp. 579-585.

Wildt, D. E.; Platz, C. C.; Chakraborty, P. K. & Seager, S. W. J. (1979). Oestrus and ovarian activity in a female jaguar (*Panthera onca*). *Journal of Reproduction and Fertility*, Vol.56, pp. 555-58.

Wildt, D. E.; Platz, C. G.; Seager, S. W. J. & Bush, M. (1981). Induction of ovarian activity in the cheetah (*Acinonyx jubatus*). *Biology of Reproduction*, Vol.24, pp. 217-222.

2

Genome Banking for Vertebrates Wildlife Conservation

Joseph Saragusty
Department of Reproduction Management,
Leibniz Institute for Zoo and Wildlife Research, Berlin
Germany

1. Introduction

About 140 years ago, Charles Darwin wrote in his book *The Descent of Man, and Selection in Relation to Sex* the following prediction: "The slowest breeder of all known animals, namely the elephant, would in a few thousand years stock the whole world." (Darwin, 1871). Unfortunately, primarily due to human activities, this prediction will probably not come true. Sadly, not only elephants face the risk of extinction. The number of species listed as endangered is on the rise. The Species Survival Commission (SSC) of the International Union for Conservation of Nature and Natural Resources (IUCN) continuously monitors the planet's fauna and flora and launches the IUCN Red List of Threatened Species (http://www.iucnredlist.org). As of the end of 2010, there were 5491 species of mammals described (IUCN, 2010). Of these, 1,131 species (21%) are now listed as endangered to some degree. In addition, there are 324 species listed as near threatened and another 836 species for which data is deficient and thus could be at risk. Adding all these numbers together, about 42% of the planet's mammalian species are at some level of threat for extinction. The list also reports on 76 species (1.4%) of mammals that became extinct in recent years and two more species that are extinct in the wild and whose survival completely depend on *ex situ* breeding programs. The situation is not distinctively different in other classes of the vertebrata subphylum or in the other subphylums of the animal kingdom. If anything, it is even worse for some such as the reptiles (21% endangered), amphibians (30%), fish (21%) or among the invertebrates: insects (22%), mollusks (41%), crustaceans (28%), anthozoa (corals and sea anemones; 27%) or arachnids (58%). With each extinct species, the stability of the entire ecological system surrounding it and the food chain of which it is an integral part is shaken. Such shaking may lead to the co-extinction of dependent species (Koh et al., 2004).

In 1992 the Convention on Biological Diversity (CBD) was ratified at the United Nations Conference on Environment and Development in Rio de Janeiro. Ten years later, during the 6th meeting of the Conference of the Parties to the CBD in 2002, it was agreed "to achieve by 2010 a significant reduction of the current rate of biodiversity loss at global, regional and national level as a contribution to poverty alleviation and to the benefit of all life on Earth" (Convention on Biological Diversity, 2002). However, 2010, which was named by the United Nations as the "Year of Biodiversity", has arrived and gone and this target not only has not been met, even some of the indicators needed to measure progress (or regress) have not yet been developed or fully implemented (Walpole et al., 2009).

Based on paleontological data, of the total biota of about 10 million species, the natural or background extinction rate is approximately 1 to 10 species per year (Reid & Miller, 1989). This may be divided into species with restricted ranges for which extinction rate might be higher and those with widespread ranges for which it is considerably lower (Pimm et al., 1995). The expected extinction rate amongst all bird and mammal species is about one species every 100 to 1,000 years, yet the current extinction rate for these and other groups is about one species per year, which is 100 to 1,000 times the natural rate by some estimates (Reid & Miller, 1989; Pimm et al., 1995; Ceballos & Ehrlich, 2002; IUCN, 2004; Living Planet Report, 2008) and even as high as 10,000 times by others (Mace et al., 2005). One of the major problems behind these predictions is that we do not really know how many species are there and this is primarily true for many understudied taxonomic groups (e.g. bacteria, marine invertebrates, insects) and endemic species in many parts of the world, which by them being endemic to limited habitats face much higher risk of extinction (Pimm et al., 1995; Hunter, 2011). Earth history has witnessed 5 major events of mass extinctions in which a significant fraction of the diversity in a wide range of taxa went extinct within relatively short period (Erwin, 2001). The last, and probably the most well known episode, took place during the late Cretaceous era, approximately 65 million years ago, when the dinosaurs became extinct. The current dramatically accelerated rate of species extinction has been likened to these evens and was termed 'the sixth mass extinction event in the history of life on Earth' (Chapin et al., 2000; Wake & Vredenburg, 2008). Various studies have demonstrated the severity of this accelerated extinction process on both the population level (Ceballos & Ehrlich, 2002) and the global biodiversity level (Living Planet Report, 2008; Rockstrom et al., 2009). Not all researchers agree with the definition of mass extinction (Barnosky et al., 2011) but all agree that the current extinction rate is far too fast. Whether we call it mass extinction or not, the cause for the current accelerated extinction rate is anthropogenic in essence, resulting from six major human interference categories: (i) habitat loss or fragmentation, (ii) over exploitation, (iii) species introduction (exotic species and diseases), (iv) pollution of water, soil and air, (v) global warming, and (vi) increasing atmospheric carbon dioxide level and the consequential acidification of the oceans. Based on different projections such as climate change, human population growth or deforestation rate, predictions suggest that large chunks of the world's biodiversity is destined to disappear (Reid & Miller, 1989; Ehrlich & Wilson, 1991; Thomas et al., 2004). For the sake of the entire ecosystem stability and for our and future generations' well being, and because we are the leading driver behind this accelerated decline in biodiversity, it is our obligation to try and slow down the current extinction rate. This, however, is not going to happen overnight, probably not even over a single generation time. So as to "buy time", the establishment of genome resource banks (GRB), which will store and manage collections of gametes (sperm and oocytes), embryos, tissues and organs of endangered species, has been proposed (Veprintsev & Rott, 1979; Benirschke, 1984; Wildt, 1992). By gathering such collections, at the moment primarily through cryopreservation, these institutions, among other services, fulfill their function as a mean to extend the reproductive lifespan of individuals beyond their biological life and prevent the loss of valuable individuals to the gene pool. Several such GRBs are already in existence. These include for example the

Frozen Ark Consortium (http://www.frozenark.org/; Clarke, 2009), the Amphibian Ark (http://www.amphibianark.org/) and the Biological Resource Bank of Southern Africa's Wildlife (Bartels & Kotze, 2006). Long-term preservation of such biological material is almost entirely a matter of how water therein is dealt with. Plant seeds, whose water content is very low, can easily be preserved at relatively high subzero temperatures of -20°C to -30°C (Ruttimann, 2006) whereas water content in animal tissues and cells is generally very high, in the range of 80%, thus requiring special handling.

About 70 years ago, the late Ernst W. Mayr coined one of the currently leading definitions for a species: "Species are groups of actually or potentially interbreeding natural populations, which are reproductively isolated from other such groups" (de Queiroz, 2005). The uniqueness of reproduction is thus central to the definition of species. It is therefore only natural that we find a wide variety of unique reproductive traits across species. These variations can come in a range of different forms, be it the anatomy of the genital system, morphology of gametes, presence or absence of accessory glands, mechanisms of ovulation, variations in the active hormones, duration of reproductive cycle and gestation and many other aspects of reproductive biology. It is also not surprising that great differences between species are found when it comes to the reaction of their gametes, embryos and tissues to the process of cryopreservation. Thus, when a successful cryopreservation protocol has been devised for a certain species, it will not necessarily be successful in other members of the same family, not to mention species that are phylogenetically further apart. In addition, to date the process of cryopreservation and the mechanisms that cause chilling- and cryodamages are not fully understood. Each new species we approach is thus a *terra incognita* and should be thoroughly studied before a successful protocol can be developed, if at all. Model species are often used for the development of basic techniques but in the end vast experimentation should be conducted in the target species. While this is relatively simple in domestic and laboratory animals, when a rare and endangered species is the target, opportunities to obtain gametes and other relevant cells and tissues are rare and far apart in terms of time and space, making progress extremely slow or practically impossible.

2. The male

The male's gametes are produced in very large numbers and are relatively easy to obtain. A wide variety of collection methods have been devised, including: 1) post-coital vaginal collection, either directly (e.g. O'Brien & Roth, 2000) or with the aid of intra-vaginal condom or vaginal sponge (e.g. Bravo et al., 2000), 2) artificial vagina (e.g. Gastal et al., 1996; Asher et al., 2000), 3) manual stimulation of either the rectum (e.g. Schmitt & Hildebrandt, 1998; Schmitt & Hildebrandt, 2000), the abdomen (Burrows & Quinn, 1937), or through stimulation of the penis (e.g. Schneiders et al., 2004; Melville et al., 2008), 4) electroejaculation (e.g. Hermes et al., 2009b), 5) pharmacologically-induced ejaculation by oral imipramine and intravenous xylazine (McDonnell, 2001) or through urethral catheterization after medetomidine administration (Zambelli et al., 2008), 6) aspiration from the cauda epididymis (e.g. Moghadam et al., 2005), and 7) semen retrieval from the cauda epididymis and proximal portion of the vas deference following castration or post mortem (e.g. Jewgenow et al., 1997; Saragusty et al., 2006; Keeley et al., 2011). Whereas techniques one to three above are relatively close to natural ejaculation, they require easy access to the animal and excessive training (e.g. Robeck & O'Brien, 2004) and are thus limited to only a handful of species and individuals. The pharmacological techniques (5) and aspiration from the epididymis (6) are too invasive to be frequently used, and extraction from the epididymis (7) is a one-time technique, which is often used as a gamete rescue procedure. Epididymal sperm extraction and preservation is a well-documented collection technique. Probably the main advantage of this method is that it enables us to collect sperm post mortem and, if stored, it can be used to extend the reproductive "life span" of that individual. When dealing with endangered species, this may enable us to preserve the spermatozoa of wild and genetically valuable captive males who die in an accident or otherwise. The spermatozoa accumulated in the cauda epididymis is already mature and fertile (Foote, 2000) making it a useful source. Several methods were described as to how to extract the sperm out of the cauda epididymis. These include squeezing the cauda epididymis (Krzywinski, 1981), making cuts in the cauda epididymis (Krzywinski, 1981; Hishinuma et al., 2003; Martinez-Pastor et al., 2006; Saragusty et al., 2006), cutting and squeezing (Quinn & White, 1967), extrusion by air pressure (Kikuchi et al., 1998; Ikeda et al., 2002) and flushing the vas deferens (Martinez-Pastor et al., 2006). Flushing the vas deference, when compared with the cutting method (Martinez-Pastor et al., 2006) , was showed to be superior, yet it seems to be less suitable for field work. For epididymal sperm extraction, spermatozoa stored chilled within the epididymis seem to survive better and for longer periods than those stored in an extender (Ringleb et al., 2011). Still, for *in vivo* sperm collection, electroejaculation became by far the most frequently used method in wildlife species. To be successful, one would need a suitable probe, which often needs to be specifically designed for the animal to be collected based on preliminary knowledge of its anatomy (Hildebrandt et al., 2000; Roth et al., 2005). Even so, ejaculates often come with urine contamination (e.g. Anel et al., 2008) or they may come with or without the relevant secretions from all accessory glands. In elephants this is manifested by occasional ejaculates with very sticky consistency indicating that high level of secretions from the bulbourethral gland are present (personal observation) whereas in rhinoceros it is manifested by high viscosity of the ejaculate (Behr et al., 2009b). In rhinoceros, measuring alkaline phosphatase in the ejaculate was suggested as a mean to identify true ejaculates (Roth et al., 2010). Despite its wide use and success in many species, there is one major drawback to electroejaculation that limits its use on a frequent basis. To conduct electroejaculation, the animal needs to be anesthetized, something many zoos would rather avoid when possible. The need to anesthetize the animal makes it impossible to collect from the same individual on a regular, frequent basis. Anesthesia may also affect the collection procedure (Santiago-Moreno et al., 2010) and the quality of the collected sample (Campion et al., 2011). One should also keep in mind that the collection technique itself may effect the composition of the ejaculate and therefore its quality (Christensen et al., 2011). Thus, the development of preservation protocols for wildlife species progress slowly and often rely on relatively small number of individuals, repeats, and/or ejaculates.

Sperm evaluation also requires understanding of the species under study as sperm competition, for instance, is a major driver behind the wide variety of sperm traits, morphologies and behaviors found in nature (Tourmente et al., 2011). In primates, semen can be as thick as paste, which requires liquefaction and extraction of the cells into a diluent (e.g. Oliveira et al., 2010). In camelids, possibly due to the absence of vesicular glands, sperm is also fairly viscous but it can be enzymatically liquefied (Bravo et al., 2000). Similar enzymatic liquefaction was also helpful when attempting to separate rhinoceros sperm from the seminal plasma, something that cannot be done efficiently with centrifugation alone in some of the ejaculates (Behr et al., 2009b). The volume and concentration also vary by several orders of magnitude among species. In the naked mole rat (*Heterocephalus glaber*) only 5 to 10 µL of sperm can be collected with cells in the hundreds to thousands at the most, many of which are morphologically abnormal (unpublished data). In the European brown hare (*Lepus europaeus*) or the Asiatic black bear (*Ursus thibetanys*), volume of semen collected by electroejaculation is often in the range of 1 mL or less with concentrations that at times can exceed 10^9 cells/mL (personal observations; Chen et al., 2007). Low volume of up to a few mL and low concentration of few millions per mL is often the case in felids both in captivity and in the wild (Barone et al., 1994; Morato et al., 2001). In the pygmy hippopotamus (*Choeropsis liberiensis*) the sperm-rich fraction can be extremely concentrated. In one case we found as much as 9.85×10^9 spermatozoa per mL (Saragusty et al., 2010a). In some other animals volumes can be very large. In boar, donkey or elephant semen can exceed 100 mL with concentrations of several hundred million cells per mL (unpublished data and e.g. Saragusty et al., 2009e; Contri et al., 2010). Initial motility is expected to be low in sperm collected from the epididymis, as epididymal sperm is immotile in most mammals. This is likely to change after a short incubation time in a suitable media. As many of the cells in the epididymis did not complete their maturation process at the time of extraction, cytoplasmic droplets can be highly prevalent (Saragusty et al., 2010b). Some specific characteristics were also noted in certain species. For example the seminal plasma pH of the black flying fox (*Pteropus alecto*) or the snow leopard (*Panthera uncial*) is high (8.2 and 8.4, respectively) (Roth et al., 1996; Melville et al., 2008) or in the Asian elephant (*Elephas maximus*) osmolarity of the seminal plasma is low, at around 270 mOsm/kg (Saragusty et al., 2009e). Such characteristics demonstrate the need to verify multiple aspects of the semen so that suitable diluents can be made. One should always keep in mind that when dealing with endangered species, many were pushed into a bottleneck situation, resulting in highly inbred populations. Inbreeding comes with a very high price with respect to the soundness of the reproductive system. This can be manifested in sperm quality (Roldan et al., 1998; Gomendio et al., 2000; Ruiz-Lopez et al., 2010) and in the outcome in term of litter size and survival (Rabon & Waddell, 2010). Once proper sample of sufficiently good quality is in hand, there are several options for its preservation.

2.1 Semen freezing

Probably the most popular preservation technique is slow freezing of semen or the cells therein. Spermatozoa are generally small in size and thus have low surface to volume ratio, an important factor in cryopreservation, which influences the movement of cryoprotectants and water in and out of the cells. They also have highly condensed and thus stable nucleus and little cytoplasm, making them relatively easy to freeze. Although problems are still numerous and even after more than 60 years of extensive research, propelled primarily by that related to human infertility and livestock and laboratory animal production, our knowledge about the exact mechanisms that eventually lead to success, failure or anywhere in-between is still very limited (Saragusty et al., 2009a). Thus, much of the progress in this field has been primarily empirical in nature (e.g. Saragusty et al., 2009e). Evolution made each species unique in many respects, one of which is the sperm that comes in different shapes, sizes, membrane composition, and sensitivity to chilling, osmotic pressure, pH and more. This means that any new species is an enigma and the specific characteristics of its spermatozoa and seminal plasma and their interaction with various components of freezing extenders and stages of the freezing and thawing process should all be verified. While this can be done relatively easy in domestic and laboratory animals where samples are ample and easy to get, conducting such studies in endangered species is very difficult. Opportunities to obtain samples are rare and often far apart in terms of time and space. Such samples are thus very valuable and using them for experiments rather than for banking would be a waste of important genetic material. Still, to date, semen from probably upward of 200 species from all five major classes of the Vertebrata subphylum (mammals, birds, fish, reptiles and amphibians) have been cryopreserved. When approaching a new species, several hurdles must be overcome before a successful cryopreservation protocol can be developed. The first step is to determine the specific characteristics of its spermatozoa and seminal plasma mentioned above. The next step would be to determine the composition of the freezing extender. Sensitivity to chilling-, freezing- and thawing-associated damages and cryoprotectant-associated toxic or osmotic damages is species-specific and often even individual-specific (Thurston et al., 2002). Similarly, sensitivity to various aspects of sperm handling in preparation for cryopreservation should also be taken into account. In some species it is better to remove the seminal plasma by centrifugation before freezing [e.g. goat, boar, elephant (Saragusty et al., 2009e; unpublished data)] while in others this is not required [e.g. hare or cattle (Hildebrandt et al., 2009; Saragusty et al., 2009c)]. When the seminal plasma is removed, at times adding at least some back after thawing is needed to facilitate fertilization [e.g. camels (Pan et al., 2001)]. Centrifugation is also used for selection of live, morphologically normal cells. When doing so, one needs to understand the basic species-specific sperm characteristics. For example, in opossum (*Monodelphis domestica*) sperm tend to team into pairs to enhance swimming speed (Moore & Taggart, 1995) or in deer mice (genus Peromyscus) sperm form large aggregates (Fisher & Hoekstra, 2010). One should also keep in mind that the fast forward moving population is not necessarily the right one to choose because in some species the slow and steady ones are the cells to eventually win the race (Dziminski et al., 2009). Some species are highly sensitive to glycerol (e.g. mice, boar) while others require concentrations as high as 28% for freezing to be successful, with even higher concentrations to maintain high DNA integrity (e.g. in marsupials: Johnston et al., 1993; Czarny et al., 2009a). In some cases insemination can be done with the thawed sample [e.g. cattle, rhinoceros (Hermes et al., 2009b)] while in others the glycerol should be removed or else fertilization does not occur (e.g. Poitou donkey: Trimeche et al., 1998). So, on the way to developing a successful cryopreservation protocol, species-specific characteristics should be identified and techniques to protect the cells from all these damaging mechanisms should be devised (Zeron et al., 2002; Saragusty et al., 2005; Pribenszky et al., 2006; Saragusty et al., 2009b; Pribenszky & Vajta, 2010). Several cryopreservation techniques were described. These can be divided into field-friendly and - unfriendly ones. The field-friendly techniques include the pellet method [placing a sample drop of ~200 μL directly on carbon dioxide ice ("dry ice")] (Gibson & Graham, 1969), the dry-shipping container technique (Roth et al., 1999), freezing in cold ethanol (Saroff & Mixner, 1955) or in liquid nitrogen vapor (Sherman, 1963; Roussel et al., 1964). The last two being a bit less field-friendly as they require bringing the cold ethanol or liquid nitrogen to the site of work. Still, freezing in liquid nitrogen vapor is currently the most popular one amongst the low-tech, equipment-free freezing techniques. The more sophisticated and more laboratory-bound techniques include the controlled-rate freezing machines (Landa & Almquist, 1979) and the directional freezing machine (Arav, 1999; O'Brien & Robeck, 2006; Si et al., 2006; Saragusty et al., 2007; Reid et al., 2009). When the initial sample is of very poor quality or with very small number of cells, small cell-number or single cell cryopreservation techniques may become useful. Starting in the late 1990's (Cohen & Garrisi, 1997; Cohen et al., 1997), reports on several single sperm cryopreservation techniques showed up in the scientific literature (Walmsley et al., 1998; Gil-Salom et al., 2000; Gvakharia & Adamson, 2001; Just et al., 2004; Herrler et al., 2006; Isaev et al., 2007; Koscinski et al., 2007; Woods et al., 2010). Using these various techniques, researchers reported a wide range of outcomes and recovery efficiency. Time will tell which of these technologies, or others that are currently under development, will emerge as the leading technique that will gain a foothold in sperm banks. Naturally, when sperm banking is considered, single sperm cryopreservation is an option to be considered only if all other possibilities were exhausted. When banking sperm from wildlife, the aim is to bank large number of cells from large number of individuals to ensure availability and variability.

About three decades ago, the thus far only technique that has reached commercial level made it possible to sort sperm according to the sex chromosome they carry (Johnson et al., 1987a; Johnson et al., 1987b). This technique has been tested in various wildlife species such as elephants, rhinoceros, dolphins and non-human primates (O'Brien et al., 2004; O'Brien & Robeck, 2006; Behr et al., 2009a; Behr et al., 2009b; Hermes et al., 2009a). Sperm sorting machines, however, are very expensive, scarce and usually situated far away from where the sperm donor and recipient are located. For this, the double freezing technique has been developed (Arav et al., 2002; Hollinshead et al., 2004; Maxwell et al., 2007; Saragusty et al., 2009c; Montano et al., 2010). This technique allows collection and cryopreservation of sperm sample near the donor, transportation of the frozen sample to the sorting center, thawing it for sorting and then freezing the sorted sample for transportation to the recipient. In this respect the advantage of large volume freezing at the sperm donor site is clear.

Although advances were made over the six decades of sperm cryopreservation history, the basic model that will predict behavior of spermatozoa during cryopreservation is still to be devised. Current knowledge is lacking in many respects and thus, when approaching a new species, much empirical work, often based on trial and error, should be conducted. Thanks to the large number of cells in each ejaculate, these can be split into several treatment modalities, thus speeding up the freezing protocol development process. Once better understanding is attained, and predictions can be made for sperm behavior under various freezing-associated conditions, probably the right course to be taken will be tailor-made, individual-based cryopreservation. This will help overcoming considerable differences between males in response to cryopreservation (Thurston et al., 2002; Saragusty et al., 2007; Loomis & Graham, 2008). However, despite all hurdles, and certainly since ICSI made sperm motility and membrane integrity obsolete, sperm banking under liquid nitrogen is probably the most widely used technique in gametes and tissue banking for reproduction preservation. Success in post-thaw survival, and often also in offspring production, has been demonstrated in many vertebrate species. And yet, there are other options to preserve male fertility.

2.2 Semen vitrification

Ice crystals, both outside and even more so – inside, can be very damaging to any frozen cell or tissue. To avoid ice formation and to minimize the pre-freezing chilling damages, vitrification can be used. Vitrification, also known as ice-free cryopreservation, is a process in which liquid is transformed into an amorphous, glass-like solid, free of any crystalline structures (Luyet, 1937). A major advantage of vitrification over slow freezing is its low-tech, low cost, simple to use, suitable for the field character. For vitrification to be successful, however, much experience in sample handling before cooling and after warming and in loading the sample into or onto the carrier system, are needed. Probability of vitrification depends on the interaction between three factors – cooling rate, sample volume and its viscosity, according to the following general relationships (Saragusty & Arav, 2011):

$$\text{Probability of Vitrification} = \frac{\text{Cooling rate} \times \text{Viscosity}}{\text{Volume}} \tag{1}$$

Thus, to achieve the state of vitrification, very high viscosity (usually attained through high concentrations of cryoprotectants or low water content), and/or very high cooling rates and/or very small volumes are needed. Since the high cryoprotectant concentrations needed are beyond what spermatozoa from most species can tolerate, the vitrified volume is usually being considerably reduced and techniques to achieve high cooling rate with adequate heat transfer throughout the sample are devised. For example, using the cryo-loop vitrification technique, it was calculated that cooling rate as high as 720,000°C/min has been achieved (Isachenko et al., 2004) or with quartz capillaries, cooling rates of around 250,000°C/min were reported (Risco et al., 2007; Lee et al., 2010). The technique, though, have several major drawbacks when sperm banking is considered: 1) The small volume that can be vitrified (at best, presently only a few microliters of semen suspended in vitrification solution) is way too small for banking for species conservation purposes and vitrifying large number of samples from each individual is not practical. 2) The small volume, and thus the small sperm number, make vitrified samples impractical for use in artificial insemination or even in standard IVF. Its optimal utilization is through ICSI, a technique that requires specialized equipment and expertise not available in most laboratories dealing with wildlife, and ICSI has not yet been developed for most species. 3) The risk of contamination through the liquid nitrogen prevails in many of the currently available vitrification carrier devices, which are open systems. 4) High permeable cryoprotectant concentrations (up to 50% compared to 3-7% in slow freezing in most species) are still needed in many of the vitrification protocols despite the reduction in volume. Such concentrations are both toxic and cause osmotic damages to the cells. To overcome this, permeating cryoprotectant-free vitrification techniques were developed through a sizable increase in cooling rate (Nawroth et al., 2002; Isachenko et al., 2003; Merino et al., 2011). Sperm vitrified this way can maintain motility (Isachenko et al., 2004) and resulted recently in human live birth (Sanchez et al., 2011).

2.3 Sperm drying

Storage of cryopreserved samples under liquid nitrogen is very demanding in terms of maintenance, storage space, storage equipment, specially trained personnel and associated costs. Resulting from the need for constant liquid nitrogen supply in large quantities, such storage facilities have very high carbon footprint. The possibility of discontinuation of the liquid nitrogen supply due to human (e.g. conflict, strike) or natural (e.g. earthquake, hurricane) put these facilities at a constant risk. An alternative that would minimize all these is the dry storage. Drying of cells can be done by either freeze-drying or convective-drying. Freeze-drying is achieved by sublimation of the ice after freezing the sample to subzero temperatures. Convective drying, on the other hand, is achieved by placing the sample in a vacuum oven at ambient temperatures. Sperm drying is, however, damaging to cellular membrane and rehydrated cells are often devoid of biological activity, motility and viability. Some degree of chromosomal damage may also take place due to endogenous nucleases. Attempts to freeze-dry spermatozoa were first reported about six decades ago on animal (Polge et al., 1949; Sherman, 1957; Yushchenko, 1957; Meryman & Kafig, 1959) and human (Sherman, 1954) sperm. Most researchers, however, consider all these early reports, dubious. The definitive proof that freeze-dried spermatozoa retain genetic integrity was established only when microsurgical procedures for bypassing the lack of motility of freeze-dried spermatozoa were developed, and normal mice were produced by intracytoplasmic sperm injection (ICSI) of freeze-dried sperm (Wakayama & Yanagimachi, 1998). To date, embryonic development after ICSI with freeze-dried sperm heads has been reported in humans (Katayose et al., 1992; Kusakabe et al., 2008), hamster (Katayose et al., 1992), cattle (Keskintepe et al., 2002; Martins et al., 2007), pigs (Kwon et al., 2004), rhesus macaque (Sanchez-Partida et al., 2008), cats (Moisan et al., 2005; Ringleb et al., 2011) and fish (Poleo et al., 2005), and live offspring were reported in mice (Wakayama & Yanagimachi, 1998; Kaneko et al., 2003; Ward et al., 2003), rabbits (Yushchenko, 1957; Liu et al., 2004), rat (Hirabayashi et al., 2005; Hochi et al., 2008), fish (Poleo et al., 2005) and horses (Choi et al., 2011). Storage at room temperature would be ideal, and at least for mid-range duration it appear to be fine (3 years storage of somatic cells; Loi et al., 2008a). High-temperature storage, however, might be damaging to DNA integrity according to some (Kaneko & Nakagata, 2005; Hochi et al., 2008) but not all (Li et al., 2007; Klooster et al., 2011) researchers. These differences may be due to differences in the drying technique or related to differences between species (Li et al., 2007; Klooster et al., 2011; Kusakabe & Tateno, 2011).

While there are many reports on freeze-drying of sperm and other relevant cells, those on convective drying are scarce. Some researchers, however, consider convective drying to be the better option for the fact that it does not involve the freezing step, thus avoiding freezing-associated damages. This technique has been used to dry fibroblasts, and spermatogonial and hematopoietic stem cells (Katkov et al., 2006; Meyers, 2006).

Regardless of the drying process used, for now sperm drying will usually be placed way behind sperm freezing or vitrification because of the loss of motility and viability, and the need for ICSI. Thus, sperm drying is still to be demonstrated in true wildlife species.

2.4 Liquid phase semen short- to mid-term storage

In many cases, sperm can be collected in the field, away from any fully equipped andrology and cryobiology laboratory or a source for liquid nitrogen, and transferring the samples to a facility for processing may take time. For such cases, or when the sample is destined to be used but not immediately, short- and mid-term supra-zero preservation techniques may help. Nature regularly preserves sperm for months to years in a wide variety of species including members of all vertebrate classes (Holt & Lloyd, 2010; Holt, 2011). The location nature has elected is within the female's reproductive tract. This ability has been described in many species and has been investigated in a few. To date the mechanism has not been discovered although a possible direction has recently emerged. In the greater Asiatic yellow bat (*Scotophilus heathii*), with a regular gap of several months between mating and fertilization, it was shown recently that sperm storage is regulated by androgens (Roy & Krishna, 2011). Administration of flutamine, an androgen antagonist, resulted in loss of sperm storage ability in treated females. It was also suggested that sperm storage duration and survival is the outcome of interplay between expression of B-cell lymphoma 2 (Bcl-2) – an anti-apoptotic factor, and caspase-3 – a promoter of apoptosis. In the absence, as yet, of clear knowledge on how nature does it, *in vitro* techniques were devised in an attempt to achieve this long-term fresh storage goal. In some species, such as the pig, chilled storage is the most widespread method of preservation as thus far sperm cryopreservation has provided only mediocre post-thaw results. When planning on extended chilled storage, several sperm energy-metabolism aspects should be taken into consideration. Both glycolysis and the Krebs cycle play an important role in sperm energy metabolism. Sperm from various species stored in a range of solutions, osmolarities and storage temperatures, were shown to be functional when inject into oocytes after storage of weeks to several months (Kanno et al., 1998; Van Thuan et al., 2005; Riel et al., 2007; Riel et al., 2011). An alternative is to simply leave the spermatozoa inside the epididymides and keep these at 4°C. This epididymal preservation option was demonstrated to produce good results in dogs (Yu & Leibo, 2002), bovine (Martins et al., 2009), gazelles (Saragusty et al., 2006), ram (Tamayo-Canul et al., 2011) and many other species. Short-term epididymal preservation has many advantages when dealing with wildlife. Animals usually have the "tendency" to die at inconvenient time or location. The ability to preserve spermatozoa within the epididymis, till it is transported to a laboratory for processing, helps us buy time for rescue procedures. This can easily be done by non-experts (zoo or park employees for example) by simply cutting off the testicles, putting them in 0.9% saline and keeping them in the refrigerator. Motility preservation for several days can also be done with ejaculated sperm in egg yolk based extenders. For instance, we have recently showed that pygmy hippopotamus (*Choeropsis liberiensis*) spermatozoa preserved some motility for 3 weeks when suspended in the Berliner Cryomedium basic solution (a TEST-egg yolk based extender) (Saragusty et al., 2010a) or, in humans, sperm suspended in PBS supplemented with salts, BSA, antibiotics and glucose had about 15% motility and over 40% viability after 10 days at room temperature (Amaral et al., 2011). During such storage, the reduced metabolism and biological activity, the disintegration of dead cells or the presence of leukocytes in the suspension, all result in the release of reactive oxygen species (ROS) and other damaging components into the solution (Whittington & Ford, 1999). Removal of the leukocytes and periodic exchange of solution should thus be beneficial to the stored cells and extend their life.

2.5 Preservation of other male reproductive-related cells

Spermatozoa, however, can only be potentially retrieved from adult, relatively healthy, individuals but not from sick, azoospermic, or prepubertal ones, and often these carry valuable genetic material that, if not preserved, will be lost for the population. Thanks to

ICSI, even early developmental stages such as elongating or elongated spermatids can be utilized for fertilization. Such cells as testicular spermatozoa and earlier developmental stages can be extracted using testicular sperm extraction (TESE) techniques, and then used through ICSI to fertilize oocytes (Schoysman et al., 1993; Devroey et al., 1995; Kimura & Yanagimachi, 1995; Hewitson et al., 2002). These early-stage cells can be used fresh but they can also be cryopreserved and used at a later stage when needed (Hirabayashi et al., 2008).

Cells of even an earlier developmental stage than the spermatocytes and spermatids are the spermatogonium or spermatogonial stem cells, which can be collected from any male, including infants and juveniles. Infant mortality rate is known to be relatively high in many populations (e.g. Howell-Stephens et al., 2009; Saragusty et al., 2009d) so methods to preserve germ cells from valuable individuals in certainly called for. Spermatogonial stem cells transplantation was first reported in mice (Brinster & Zimmermann, 1994) when it was demonstrated that such transplantation can lead to spermatogenesis. The transplantation technique was later extended to other species such as pigs (Honaramooz et al., 2002), bovine (Izadyar et al., 2003), goats (Honaramooz et al., 2003a; Honaramooz et al., 2003b), cynomolgus monkeys (Schlatt et al., 2002a), and recently to felids as well (Silva et al., 2011). Xenogeneic transplantation, usually from other mammals to nude, immune-deficient mice, has also been reported. However, the further apart (phylogenetically) the donor and recipient species are, the more difficult it becomes. Using this technique, isolated donor testis cells are infused into the seminiferous tubules of the recipient whose testes have been depleted of all germ cells (by irradiation or chemotherapy). The spermatogonial stem cells establish themselves in the testis and through spermatogenesis, produce spermatozoa carrying the donor genetic material. In 2006 the proof that such xenotransplanted cells can actually produce normal, functioning spermatozoa was reported (Shinohara et al., 2006). In their study, spermatogonial stem cells collected from immature rats were transplanted into chemically sterilized mice and the spermatozoa or spermatids collected from the recipient mice produced normal, fertile rat offspring, both when freshly used and following cryopreservation. The donor stem cells can also be grown in culture to generate more cells for transplantation (Nagano et al., 1998) and they can be cryopreserved for future use (Avarbock et al., 1996). Under very complex *in vitro* culture conditions, and with very low efficiency, morphologically normal and even motile spermatozoa were generated from spermatogonial stem cells (Feng et al., 2002; Hong et al., 2004; Stukenborg et al., 2009).

2.6 Testicular tissue cryopreservation

Tissue cryopreservation is more complex than cellular preservation because tissue is composed of more than one cell type and thus of different water and cryoprotectant permeability coefficient values and different sensitivities to chilling and osmotic challenges. Tissue is also larger in volume and thus cryoprotectant penetration is difficult and heat transfer is not uniform, putting the center of the sample at greater risk of intracellular ice formation and death. This is true for testicular tissue, ovarian tissues and many other types of tissues and whole organs. Testicular tissue preservation can be done in one of three basic forms. The tissue can be cryopreserved for future use, it can be cultured *in vitro* for short to mid-term preservation or it can be transplanted. When preserved in the cryopreserved form, one can freeze the whole organ or even the entire animal. Recently it was demonstrated that spermatozoa or spermatids retrieved from reproductive tissues (whole testes or epididymides) frozen for up to one year at -80°C or from whole mice frozen at -20°C for up to 15 years, can produce normal offspring when used, through ICSI, to fertilize mature oocytes (Ogonuki et al., 2006). This success followed a previous, failed, attempt to cryopreserve the entire testis (Yin et al., 2003). The other option is to cryopreserve testicular tissue slices. This technique is widely used today in both adult and pediatric human medicine as a mean to preserve fertility of patients undergoing cancer treatments. To cryopreserve the tissue, it is cut into tiny pieces, usually in the range of 1-2 mm^3 to ensure cryoprotectant penetration, efficient heat transfer and eventual successful grafting. Other alternatives that have been proposed are to mince the tissue and then suspend it in freezing extender to achieve better cryoprotection (Crabbe et al., 1999) or to cut the testicular tissue into thin stripes (e.g. ~9×5×1 mm in sheep) to increase the total number of seminiferous tubules in each graft (Rodriguez-Sosa et al., 2010). Although such tissue samples can be obtained from every individual, infant, juvenile or adult, almost all successful studies to date used immature tissue (Ehmcke & Schlatt, 2008). Like in semen cryopreservation, there are differences between species in the reaction of their testicular tissue to cryoprotectants, chilling and cryopreservation (Schlatt et al., 2002b). The preserved testicular tissue can be handled in several ways. From these tissues, spermatozoa, spermatocytes and round and elongated spermatids can all be retrieved and used to fertilize oocytes through ICSI (Hovatta et al., 1996; Gianaroli et al., 1999). Testicular tissue can also be transplanted back to the donating individual (autografting), to another individual of the same species (allografting) or to individual of a different species, usually to nude or immunodeficient mice (xenografting). After transplantation, the graft may be lost due to tissue rejection or ischemia. If it manages to survive the critical first few days, blood supply will reach the graft, it will be supported by the recipient system and, after some time, will start producing spermatozoa, which can be harvested by surgical excision of all or part of the graft (Schlatt et al., 2002b). Although dependent on the recipient system for support, the spermatogenesis cycle length is assumed to be inherent to the spermatogonial stem cells, which are expected to preserve the donating species spermatogenesis length (Zeng et al., 2006). However other studies showed that in some species, the process is accelerated when their testicular tissue was xenografted into mice (rhesus monkeys; Honaramooz et al., 2004) while in others it is not (domestic cat; Snedaker et al., 2004). Acceleration, when identified, bears special interest for species preservation as it can shorten generation time and thus speed up population growth. This acceleration, however, may also mean abnormal spermatogenesis process that produces abnormal gametes. The sperm produced this way does not go through epididymal maturation process so the only way it can be utilized is by ICSI (Shinohara et al., 2002). One should also keep in mind that it is very costly to keep immunodeficient mice and handle them under germ-free conditions and, of course, repeated transplantations from one mouse to another are required to maintain viable tissue for many years. Still, testicular tissue cryopreservation was done in several species and pregnancies were achieved in mice (Schlatt et al., 2002b; Shinohara et al., 2002), rabbit (Shinohara et al., 2002), human (Hovatta et al., 1996), Djungarian hamsters (Schlatt et al., 2002b) and marmoset monkeys (Schlatt et al., 2002b), to name a few. Testicular tissue can also be cultured *in vitro* to give rise to mature and competent cells. Culture conditions, however, are very complex and, until recently, attempts were encouraging but still unsuccessful (Gohbara et al., 2010). Earlier this year, generation of offspring from such tissues was demonstrated (Sato et al., 2011). In that study, neonatal mouse testicular tissue cultured *in vitro* for over two months (with or without being previously cryopreserved), generated fully competent spermatids and spermatozoa, which led to embryonic development and healthy and reproductive-active offspring production. This exciting development still needs to be evaluated in terms of accuracy of the genetic profile and absence of aneuploidy in haploid cells (Cheung & Rennert, 2011) as well as its applicability to other species. However, the fact that healthy and fertile offspring were produced is very encouraging.

3. The female

In comparison to the male, females' gametes pose several difficulties when it comes to preservation (Table 1). Very small number of gametes is progressing to the more advanced developmental stages during each cycle, and at best only a handful mature and ovulate. When dealing with rare and endangered species in which the number of available individuals for research is extremely limited and often spatially and temporally far apart, progress is very slow and limited by the small numbers. Oocytes and embryos are orders of magnitude larger than spermatozoa, thus bringing down the ratio between surface area and volume. The outcome is slower movement of water and cryoprotectants across the cellular membrane and elevated risk for intracellular ice formation. Unlike in males, *in vivo* collection of oocytes, and to a lesser extent - embryos, is an invasive procedure requiring anesthesia or sedation. Although production of new oocytes exists even in adulthood (Niikura et al., 2009; Tilly et al., 2009), it is very minimal. So, in general terms, the female can be considered as if it is born with a limited life-long supply. Males on the other hand produce sperm continuously, throughout their adult life, sperm that can be collected relatively easy almost any time. All these differences contribute to the fact that while the number of species in which sperm was cryopreserved is in the hundreds, the number of species in which embryo cryopreservation was reported (not all successful) is currently less than 50 and the number of species in which oocyte cryopreservation was attempted is far less than that.

	Male	Female
Gamete size	<10μm (head)	10s-100s μm (species specific)
Numbers	Millions to billions	Few at a time
Production	Continuous	Very limited new production
Accessibility	Easy to collect	Invasive procedure
Collection	Almost any time	In estrous (need monitoring)

Table 1. comparison between male and female gametes in relation to gamete cryopreservation.

Female gametes can be collected at different time points in their maturation process: 1) as mature oocytes, following ovulation (natural or chemically induced), 2) as mature and immature oocytes, by ovum pick up, either transabdominally, transvaginally or transrectally. This can be done during natural estrus cycle or following chemical stimulation, 3) at all developmental stages, mostly immature, following ovariectomy, either when neutering the animal or post mortem, a possibility with time constraint because deterioration is fast *in vitro* and even faster *in vivo* – reasonable quality oocytes can be harvested only up to ~24h after the removal of the ovaries if they were held at 4°C (Wood et al., 1997; Cleary et al., 2001; Personal experience), 4) after fertilization (natural or by artificial insemination), as embryos. This can be done at any stage prior to implantation. The collected oocytes can be at any level of maturation including oocytes found in primordial, preantral or antral follicles, each presenting its own special requirements and sensitivities. Harvesting and preserving oocytes is almost pointless if all other associated assisted reproductive technologies – *in vitro* maturation, *in vitro* fertilization, *in vitro* culture and embryo transfer, are not mastered (at present or in the future) to support it. Female fertility preservation can be done through preservation of oocytes and/or embryos at various developmental stages, as well as by preservation of ovarian tissue or entire ovaries, all of which will be discussed in details in the following sections.

3.1 Oocyte cryopreservation

For decades it was believed that females are born with their life supply of oocytes in their ovaries, all dormant at a very early maturation stage (Zuckerman, 1951). This dogma, however, was recently challenged by a number of studies suggesting that the female gonads retain the ability to regenerate oocytes throughout adulthood, albeit at a very limited number (e.g. Niikura et al., 2009; and reviewed in Tilly et al., 2009). The vast majority of oocytes, however, is already in the ovaries at birth and remains dormant at a very early stage of maturation to adulthood and beyond. Once the female reaches puberty, one or more cohorts of oocytes are selected at each estrus cycle to progress in the maturation process and, depending on the species, one or several oocytes are ovulated. The remaining oocytes in these selected cohorts degenerate or luteinize to form accessory corpora lutea. To be fertilized, an oocyte needs to overcome the meiotic block and progress to the metaphase II (MII) stage of maturation or else only very few oocytes will fertilize (Luvoni & Pellizzari, 2000). Thus, an *in vitro* maturation procedure should be in hand to handle immature oocytes. This process is currently developed for only a handful of species and even for these success is often fairly limited (Krisher, 2004). Furthermore, collection of immature oocytes disrupts the natural maturation process and thus compromises the quality of the oocytes even if they are later matured *in vitro*. During oocyte maturation and follicular growth, the oocyte accumulates large quantities of mRNA and proteins needed for the continuation of meiosis, fertilization and embryonic development. In the absence of the entire supporting system in the *in vitro* culture, production of some of these needed components is hampered. The resulting mature oocytes are therefore of inferior quality when compared to *in vivo* matured oocytes. In seasonal animals, oocytes collected out of the season may show resistance to IVM and IVF (Spindler et al., 2000; Berg & Asher, 2003; Comizzoli et al., 2003). In red deer for example, while about 15% of cleaved oocytes collected during the season (April-July) developed *in vitro* to blastocysts, none have developed if collected after July (Berg & Asher, 2003). Comizzoli et al. (2003) showed that anti-oxidants and FSH in the culture media can overcome this problem in the domestic cat model they have studied. Naturally, *in vitro* fertilization and culture should also be developed so that embryos can be generated for transfer. During the development of such techniques, as well as in those cases when conspecific oocytes are not available, interspecific IVF can be considered. This was done, for example between the mouflon (*Ovis orientalis musimon*) and the domestic sheep

(Ptak et al., 2002) or between some small cat species and the domestic cat (Herrick et al., 2010) or even between a cat and a mouse (Xu et al., 2011). To enhance the number of oocytes collected at any ovum pick-up procedure, hormonal stimulation can be used. This, however, will result in both mature and immature oocytes and the quality of both may be compromised (Blondin et al., 1996; Moor et al., 1998; Takagi et al., 2001). Although to date no morphological or other method is able to accurately predict which oocytes have optimal developmental potential (Coticchio et al., 2004), it is clear that oocyte quality is a major determining factor in the success of IVF (Coticchio et al., 2004; Krisher, 2004; Combelles & Racowsky, 2005), early embryonic survival, the establishment and maintenance of pregnancy, fetal development, and even adult disease (reviewed in Krisher, 2004). Once all these hurdles have been overcome and while keeping in mind the importance of oocyte quality, the next major hurdle to overcome is oocyte cryopreservation.

Oocytes are very different from sperm or embryos with respect to cryopreservation. Oocytes (and embryos) are in the range of three to four orders of magnitude larger than spermatozoa, thus substantially increasing their surface-to-volume ratio and making them sensitive to chilling and susceptible to intracellular ice formation (Arav et al., 1996; Zeron et al., 1999; Chen & Yang, 2009). Oocytes at the MII stage also have a formed fuse that is chilling-sensitive (Chen & Yang, 2009) and their plasma membrane has low (temperature dependent) permeability coefficient, thus making the movement of cryoprotectants and water slower (Jackowski et al., 1980; Ruffing et al., 1993). This, however, may vary between species. Membrane permeability increases after fertilization (Jackowski et al., 1980) and seem to be higher in morula/blastocyst stages as compared to earlier embryonic stages (Jin et al., 2011), thus contributing to the fact that embryos are easier to cryopreserve. The oocyte cytoskeleton is highly sensitive to chilling and gets disorganized at suboptimal temperatures (Trounson & Kirby, 1989). Oocytes also have high cytoplasmic lipid content which increases chilling sensitivity (Ruffing et al., 1993). They have less submembranous actin microtubules (Gook et al., 1993) making their membrane less robust. The meiotic spindle, which has formed by the MII stage, is very sensitive to chilling and may be compromised as well (Ciotti et al., 2009) resulting in uneuploidy (Sathananthan et al., 1988) and oocytes are more susceptible to the damaging effects of reactive oxygen species (Gupta et al., 2010). Many of these parameters change after fertilization, making embryos less chilling sensitive and easier to cryopreserve (Jackowski et al., 1980; Gook et al., 1993; Fabbri et al., 2000). Despite many advances in the field of cryopreservation, oocyte (ovulated, mature or immature) cryopreservation still has a long way to go before it can be routinely utilized in many species. Even in human medicine, fewer than 200 births resulting from cryopreserved oocytes were reported as of 2007 (Edgar & Gook), a number that went up to around 500 by 2009 (Nagy et al.). Yet, despite all these difficulties, some success in oocyte cryopreservation has been reported.

Two main cryopreservation techniques are used for oocyte cryopreservation – slow (equilibrium) freezing and vitrification. In slow freezing, oocytes are exposed to permeating cryoprotectants in the range of 1.0-1.5 M and are frozen, following equilibration and seeding, at a rate of 0.3°C to 0.5°C per minute down to -30°C or lower. Once at the desired temperature they are plunged into liquid nitrogen to vitrify the intra- and extracellular still unfrozen compartments and for storage. Attempts to improve outcome by altering the components of the freezing extender (e.g. replacing sodium chloride with choline chloride; Stachecki et al., 1998a; Stachecki et al., 1998b; Quintans et al., 2002) suggest that there is still some room for improvements in the standard techniques widely in use. Vitrification usually exposes the oocytes to substantially higher concentration of cryoprotectants, in the range of

5.0 to 7.0 M, and cryopreservation is done at cooling rates of 2,500°C per minute or more, depending on the technique used. Vitrification can, however, be achieved even at cryoprotectant concentrations similar to those used for slow freezing if sample volume is small enough and/or cooling rate is high enough to achieve vitrification. One advantage of vitrification over slow freezing, when oocytes are concerned, is the higher survival rate that the fast cooling facilitates. To achieve very high cooling rates, a wide variety of carrier systems were developed (reviewed by Saragusty & Arav, 2011). The small-volume sample, with the carrier, is plunged directly into liquid nitrogen or nitrogen slush. For vitrification to be successful, one should be highly experienced in handling the oocytes throughout the dilution process and in loading them onto or into the carrier system. By cooling liquid nitrogen from its boiling temperature (-196°C) to close to its freezing temperature (-210°C), nitrogen slush is formed. Vitrification in slush gives at least two major advantages. When a sample is inserted into liquid nitrogen, the nitrogen boils and forms an insulation vapor layer around the sample (the Leidenfrost effect). Boiling is considerably reduced when slush is used. Slush also significantly increases the cooling rate. Several studies have demonstrated the superiority of slush over liquid nitrogen (Arav & Zeron, 1997; Isachenko et al., 2001; Beebe et al., 2005; Santos et al., 2006; Lee et al., 2007; Criado et al., 2010) but some found little or no difference (Martino et al., 1996; Cuello et al., 2004; Cai et al., 2005) (Table 2). When cooled at such high cooling rates, oocytes spend very short interval at their lipid phase transition temperature, thus avoiding, or at least minimizing, chilling injury (Arav et al., 1996). Vitrification also reduces the loss of mRNA from the cryopreserved oocytes (Chamayou et al., 2011), mRNA that is crucial for embryonic development and beyond.

The first human pregnancy from cryopreserved (by slow freezing), *in vitro* fertilized oocyte was reported in 1986 (Chen, 1986) following success in other (laboratory) species that came a few years earlier, such as mice (Whittingham, 1977) and rat (Kasai et al., 1979) oocytes cryopreserved to -196°C or mice oocytes frozen to -75°C (Parkening et al., 1976). Still, despite several decades of research and many advances in the field, success is very limited and oocyte cryopreservation is still labeled as experimental even in human medicine (Noyes et al., 2010). Cryopreservation can cause cytoskeleton disorganization (Trounson & Kirby, 1989), chromosome and DNA abnormalities (Van Blerkom, 1989), spindle disintegration (Pickering & Johnson, 1987), plasma membrane disruption (Van Blerkom, 1989) and premature cortical granule exocytosis with its related zona pellucida hardening, making it impermeable to spermatozoa (Johnson et al., 1988). It also hamper, at least to some extant, the ability of oocytes to mature *in vitro* after thawing/warming (Rao et al., 2011). When comparing these parameters, in addition to survival rate, oocyte cryopreservation by vitrification seem to be superior to slow freezing, which explains why oocyte vitrification is gradually replacing slow freezing as the leading technique of preservation. Either open or closed carrier systems are used for vitrification. The closed systems are more secure while the open systems can provide higher cooling rates by direct exposure of the sample to liquid nitrogen. The large number of carrier systems (see Saragusty & Arav, 2011 for a most current list) suggests that the field is still developing and even decision if the open or the closed system is better is still under debate. While in most carrier systems, the volume that enables vitrification limits the number of oocytes that can be contained in it to just a few, some carrier systems such as the electron microscope grid (Steponkus et al., 1990) or nylon mesh (Matsumoto et al., 2001) allow simultaneous vitrification of a large number (as many as 65 in one study) of oocytes. Most reports on oocyte vitrification are, however, sporadic in nature and usually on small number of oocytes. The open system [Cryotop and Open Pulled Straw (OPS)] was used to vitrify germinal vesicle-stage oocytes of the minke whale

(*Balaenoptera bonaerensis*) with the Cryotop producing better results in post-warming morphology and rate of maturation (Iwayama et al., 2005) and both carrier systems produced better results compared to an earlier attempt to cryopreserve minke whale oocytes by slow freezing (Asada et al., 2000). Oocytes of the Mexican gray wolf (*Canis lupus baileyi*) and the domestic dog were also vitrified recently using the Cryotop carrier system (Boutelle et al., 2011). Post warming viability was 61% of intact dog oocytes and 57% of intact wolf cells. Open systems were also used to vitrify granulosa-oocyte complexes (GOC) from primary follicles of marsupials. In two different studies the fat-tailed dunnart (*Sminthopsis crassicaudata*) (Czarny et al., 2009b) and the Tasmanian devil (*Sarcophilus harrisii*) (Czarny & Rodger, 2010) GOC were vitrified in self-made OPS. Post-warming viability was about 70% in both studies. Immature oocytes of the lowland gorilla (*Gorilla gorilla gorilla*) were also cryopreserved, using slow freezing. Of the thawed oocytes, 4/6 were morphologically degenerated, one arrested at the GV stage and the other progressed to the MI stage and then arrested (Lanzendorf, 1992). Immature oocytes of chousingha (*Tetracerus quadricorni*) were also vitrified using the OPS as a carrier system but post warming maturation rate (29.4%) was considerably lower than that of fresh oocytes (69.3%) (Rao et al., 2011). What unifies all these studies is the fact that only small number of oocytes were cryopreserved and only *in vitro* post thaw / warming evaluations were conducted.

Species	Model	Liquid nitrogen	Nitrogen slush	Sig.	Refernce
Bovine	MII oocytes	40% cleavage	25% cleavage	NS	(Martino et al., 1996)
Bovine	MII oocytes	28% cleavage	48% cleavage	P<0.05	(Arav & Zeron, 1997)
Ovine	GV-oocytes	25% survival	5% survival	P<0.05	(Isachenko et al., 2001)
Porcine	Early blastocysts	77% survive	95% survive	NS	(Cuello et al., 2004)
Rabbit	MII oocytes	83% survive	82% survive	NS	(Cai et al., 2005)
Porcine	Blastocysts	62% survive	83% survive	P<0.05	
Bovine	MII oocytes	39% survive	48% survive	P<0.05	(Santos et al., 2006)
Mouse	4-cell embryo with biopsy	50% survive	87% survive	P<0.05	(Lee et al., 2007)
Mouse	Blastocysts	10% survive	54% survive	P<0.05	(Yavin et al., 2009)
Rabbit	Morulae	83% develop	92% develop	P<0.05	(Papis et al., 2009)
Mouse	MII oocytes	45% survive	90% survive	P<0.001	(Lee et al., 2010)
Human	MII oocyte	56% survive	92% survive	P<0.05	(Criado et al., 2010)

NS = not significant.

Table 2. When oocytes or embryos from various species were vitrified in liquid nitrogen slush in comparison to regular vitrification, results either showed no difference or, more frequently, that slush was superior.

Immature oocytes seem to be less prone to damages caused by the chilling, freezing and thawing or warming procedures (Arav et al., 1996) and they, too, can be cryopreserved by slow freezing (Luvoni et al., 1997) or vitrification (Arav et al., 1993; Czarny et al., 2009b). Preantral oocytes can be preserved inside the follicle and about 10% seem to be physiologically active after thawing and one week of culture. Of over 16,000 small preantral oocytes recovered from the ovaries of 25 cats, 66.3% were intact after thawing (Jewgenow et al., 1998). Before freezing 33.9% of the follicles contained viable oocytes while after thawing there were 19.3% if frozen in Me_2SO and 18.5% if frozen in 1,2-propanediol. However, culture conditions that will allow these oocytes to grow and reach full maturation are still largely unknown despite attempts in several species (Jewgenow et al., 1998; Nayudu et al., 2003). For example, in the marmoset monkey, oocytes collected from secondary pre-antral follicles of either mature or pre-pubertal females were able to develop *in vitro* to the polar body stage but could not complete the maturation process (Nayudu et al., 2003). The exception is the mouse, in which this was done and embryos were produced following IVF of frozen-thawed primary follicles matured *in vitro* and live young were born after embryo transfer (Carroll et al., 1990). Some, very limited, success was also reported in cats, where following vitrification in 40% ethylene glycol, 3.7% of the *in vitro* matured oocytes were able to develop to the blastocyst stage following IVF (Murakami et al., 2004). The problems associated with maturation of early-stage oocytes *in vitro* are the need to develop the complex endocrine system that support the development at different stages, other culture conditions that will ensure survival (oxygen pressure for example) and, in many species, the duration of time required to keep the follicles in culture – 6 months or more. An alternative to isolated oocyte cryopreservation is cryopreservation of individual primordial follicles and later transplanting them to the ovarian bursa, where they can mature and eventually produce young offspring following natural mating as was shown in mice (Carroll & Gosden, 1993).

Liquid-phase sperm preservation is relatively simple. Doing the same with oocytes was, until recently, much more challenging. A recent report on pig oocytes, however, has demonstrated ambient-temperature (27.5°C) preservation for 3 days with as many as 65% of the GV oocytes maintaining viability and developmental competence (Yang et al., 2010). This study demonstrated that oocyte preservation without freezing for several days is possible and relatively simple. This is of great importance for wildlife as cryopreservation or IVF of oocytes collected from dead animals in the field often cannot be done on the spot. The ability to keep oocytes alive while transporting them to the laboratory will considerably increase the number of possibilities.

3.2 Embryo cryopreservation

As discussed earlier with regards to oocytes, the vast difference in size, components and associated structures between spermatozoa on the one hand and oocytes and embryos on the other make cryopreservation of the latter much more complex. The issue of intracellular ice formation becomes a major concern, even at relatively slow cooling rates. To avoid this from happening, small volume cryopreservation and either high cryoprotectant concentration coupled with very fast cooling rate to achieve a state of vitrification or lower cryoprotectant concentration and slow cooling rate (slow freezing) are utilized. The first report on fertilized eggs cryopreservation was on rabbit fertilized ova frozen to -79°C (Ferdows et al., 1958). Some of these cryopreserved ova resulted in pregnancies after thawing and transfer. This abstract, however, seem not to have been followed by a full peer-reviewed manuscript so it is not clear if those zygotes really froze and resulted in pregnancies. In 1971 another report on successful mouse embryo cryopreservation to -79°C, using 7.5% polyvinylpyrolidine (PVP) as cryoprotectant, was published, reporting post-thaw *in vitro* development to blastocysts and *in vivo* development to day-18 fetuses (Whittingham, 1971). Several researchers tried to repeat these results but none was successful (Whittingham et al., 1972; Wilmut, 1972; Ashwood-Smith, 1986; Leibo & Oda, 1993). The real start of the embryo cryopreservation era can therefore be considered as the year 1972. During that year two groups reported successful cryopreservation of mouse embryos to -196°C (Whittingham et al., 1972; Wilmut, 1972). These reports came more than two decades after Polge et al. (1949) reported their chance observation that led to successful freezing of spermatozoa and opened a new era in cryobiology and assisted reproduction. Despite the decades that went by and numerous studies attempting a plethora of protocols and combinations of cryoprotectants, it is amazing to note that besides modification to cooling rate that came a few years later (Willadsen et al., 1976; Willadsen et al., 1978), the same basic protocol is still in vast use today. From conservation standpoint, embryo cryopreservation has the advantage of preserving the entire genetic complement of both parents. Naturally, a number of both male and female embryos should be stored to ensure representation of both sexes and a wide genetic diversity. Since sexing each embryo before cryopreservation is not practical, a large number of embryos should be preserved to increase the probability for sufficient representation of embryos from both sexes. Cryobanking of embryos can thus help establishing founder population with the aim of eventual reintroduction into the wild (Ptak et al., 2002) or revive isolated small population. However, while millions of offspring were born following the transfer of cryopreserved embryos in humans, cattle, sheep and mice, success is very limited in many other, even closely related species. To date the number of species in which embryo cryopreservation has been reported stands at less than 50 mammals (human, domestic and laboratory animals included), with live birth achieved in only about half of them (Table 3). There are also a few reports on non-mammalian embryo cryopreservation, all of them in fish (Table 3). Looking through the table, one can see that the majority of species in which embryo cryopreservation led eventually to pregnancy and live birth are domestic, companion, and laboratory species and species of commercial value. Only very few are truly wildlife species. Much of the knowledge gained came from studies on model animals since endangered species are too rare and studying them directly is often too difficult or practically impossible. By definition, however, each species has a unique reproductive specialization so, no matter how close we get with the aid of model animals, we must in the end gain access to the target species and verify that what worked in the model also works in the target. For example, studies on the domestic cat helped develop various technologies, which were later used in non-domestic cats (Dresser et al., 1988; Pope et al., 1994; Pope, 2000), or cattle served as a model for other ungulates (Dixon et al., 1991; Loskutoff et al., 1995). Too often direct adaptation is not possible and either adjustments to protocols or complete revision are required, forcing researchers to settle for small animal study population, at times comprised of a single animal (e.g. Robeck et al., 2011), and samples that are hard to come by. As in the case of oocytes, slow freezing and vitrification are currently used for embryo cryopreservation. Unlike oocytes, however, slow freezing has been producing good results so vitrification does not occupy as important a role in embryo cryopreservation as it does with oocytes. Two main sources of embryos can be considered – *in vivo* produced embryos and those produced *in vitro*. These two embryo groups can develop *in vivo* to produce live offspring but the *in vivo* produced embryos seem to be superior to the *in vitro* ones in many respects, including their sturdiness and ability to survive cryopreservation (Rizos et al., 2002). Obtaining *in vivo*-produced embryos from an endangered species for cryopreservation is a difficult ethical question. If pregnancy has already occurred, shouldn't we let it proceed? Still, because of their superiority, *in vivo*-produced embryos were used in many of the studies on embryo cryopreservation in wildlife.

Species	Procedure	Outcome	References
Primates			
Human (*Homo sapiens*)	4 to 8-cell, freezing, 4 to 16-cell freezing	Pregnancy, Pregnancy to term	(Trounson & Mohr, 1983; Zeilmaker et al., 1984)
Baboon (*Papio* sp.)	*In vivo*-produced 6-cell to blastocyst, freezing	Pregnancy to term	(Pope et al., 1984)
Marmoset monkey (*Callithrix jacchus*)	*In vivo*-produced 4 to 10-cells and morulae freezing	Pregnancy to term	(Hearn & Summers, 1986; Summers et al., 1987)
Cynomolgus monkey (*Macaca fascicularis*)	IVF, 4 to 8-cell freezing 2 to 8-cell vitrification	Pregnancy *In vitro* survival	(Balmaceda et al., 1986; Curnow et al., 2002)
Rhesus macaque (*Macaca mulatta*)	IVF, early-stage freezing ICSI blastocysts vitrification	Pregnancy to term Pregnancy to term	(Wolf et al., 1989; Yeoman et al., 2001)
Hybrid macaque [pig-tailed (*Macaca nemestrina*) & lion-tailed (*M. silenus*)]	IVF, 2-cell freezing	Pregnancy to term	(Cranfield et al., 1992)
Western lowland gorilla (*Gorilla gorilla gorilla*)	IVF, 2-cell freezing	Not reported	(Pope et al., 1997a)
Ungulates			
Bovine (*Bos taurus*)	*In vivo*-produced blastocysts freezing	Pregnancy to term	(Wilmut & Rowson, 1973; Willadsen et al., 1978)
Sheep (*Ovis aries*)	*In vivo*-produced morula and blastocyst freezing	Pregnancy to term	(Willadsen et al., 1974, 1976)
Goat (*Capra aegagrus*)	*In vivo*-produced morula and blastocyst freezing	Pregnancy to term	(Bilton & Moore, 1976)

Species	Procedure	Outcome	References
Horse (*Equus caballus*)	*In vivo*-produced blastocysts freezing	Pregnancy to term	(Yamamoto et al., 1982; Slade et al., 1985)
African eland antelope (*Taurotragus oryx*)	Details not provided	Stillbirth	(Kramer et al., 1983; Dresser et al., 1984; both cited in Schiewe, 1991)
Arabian Oryx (*Oryx leucoryx*)	*In vivo*-produced morula freezing	Transferred but no pregnancy	(Durrant, 1983)
Gaur (*Bos gaurus*)	*In vivo*-produced blastocysts freezing	Transferred to both cow and gaur. Pregnancy at day 135 in cow	(Stover & Evans, 1984; Armstrong et al., 1995)
Bongo (*Tragelphus euryceros*)	*In vivo*-produced blastocysts freezing	Transferred but outcome not reported	(Dresser et al., 1985)
Swine (*Sus domestica*)	*In vivo*-produced blastocysts freezing to -35°C and -196°C	Pregnancy to term from -35°C, no pregnancy from -196°C	(Hayashi et al., 1989)
Scimitar-horned Oryx (*Oryx dammah*)	*In vivo*-produced morula and blastocysts freezing	Transferred but no pregnancy	(Schiewe et al., 1991a)
Red deer (*Cervus elaphus*)	*In vivo*-produced blastocysts freezing	Pregnancy by ultrasound	(Dixon et al., 1991)
Suni Antelope (*Neotragus moschatus zuluensis*)	8-cell freezing	Transferred but no pregnancy	(Schiewe, 1991)
Water buffalo (*Bubalis bubalis*)	*In vivo*-produced morula and blastocysts freezing	Pregnancy to term	(Kasiraj et al., 1993)
Fallow deer (*Dama dama*)	*In vivo*-produced morula and blastocysts freezing	Pregnancy by ultrasound at day 45.	(Morrow et al., 1994)
Domestic donkey (*Equus acinus*)	*In vivo*-produced blastocysts freezing	Outcome not reported	(Vendramini et al., 1997)
Dromedary camel (*Camelus dromedarius*)	*In vivo*-produced blastocysts freezing and vitrification	Freezing – pregnancy by ultrasound, vitrification – pregnancy to term	(Skidmore & Loskutoff, 1999; Nowshari et al., 2005)

Species	Procedure	Outcome	References
Wapiti (*Cervus canadensis*)	Details not mentioned	Pregnancy to term	(cited in Rall, 2001)
European mouflon (*Ovis orientalis musimon*)	IVF blastocysts vitrification	Outcome not reported	(Ptak et al., 2002)
Llama (*Lama glama*)	*In vivo*-produced blastocysts freezing and vitrification	Pregnancy by ultrasound after vitrification	(Aller et al., 2002; Lattanzi et al., 2002)
Wood bison (*Bison bison athabascae*)	IVF morula and blastocysts vitrification	Not evaluated	(Thundathil et al., 2007)
Sika deer (*Cervus nippon nippon*)	IVF blastocysts freezing	Pregnancy to term in red deer surrogate hind	(Locatelli et al., 2008)
Carnivores			
Domestic cat (*Felis catus*)	*In vivo*-produced blastocysts freezing	Pregnancy to term	(Dresser et al., 1988)
African wildcat (*Felis silvestris*)	IVF morula and blastocysts freezing	Pregnancy to term in domestic cat	(Pope et al., 2000)
Siberian Tiger (*Panthera tigris altaica*)	IVF 2 to 4-cell freezing and vitrification	*In vitro* development of vitrified only	(Crichton et al., 2000; Crichton et al., 2003)
Blue fox (*Alopex lagopus*)	Frozen and vitrified embryos, stage and source not mentioned	Both transferred and implanted but not carried to term	(cited in Farstad, 2000a)
Ocelot (*Leopardus pardalis*)	IVF (stage not reported) freezing	Pregnancy to term	(Swanson, 2001, 2003)
Tigrina (*Leopardus tigrinus*)	IVF 2 to 8-cell freezing	Not evaluated	(Swanson et al., 2002)
Bobcat (*Lynx rufus*)	*In vivo*-produced blastocyst freezing	Transferred but no pregnancy	(Miller et al., 2002)
European Polecat (*Mustela putorius*)	*In vivo*-produced morula and blastocysts freezing and vitrification.	Pregnancy to term in both cryopreservation techniques	(Lindeberg et al., 2003; Piltti et al., 2004)
Caracal (*Felis caracal* or *Caracal caracal*)	IVF day 5 to 6 freezing	Pregnancy to term	(cited in Swanson, 2003; Pope et al., 2006)
Geoffroy's cat (*Felis geoffroyi*)	Source and technique not mentioned	Outcome not mentioned	(Swanson & Brown, 2004)

Species	Procedure	Outcome	References
Serval (*Leptailurus serval*)	IVF morula and blastocysts freezing	Transferred but no pregnancy	(Pope et al., 2005)
Dog (*Canis lupus familiaris*)	*In vivo*-produced 1-cell to blastocyst vitrification	Pregnancy to term from 8 to 16-cell embryos	(Suzuki et al., 2009)
Clouded leopard (*Neofelis nebulosa*)	IVF and ICSI day-five freezing	Transferred but no pregnancy	(Pope et al., 2009)
Glires			
European rabbit (*Oryctolagus cuniculus*)	Fertilized ova frozen to -79°C, later 4 to 16-cell and morula freezing to -196°C and vitrification.	Confirmed pregnancy and later pregnancy to term by freezing and *in vitro* survival for vitrification	(Ferdows et al., 1958; Bank & Maurer, 1974; Whittingham & Adams, 1974, 1976; Popelkova et al., 2009)
Mouse (*Mus musculus*)	*In vivo*-produced 8-cell freezing	Live fetuses to term	(Whittingham et al., 1972; Wilmut, 1972)
Rat (*Rattus norvegicus*)	*In vivo*-produced 2 to 8-cell freezing and blastocysts vitrification	Confirmed pregnancy on day 18 for freezing, pregnancy to term for vitrification	(Whittingham, 1975; Kono et al., 1988)
Syrian hamster (*Mesocricetus auratus*)	*In vivo*-produced 1-cell to morula freezing and 1 to 2-cell vitrification	Confirmed pregnancy on day 14 for freezing and pregnancy to term for vitrification	(Ridha & Dukelow, 1985; Lane et al., 1999)
Mongolian gerbil (*Moriones unguieulatus*)	*In vivo*-produced 2-cell, morula and blastocyst vitrification	Pregnancy to term	(Mochida et al., 2005)
Marsupials			
Fat-tailed dunnart (*Sminthopsis crassicaudata*)	*In vivo*-produced day 2 to 4 freezing and vitrification	*In vitro* survival in both systems	(Breed et al., 1994)

Species	Procedure	Outcome	References
Others (fish)			
Zebrafish (*Danio rerio*)	6-somite and heartbeat stage vitrification	No viability	(Zhang & Rawson, 1996)
Turbot (*Psetta maxima*)	Tail bud and tail bud free stages vitrification	No viability	(Robles et al., 2003)
Flounder (*Paralichthys olivaceus*)	14-somite to pre-hatching stage vitrification	Post-warming hatching	(Chen & Tian, 2005)
Gilthead seabream (*Sparus aurata*)	Tail bud and tail bud free stages vitrification	No viability	(Cabrita et al., 2006)
Red seabream (*Pagrus major*)	Heart beat stage vitrification	No viability	(Ding et al., 2007)
Cascudo preto (*Rhinelepis aspera*)	Blastoporous closing stage freezing to -8°C for 6h	Post-thaw hatching	(Fornari et al., 2011)

Table 3. Embryo cryopreservation in vertebrates. The table make it clear that attempts were made almost only in mammals and success in terms of pregnancy carried to term was achieved almost only in domestic, laboratory or companion species and species of commercial value.

3.2.1 Mammals

3.2.1.1 Non-human primates

The number of cryopreserved human embryos successfully transferred since the first report on birth resulting from a transfer of a frozen-thawed embryo (Trounson & Mohr, 1983) is probably over half a million. Yet, despite the fact that non-human primates are used as laboratory models for humans in many studies, progress in primate embryo cryopreservation has been very limited (Mazur et al., 2008) and reports are scarce but with promising results. Observing the progress in non-human primate embryo cryopreservation, it seems that in this field humans act as models for other primates rather than the other way around. The first report on the birth of a non-human primate (baboon; *Papio* sp.) following transfer of frozen-thawed embryo came in 1984, about a year after similar report in humans (Pope et al., 1984). Six *in vivo* produced embryos were retrieved and frozen using glycerol as cryoprotectant. All six embryos survived the freeze-thaw procedure and resulted in two pregnancies (33.3%) after being transferred to six recipients. A similar report on cryopreservation of *in vivo* produced embryos in marmoset monkey (*Callithrix jacchus*) showed higher pregnancy rates (Hearn & Summers, 1986; Summers et al., 1987). In one of these studies, for example, 70% (7/10) of cryopreserved four- to 10-cell embryos and 56% (5/9) of cryopreserved morulae resulted in pregnancies (Summers et al., 1987). Five pregnancies of the first and four of the latter were carried to term resulting in six babies in each group. These authors noted that 1.5M Me$_2$SO was superior to 1.0M glycerol; the latter causing severe osmotic damage. Relying on success in IVF followed by embryo transfer (Balmaceda et al., 1984), pregnancies resulting from frozen-thawed IVF-produced embryos in cynomolgus monkeys (*Macaca fascicularis*) were reported (Balmaceda et al., 1986). Fifty-six cynomolgus macaque embryos were cryopreserved at the four- to eight-cell stage using 1.5 M Me$_2$SO as cryoprotectant and the slow-freezing technique. After thawing, 39 embryos (70%) were still viable. Of these, 25 were transferred to nine synchronized recipients 24 to 48 h after ovulation, resulting in three pregnancies. Report on pregnancy carried to term from frozen-thawed transferred embryo in the rhesus macaque (*Macaca mulatta*) came not too long after that (Wolf et al., 1989). Using hormonal stimulation to achieve superovulation, oocytes (68% mature) were retrieved and inseminated *in vitro*. Embryos were then cryopreserved at the three- to six-cell stage following a propanediol-based freezing protocol, originally developed for humans. Embryo post-thaw survival was high (100%; 11/11). After transferring two embryos to each of three recipients during the early luteal phase of spontaneous menstrual cycles, one pregnancy was achieved and was carried to term. The same group also attempted *in vitro* maturation (IVM) of oocytes prior to IVF, freezing and transfer (Lanzendorf et al., 1990). Oocytes collected at the germinal vesicle (GV) stage did not fertilize *in vitro* and fertilization rate of those collected at the metaphase I (MI) stage was low (32%), even if these were matured *in vitro* to the metaphase II (MII) stage. Fertilization rate of oocytes collected at the MII stage was high (93%) and eight embryos frozen and transferred at the two- to six-cell stage to four recipients (two embryos to each) resulted in three pregnancies culminating in the delivery of three twins. Cross-species IVF was also attempted using *in vitro*-matured oocytes from the non-endangered pig-tailed macaque (*Macaca nemestrina*) and sperm from the endangered lion-tailed macaque (*M. silenus*) (Cranfield et al., 1992). Of the 65 oocytes collected, 25 (38%) were fertilized and 15 (24%) have developed to good quality embryos. These embryos were cryopreserved in propandiol-based extender and the slow freezing technique. Nine embryos were transferred to naturally cycling *M. nemestrina* foster mothers, one of which delivered a healthy hybrind male infant. In Western lowland gorilla (*Gorilla gorilla gorilla*), associated *in vitro* techniques (IVM, IVF, IVC) were adopted successfully from humans (Pope et al., 1997a). Of eight embryos at the two-cell stage produced *in vitro*, three were transferred to a single female, leading to a pregnancy and birth of a female infant. The other five embryos were cryopreserved in 1.5 M 1,2-propanediol containing cryoprotectant. Regrettably, cryopreservation outcome was not reported.

Vitrification is a good alternative to the slow freezing. Following the lead of human and laboratory and farm animals' embryo cryopreservation, the use of vitrification was attempted and compared to slow freezing in non-human primates as well (Yeoman et al., 2001; Curnow et al., 2002). Early-stage (two- to eight-cells) cynomolgus macaque embryos were used to compare vitrification using open pulled straw (OPS) as a carrier system to slow freezing (Curnow et al., 2002). Vitrification proved to be inferior to slow freezing in cell survival rate (18 to 29% vs. 82%), embryo survival (26 to 32% vs. 90%) and cleavage rate (29 to 38% vs. 83%). In another study, on rhesus monkey blastocysts cryopreservation, vitrification using the cryoloop as a carrier system was compared to slow freezing (Yeoman et al., 2001). Embryos were produced *in vitro* by ICSI into mature oocytes and then *in vitro* cultured to the blastocyst stage. Cryopreservation was carried out by either the slow freezing technique or vitrification using two different cryoprotectant combinations – 2.8M

Me$_2$SO with 3.6M EG (combination A) or 3.4M glycerol with 4.5M EG (combination B). Similar results were achieved when blastocysts were cryopreserved by slow freezing [8/22 (36.4%) embryos survived and 1/22 (4.5%) hatched following co-culture] or combination A of cryoprotectants [6/16 (37.5%) embryos survived and 1/16 (6.3%) hatched]. In comparison, using vitrification with cryoprotectant combination B 28/33 (84.8%) of the blastocysts survived and 23/33 (69.7%) hatched. This last study not only achieved high embryonic survival using vitrification, it has also demonstrated the suitability of this technique to overcome the problem of advanced-stage embryo preservation. Six embryos vitrified with cryoprotectant combination B and transferred to three recipients (two to each) resulted in a twin pregnancy carried to term.

3.2.1.2 Ungulates

Embryo cryopreservation has reached a commercial level in the cattle industry and to a lesser extant in sheep and goats. According to a report of the International Embryo Transfer Society (IETS), 297,677 in vivo-derived frozen-thawed bovine embryos were transferred in 2008 worldwide, representing 55.2% of all transferred in vivo-derived bovine embryos in that year (Thibier, 2009). There were also 26,914 frozen-thawed IVF embryos, comprising 10.6% of all transferred bovine IVF embryos in 2008. The actual numbers are most probably much larger since not all transfers are reported to IETS. Major Asian countries such as China, India, Korea and Thailand as well as some of the South American countries did not report their activities to IETS and the reports from Oceania are only partial. At least four important factors are responsible for this success: availability of almost unlimited number of oocytes for research, the possibility to collect in vivo produced embryos non-surgically and without the need for anesthesia or sedation, the availability of financial resources to finance overwhelming body of studies and the needs of the cattle industry. Because none of these factors is helping to push studies on endangered ungulates, situation is dramatically less developed in other species of this group. Statements in reviews on assisted reproductive technologies in non-domestic ungulates from only a decade ago were to the effect that by that time only one successful embryo cryopreservation has been achieved (Holt, 2001). Non-domestic ungulates usually do not show discernable signs of estrous and their receptive period is fairly short. This requires a thorough understanding of the estrus cycle endocrine activity, methods for it's monitoring in each species under study and the development of species-specific hormonal administration for ovarian stimulation. As in all other wildlife species, one should always keep in mind that what works for one species not necessarily will also work for another, even closely related species. For example, the bovine IVC protocol works well for the water buffalo (*Bubalis bubalis*) but when this protocol was used for the African buffalo (*Syncerus caffer*), embryos did not develop beyond the morula stage (Loskutoff et al., 1995). Hormonal monitoring can be achieved non-invasively through fecal or urine analysis but even developing such techniques is not always eventless and not always successful (Paris et al., 2008). Hormonal administration requires stress-afflicting activities such as repeated darting, general anesthesia or movement restriction by chute. Thus, progress in this field has been slow and efficiency in in vitro technologies (IVM, IVF, IVC) has been low. For example, in the Kudu (*Tragelaphus* sp.), of 397 oocytes collected, 79 zygotes cleaved yet only two blastocysts were achieved (0.5%) (Loskutoff et al., 1995). Another example is the Mohor gazelle (*Gazella dama mhorr*) in which embryos produced by IVF with frozen-thawed semen did not develop beyond the six- to eight-cell stage (Berlinguer et al., 2008). These studies suggest that while embryo cryopreservation is a technology worthwhile pursuing, other associated technologies should also reach a level of maturation to support it.

The European mouflon (*Ovis orientalis musimon*) is a wild sheep threatened by extinction. During the efforts to develop the necessary assisted reproductive technologies, the domestic sheep was used as a model. Using 25% glycerol and 25% ethylene glycol as cryoprotectants, *in vitro* produced embryos at the expanded blastocyst stage were vitrified (Ptak et al., 2002). Twenty blastocysts were transferred to domestic sheep foster mothers (two embryos each). At 40 days, seven of the sheep were pregnant and three carried the pregnancy to term, delivering four normal mouflon offspring. In another study, *in vivo* produced embryos were vitrified following embryo vitrification protocol developed for sheep (Naitana et al., 1997; Naitana et al., 2000). Of the five vitrified blastocysts, four survived and were transferred to four synchronized domestic sheep ewes, two of which became pregnant and one pregnancy was carried to term. The domestic sheep, and in part the cow as well, acted as a model for the scimitar-horned Oryx (*Oryx dmmah*) as well. After developing the needed methods, including embryos collection, cryopreservation and transfer, in the sheep, the gained knowledge was used in the scimitar-horned Oryx. *In vivo* produced embryos were frozen in propylene glycol or glycerol but no specific results were reported (Wildt et al., 1986). In another, later study performed on scimitar horned Oryx embryos, thirty late morula- to blastocyst-stage embryos were frozen in cryoprotectant containing Me_2SO, glycerol, or propylene glycol, 10 embryos in each (Schiewe et al., 1991a). Survival was higher in the Me_2SO and glycerol groups. Although the majority (67%) of *in vitro*-cultured embryos developed into hatched blastocysts after 48 h, no pregnancies were established following nonsurgical (n = 8) or laparoscopic (n = 1) transfer of the remaining transferable embryos. Another Oryx species in which an attempt to cryopreserve embryos was made is the Arabian Oryx (*Oryx leucoryx*). Morula-stage *in vivo*-produced embryos were collected and one was frozen in 1.5M Me_2SO. After thawing, the embryo was rated as having a good quality grade. It was transferred to a scimitar-horned Oryx foster female but failed to produce a pregnancy following surgical transfer (Durrant, 1983). Another failed attempt concerns cryopreservation of suni antelope (*Neotragus moschatus zuluensis*) eight-cell stage embryos (N. Loskutoff, personal communication cited in Schiewe, 1991). Of the 18 embryos frozen, nine completely degenerated after thawing. The other nine embryos were transferred by laparoscopy despite the fact that all of them exhibited partial blastomere degradation. No pregnancies were achieved. Attempts were also carried out to freeze *in vivo* produced embryos of African eland antelope (*Taurotragus oryx*) and bongo (*Tragelaphus euryceros*) using glycerol as cryoprotectant. Post-thaw evaluations indicated that six of seven eland (Dresser et al., 1984) and bongo (Dresser et al., 1985) embryos were considered viable and of good enough quality for transfer. Damage to the zona pellucida was noted in one of the eland embryos. Only one pregnancy was carried to term but resulted in a stillborn eland offspring due to dystocia. This attempt was followed by subsequent transfer attempts that resulted in a live eland offspring (B.L. Dressen, personal communication cited in Schiewe, 1991).

The red deer (*Cervus elaphus*), an animal of commercial value in various parts of the world, can also act as a model animal for other closely related species. Slow freezing of red deer *in vivo*-produced embryos in 1.4M glycerol followed by embryo transfer in another country resulted in pregnancy rate ranging between 50 and 72% in different farms, with an average pregnancy rate of 61.2% (153/247) (Dixon et al., 1991). In another study, slow freezing was compared to vitrification by the OPS technique and fresh embryos as control (Soler et al., 2007). Pregnancy rates were 64.3% (18/28), 53.3% (8/15) and 70.0% (7/10) for fresh, vitrified and frozen embryos, respectively. The knowledge accumulated through experiments on red deer was used to freeze embryos from fallow deer (*Dama dama*) (Morrow et al., 1994). *In vivo*-produced embryos resulting from AI were collected surgically from fallow deer and transferred either fresh or following cryopreservation to recipients. Pregnancy rate of frozen-thawed embryos was half that of fresh (26% vs. 53%) and the overall efficiency of the program was low (0.9 to 1.0 surrogate pregnancy per donor). Another deer species in which embryos cryopreservation was attempted is sika deer (*Cervus nippon nippon*). Here, too, the protocol developed for the red deer (Dixon et al., 1991) was used. Of 142 oocytes collected following chemical synchronization, 57 (40.1%) cleaved after IVF and 14 of them reached the blastocyst stage. These embryos were cryopreserved by slow freezing and were later transferred (two per recipient) to synchronized red deer hinds. One of the seven recipients delivered a healthy young sika deer fawn after 224 days of pregnancy (Locatelli et al., 2008).

The domestic cow has acted as a model for other members of the Bovinae subfamily. The gaur (*Bos gaurus*), a member of this subfamily living in the forested areas of South and South East Asia is classified in the IUCN red list as vulnerable. Following protocols developed for the cow, nine *in vitro* produced blastocysts were cryopreserved. One embryo was transferred to a domestic cow which was confirmed pregnant on day 135 (Armstrong et al., 1995). Cryopreservation of gaur embryos was reported more than a decade earlier (Stover & Evans, 1984) however that report did not elaborate on the freezing protocol nor was any information provided as to the outcome of the procedure. Another member of this subfamily is the wood bison (*Bison bison athabascae*), a sub species of the North American bison. Using IVM, IVF, IVC and vitrification protocols developed for bovine, *in vitro*-produced embryos were vitrified (Thundathil et al., 2007). Regettably, protocols that works very well for cattle, gave fairly poor results in wood bison. Only 6.9% (11/160) of the embryos reached the blastocyst stage. Morula-stage (n=27) and blastocyst-stage (n=6) embryos were vitrified. Disappointingly, the researchers failed to report on the evaluation of the embryos after warming.

Camelids are seasonal breeders and induced ovulators. *In vivo*-produced embryos of dromedary camel (*Camelus dromedarius*), collected at the blastocyst stage, were vitrified. Post-warming survival and intact morphology were high (92%) and following transfer of 45 embryos (20 during the breeding season and 25 off-season), three pregnancy were obtained, one of which was carried to term (Nowshari et al., 2005). This report follows a previous one in which cryopreserved embryos did not lead to a pregnancy after transfer (Skidmore & Loskutoff, 1999). Among the South American camelids, attempts have reached some level of success in the Llama (*Lama glama*) whose embryos were found to be three- to five-fold larger than bovine embryos of the same stage (Lattanzi et al., 2002). In one attempt, *in vivo* produced hatched blastocysts were either vitrified or frozen slowly (Lattanzi et al., 2002). After 24 h of *in vitro* culture, 64% (21/33) of the vitrified embryos and in 63% (12/19) of the slow freezing embryos re-expanded. In another attempt to vitrify llama embryos, 10/40 embryos re-expanded after warming (von Baer et al., 2002). Three fresh-chilled and two vitrified-warmed embryos were transferred to synchronized recipients but only one of the fresh embryos resulted in a pregnancy. In yet another report from about the same time, by a different group, success was achieved. *In vivo* produced embryos were collected non-surgically and vitrified at the expanded blastocyst stage. Eight embryos were transferred after warming to four recipients (two embryos, each) and two of them became pregnant, delivering two offspring (Aller et al., 2002).

As for other members of this group, some but very modest success have been reported on cryopreservation of domestic species like the horse (Yamamoto et al., 1982; Slade et al., 1985; Barfield et al., 2009; Choi et al., 2009) and swine (Nagashima et al., 1995; Dobrinsky et al., 2000) but very little success have been reported in other species.

3.2.1.3 Carnivores

The order Carnivora includes two suborders – Caniformia (dog-like species) and Feliformia (cat-like species). Similar to cows among the ungulates, the domestic dog (*Canis lupus familiaris*) and cat (*Felis catus*) are representatives of these two suborders and are highly accessible in terms of their frequent use as laboratory animals and the availability of large number of ovaries from neutered or euthanized animals. Still, despite these similarities and their being members of the same order, embryo cryopreservation and all associated technologies are highly developed for cats but lagging far behind in dogs. The domestic cat was found to be a very suitable model for other felid species, which may partially explain why things are more advanced among felids. Felines are induced ovulators (the release of LH that leads to ovulation is induced by mating) and mostly seasonal breeders. The first report on successful IVF and IVC to the blastocyst stage in a cat came in 1977 (Bowen, 1977). Eleven years later the first in-depth study on cat IVF and the first report on birth of live kittens after embryo transfer of cryopreserved, *in vivo*-derived embryos at the morula stage were published (Dresser et al., 1988; Goodrowe et al., 1988). Cryopreservation was carried out using the slow freezing technique and glycerol as cryoprotectant. However, success rate of embryo transfer was relatively low (14.4%, 17/118), most probably because all thawed embryos were transferred, regardless of their grade. Subsequently, production of offspring after transfer of *in vitro*-derived embryos from *in vivo* and *in vitro* matured oocytes and with or without post thaw culture were described (Pope et al., 1994; Wolfe & Wildt, 1996; Pope et al., 1997b; Wood & Wildt, 1997; Pope et al., 2002). Recently it was suggested that removing some of the lipids from the embryo before cryopreservation, a process known as delipidation, result in higher survival rate and higher rates of post-thaw development to morula and blastocyst stages (Tharasanit & Techakumphu, 2010).

Differences between the domestic cat and other feline species still exist and transfer of knowledge is not entirely straightforward. Still, following the success in the domestic cat, maturation and *in vitro* fertilization of oocytes from a large number of feline species was demonstrated (Johnston et al., 1991). This included tiger (*Panthera tigris*), lion (*Panthera leo*), leopard (*Panthera pardus*), jaguar (*Panthera onca*), snow leopard (*Panthera uncia*), puma (*Felis concolor*), cheetah (*Acinonyx jubatus*), clouded leopard (*Neofelis nebulosa*), bobcat (*Lynx rufus*), serval (*Felis serval*), Geoffroy's cat (*Felis geoffroyi*), Temminck's golden cat (*Felis temmincki*), and leopard cat (*Felis bengalensis*). A total of 846 oocytes were recovered from ovaries of 35 individuals from these 13 species, 508 of them were of fair to excellent quality, yet only 4 (0.8%) cleaved – one of leopard using homologous sperm and three of puma using domestic cat sperm. Matured oocytes were achieved in all but fertilization was not achieved in jaguar, cheetah, clouded leopard, bobcat and Temminck's golden cat. In another study, on puma, 6/25 recovered oocytes fertilized and five of them cleaved (Jewgenow et al., 1994). Success in IVF came at about the same time in other species, e.g. in the tiger (Donoghue et al., 1990), the Indian desert cat (*Felis silvestris ornata*) (cited in Pope, 2000) or leopard cat (*Felis bengalensis*) (Goodrowe et al., 1989). Pope (2000) also mentions IVF/ET in African wild cat (*Felis sylvestris lybica*) but pregnancy here ended with stillbirths.

Over the past decade or so, several reports on embryo cryopreservation in felids appeared in the scientific literature. Some investigators, using the domestic cat as a surrogate mother for frozen-thawed embryos of similar-sized wild feline species, produced offspring of ocelot (*Felis pardalis*) (Swanson, 2001) and the African wild cat (Pope et al., 2000). Transfers of frozen-thawed embryos to conspecific recipients have often failed to produce live offspring. In clouded leopard, no pregnancies were achieved with either frozen-thawed or control embryos (Pope et al., 2009). Similarly, frozen-thawed morula-stage cerval embryos failed to result in pregnancies after transfer (Pope et al., 2005). In the bobcat (*Lynx rufus*), out of three transferred embryos – two fresh and one frozen-thawed, one pregnancy (from a fresh embryo) was achieved (Miller et al., 2002). Failure, however, was not a universal phenomenon. In the ocelot (*Felis pardalis*) over 80 IVF embryos, representing 15 founders of the North American population of this species were cryopreserved for safekeeping (Swanson, 2003) and two pregnancies were established following laparoscopic transfer of frozen-thawed embryos (Swanson, 2006). IVF was also carried out in tigrina (*Leopardus tigrinus*), another South American wild felid, and the resulting embryos (n=52) were cryopreserved (Swanson et al., 2002). Regrettably, the researches failed to report on post-thaw evaluation. In caracal (*Felis caracal*) from 452 recovered matured oocytes, 297 embryos were produced. Additional 16 embryos were produced following IVM of 83 oocytes. A total of 109 embryos were cryopreserved using slow freezing. Of nine recipients, three became pregnant and three kittens were delivered (Pope et al., 2006). Vitrification was also attempted in wild felids and was shown to produce superior results as compared to slow freezing. Siberian tiger (*Panthera tigris altaica*) oocytes were collected by laparoscopy from chemically stimulated ovaries. Following IVF with frozen-thawed sperm and IVC to the 2- to 4-cell stage, embryos were cryopreserved by either slow freezing or vitrified. None of the slow freezing embryos survived (0/89). From those vitrified, 46% (32/70) survived (Crichton et al., 2000; Crichton et al., 2003).

Whereas some success has been achieved in felids, situation is lagging far behind in canids and progress has been slow (Farstad, 2000a, b). Associated ART techniques such as IVM, IVF and IVC still face many difficulties and outcome is often unpredictable, most probably because *in vitro* culture media and conditions are not optimized for this group (Rodrigues & Rodrigues, 2006; Mastromonaco & King, 2007). In the vast majority of the studies, dog zygotes did not progress to the advanced embryonic developmental stages – morula and blastocyst (Rodrigues & Rodrigues, 2006). The first successful embryo cryopreservation in dogs, leading to pregnancy after ET, was reported only in 2007 (Abe et al., 2007) and pup delivery following embryo cryopreservation came two years later (Suzuki et al., 2009). This success was later repeated with *in vivo*-produced embryos using vitrification as the cryopreservation method (Abe et al., 2011). Canine females are unique in their reproductive cycle in the fact that the ovulated oocytes are still immature and their maturation may take two or more days (estimated at 48 to 60 h) while in the distal uterine horn. Also unique is the fact that luteinization and the increase in progesterone actually occur before ovulation (Reynaud et al., 2005; Chastant-Maillard et al., 2011; Concannon, 2011). The extra-follicular maturation process has proved hard to mimic and to date *in vitro* maturation and fertilization are not yet developed in dogs. The bitch anatomy makes retrieval of *in vivo*-produced embryos very difficult, leading researchers to resort to a complete surgical removal of the uterus and associated structures, a procedure that limits its application. From the same reason, embryo transfer was also done surgically until the recent development of a non-surgical technique (Abe et al., 2011). Canine oocytes and early-stage embryos also have high lipid content (Reynaud et al., 2005) making their cryopreservation challenging. These multiple factors are responsible for the slow progress in ART developments in canids. Despite extensive search, the only report on embryo cryopreservation in a non-domestic canid found in the scientific literature is a few words on a trial with blue fox (*Alopex lagopus*) embryos. These were cryopreserved by slow freezing and vitrification and were later transferred to recipients. Although no live pups were achieved, two implantation sites from each of the two cryopreservation techniques were found. (Personal communication with H. Lindeberg, cited in Farstad, 2000a).

Some progress has also been reported in other carnivore families. In the Mustilidae family, a member of the caniformia suborder, some species are of commercial value, primarily in the fur industry. These include, for example, the European polecat (*Mustela putorius*) and the American mink (*Mustela vison* or *Neovision vison*). Other members in this family are listed as endangered or critically endangered species, including the black-footed ferret (*Mustela nigripes*) and the European mink (*Mustela lutreola*). The species of commercial value can thus act as models for developing reproduction technologies and for gaining needed knowledge on specific attributes of the Mustelidae family. European polecat, for example, acted as a model for the European mink and the first successful embryo cryopreservation in this family was reported in this species (Lindeberg et al., 2003). Surgically recovered *in vivo*-produced European polecat embryos were cryopreserved by slow freezing and resulted, following surgical transfer, in 3/8 pregnancies and nine pups were delivers out of a total of 93 embryos transferred (9.7%). A second paper by the same group (Piltti et al., 2004) reported on the first successful embryo vitrification in carnivores. Out of 98 European polecat *in vivo*-produced embryos at the morula and blastocyst stages, 50 survived and were transferred to four recipients. Two of the recipients delivered a total of eight pups, a success rate similar to that of slow freezing (8/98; 8.2%). Further improvements came when a different vitrification technique, pipette tip, was used. Using this technique, 43.6% (44/101) of the embryos survived vitrification and resulted in live births (Sun et al., 2008). Vitrified embryos that were cultured for two or 16 h before transfer resulted in success rate (71.3% and 77.4% live births, respectively) similar to that of the control (79.3%) and significantly higher than in embryos cultured for 32 h (25%) and 48 h (7.8%).

3.2.1.4 Glires – rodents and lagomorphs

Mouse was the first animal in which embryo cryopreservation was reported (Whittingham et al., 1972; Wilmut, 1972). Since then work on glires has largely concentrated on mice, rats, gerbils, hamsters and rabbits – all species in extensive laboratory use. The major cryoprotectant used for freezing embryos in this group is Me_2SO. Although vitrification seem to be gradually taking the lead and many studies claim similar results to fresh controls, a recent meta-analysis found that vitrification is still inferior to fresh embryos (Manno III, 2010). It also found that a variety of covariates are associated with vitrified but not fresh embryos. These include issues such as the time lapse between hCG treatment and embryo cryopreservation, maternal age, and the time from hCG treatment to post-warming assessment. These and possibly other factors might be the result of heterogeneity of conditions of the studies included in such analysis but they can also be real factors arising from the process of cryopreservation. In rabbits, using *in vivo* produced embryos and either slow freezing (Bank & Maurer, 1974; Whittingham & Adams, 1974, 1976) or vitrification (Popelkova et al., 2009; Mocè et al., 2010), resulted in fairly high survival (up to 83%) and pregnancy (up to 92%) rates. However, rate of young born was still relatively low, in the range of 7 to 17% (Bank

& Maurer, 1974; Whittingham & Adams, 1974, 1976). In rats both slow freezing and vitrification were attempted, with considerably better results in the latter. *In vivo* produced embryos at the two-, four- and eight-cell stages were recovered and frozen with 3.0M Me₂SO. Post-thaw normal morphology recovery rate ranged between 65% and 68%. Rate of embryos carried to term, however, was low - 11% for two-cell embryos, zero for four-cell embryos and 9% for eight-cell embryos (Whittingham, 1975). In contrast, in the vitrification study, 79% (117/149) of the vitrified *in vivo* produced blastocysts were morphologically normal after warming. These were split between *in vitro* culture (n=48) and transfer to recipient rats (n=69). All cultured embryos progressed to expanded and hatched blastocysts and of the 69 embryos transferred, 41% (n=28) resulted in live pups (Kono et al., 1988). The golden hamster, also known as the Syrian hamster (*Mesocricetus auratus*), is another member of this group in frequent use as a laboratory research subject. *In vivo* produced embryos at the one- and two-cell stages were flushed and vitrified by the cryoloop technique (Lane et al., 1999). Of 216 vitrified two-cell embryos, 54.2% continued development to the morula/blastocyst stage after warming. Such embryos were transferred to two recipients who delivered 6 pups. In another study, *in vivo* produced embryos at the eight-cell stage were vitrified in 250μL straws, following the technique developed for mouse embryos (Mochida et al., 2000). This study evaluated only *in vitro* development and this was fairly poor, as only two out of 37 embryos developed to the blastocyst stage. Similar to the hamster, *in vivo*-produced Mongolian gerbil (*Mesocricetus auratus*) embryos were vitrified in 250μL straws (Mochida et al., 1999). Following vitrification, 155 embryos developed to the blastocyst stage were transferred to 10 synchronized females, 3 of which became pregnant and delivered 15 pups (9.7%). In a follow-up study by the same group it was shown that embryos at later developmental stages (four-cell, morula and blastocyst) can also be vitrified and result in very high post-warming normal morphology (ranging between 87% and 100%) (Mochida et al., 2005). In this last study, after transfer into recipient females, 3% (4/123), 1% (1/102), 5% (4/73), and 10% (15/155) of embryos developed to full-term offspring from vitrified-warmed early two-cell embryos, late two-cell embryos, morulae, and blastocysts, respectively. The general tendency in all glires seem to be the same – post-thaw/warming *in vitro* quality of the embryos is good but when transferred to recipient females, only around 10% of transferred embryos develop to term. The study by Kono and colleagues (1988) with the reported 41% pups delivered is the exception to this rule. When vitrification was attempted, it seems to result in better outcome.

3.2.1.5 Marsupials

Marsupials are very different from eutherian mammals in many respects, attributes related to their oocytes is one of them. Their oocytes are about twice as large as those of humans (about 200 to 250 μm vs. 100 to 120 μm in humans) (Rodger et al., 1992; Breed et al., 1994). The size is probably that large because of the very large yolk sac that occupies much of the cell volume. The much larger volume and the large yolk compartment make their cryopreservation even more difficult than that of the already hard-to-cryopreserve eutherian oocytes. The alternative is to cryopreserve embryos and in that direction only a single report was found (Breed et al., 1994). In that study, *in vivo*-produced embryos of the carnivorous fat-tailed dunnart (*Sminthopsis crassicaudata*) were cryopreserved by slow freezing or vitrification. Me₂SO proved to be not suitable for vitrification of embryos in this species as none of the embryos vitrified with this cryoprotectant cleaved after warming. Embryos cryopreserved by slow freezing or vitrification (with ethylene glycol as cryoprotectant) had similar cleavage rates or 17% and 18%, respectively. Even when morphological examination found embryos to be normal, examination by electron microscopy revealed multiple damages to intracellular components.

3.2.2 Cetaceans

Only very few studies have reported attempts at cryopreservation of marine mammals oocytes and the only ones I was able to locate were on the common minke whale (*Balaenoptera acutorostrata*). These include studies on both slow freezing (Asada et al., 2000; Asada et al., 2001) and vitrification (Iwayama et al., 2005; Fujihira et al., 2006). To date, no study reporting embryo cryopreservation in cetaceans has been published (O'Brien & Robeck, 2010).

3.2.3 Non-mammal vertebrates

Whereas embryo cryopreservation in mammals shows some success, at least in those extensively studied species, situation lagging far behind in all other vertebrates (fishes, birds, reptiles and amphibians). It is true that considerably less efforts have been invested in embryo cryopreservation in most members of these groups, but the more important cause is the different structure embryos in these vertebrates have, difference that complicates their cryopreservation. From the little that has been done in these vertebrates, the vast majority of studies were done on fish (primarily the zebrafish; *Dino rerio*) and to a lesser extent also in amphibians – the two classes with the smaller oocytes among the non-mammalian vertebrates. The ensuing discussion will therefore be primarily on fishes as representatives for these classes. When sex chromosomes are the determination method, as is the case in most vertebrates, either the male or the female can be the heterogametic sex. In mammals the male carry both X- and Y-chromosomes while the female carries two copies of X-chromosome. In birds, on the other hand, it is the female that carry the Z- and W-chromosomes while the male carries two copies of the Z-chromosome. In fishes and amphibians both systems can be found. To have both chromosomes represented, one should aim to at least preserve enough gametes of the heterogametic sex. In many of the non-mammal species this means preserving the female's gametes, which, as will be discussed here, is problematic. Several attributes differentiate oocytes in these classes from those of mammals. To start with, they are considerably larger, resulting in lower surface area to volume ratio. For example, while the diameter of human oocyte is ~120 μm or that of the mouse is ~80 μm, oocyte of the zebrafish is ~750 μm (Selman et al., 1993) or that of the marsh frog (*Rana ridibunda*) is ~1,400 μm (Kyriakopoulou-Sklavounou & Loumbourdis, 1990), oocytes of the American alligator (*Alligator mississippiensis*) are ~4,000 μm (Uribe & Guillette, 2000), those of the pink salmon (*Oncorhynchus gorbuscha*) in the range of 5,150 to 6,340 μm, and the sizes go even higher in snakes such as kingsnakes (genus: *Lampropeltis*) with diameter of about 22,000 μm (Tryon & Murphy, 1982), and birds like the Japanese quail (*Coturnix coturnix japonica*) – 17,000 to 19,000 μm (Callebaut, 1973) or the domestic chicken (*Gallus gallus domesticus*) with a diameter of about 35,000 to 40,000 μm (Schneider, 1992). The consequence of this is relatively poor water and cryoprotectant movement across the cellular membrane during chilling, freezing and thawing. The difference in size also means considerably larger volume of water to vitrify, thus greatly increasing the risk for intracellular ice formation and cell death. Fish embryos contain a large yolk compartment, enclosed in the yolk syncytial layer (YSL). The behavior of the yolk during freezing defer from the behavior of other embryonic compartments, making freezing very complex. These embryos have at least three membrane structures (YSL, plasma membrane of the developing embryo and the chorionic membrane which surrounds the periviteline space) (Kalicharan et al., 1998; Rawson et al., 2000). Each of these membranes has a different permeability coefficient for water and cryoprotectants, resulting, for example, in water permeability in the range of one order of magnitude lower in fish embryos compared to other animals - 0.022 to 0.1 μm $\times$ min^{-1} $\times$ atm^{-1} in zebrafish (Hagedorn et al., 1997a) compared to 0.722 in drosophila (Lin et al., 1989) or 0.43 in mice (Leibo, 1980). To complicate things even further, the different embryonic compartments have different water content and different osmotically inactive water content (Hagedorn et al., 1997b). Since the chorionic membrane can be removed enzymatically (by pronase) and its removal does not hinder embryonic development (Hagedorn et al., 1997c), Hagedorn et al. (1997a) suggested that the YSL was the primary barrier to crtyoprotectants resulting in the yolk sac reaching lower levels of cryoprotection compared to other embryonic compartments. Using magnetic resonance microscopy, they have shown that while no cryoprotectant injected into the yolk was able to leave, some cryoprotectant was able to enter the blastoderm (Hagedorn et al., 1996). Attempts to solve this permeability issue by adding aquaporin 3 water channels to the zebrafish embryonic membranes (Hagedorn et al., 2002) or inserting cryoprotectants into the yolk by microinjection (Janik et al., 2000) were unsuccessful. Efforts to test various permeating and non-permeating cryoprotectants including methanol, Me_2SO, glycerol, 1,2-propanediol, PG, EG, trehalose, and sucrose also took place. Embryos were shown to be very sensitive to glycerol and EG at a concentration of 1.5M, but less so to methanol, Me_2SO or PG (Hagedorn et al., 1997c). Studies also showed that later-stage embryos were less chilling sensitive than early-stage ones and thus probably more suitable for cryopreservation (Zhang & Rawson, 1995). However, attempts to cryopreserve fish embryos by slow freezing or vitrification generally met with lack of success. (reviewed in Robles et al., 2009). For instance, when intact embryos were cryopreserved by slow freezing, only about 2% of the cells in them survived the process (Harvey, 1983). Attempts were also carried out to cryopreserve amphibian (the frog *Xenopus*) oocytes with similar lack of success (Guenther et al., 2006; Kleinhans et al., 2006).

So, if oocytes and embryos are not an option at the moment, the alternatives are blastodermal cells and primordial germ cells. These cells can be cryopreserved by slow freezing (Naito et al., 1992; Naito et al., 1994) or vitrification (Kohara et al., 2008; Higaki et al., 2010) with good over all post-thaw/warming viability. Goose blastodermal cells, cryopreserved by slow freezing resulted in relatively low survival rate of 25% or less, depending on the cryovial used (Patakine Varkonyi et al., 2007). In another study, quail blastodermal cells were isolated, cryopreserved and the thawed viable cells were used to create quail-chicken chimeras (Naito et al., 1992). Chicken primordial germ cells had survival rate of 85.8 $\pm$ 1.2% and 91.2 $\pm$ 2.8% for vitrified-warmed and frozen-thawed cells, respectively with no significant difference between treatments and the control (Kohara et al., 2008). Blastodermal cells can be used to create chimeras, which are organisms made out of cells from two or more donors with different genetic background. Using this system, duck blastodermal cells were injected into the subgerminal cavity of same stage gamma-irradiated chicken embryo to produce duck-chicken chimeras (Li et al., 2002). These chimeras were mated with ducks to produce six duck hatchlings (out of 622 eggs collected) indicating that, albeit at low efficiency, this system can produce offspring of the blastodermal cells donor. The alternative, which seems to have higher potential from conservation point of view, is the preservation of primordial germ cells. These can later be allo- or xenotransplanted to produce viable offspring of the donor. As a demonstration of concept, primordial germ cells from pheasant (*phasianus colchicus*) were injected into the bloodstream of domestic chicken (*Gallus gallus domesticus*) embryos to produce pheasant-chicken chimeras (Kang et al., 2008). Back-crossing chimera males with pheasant females produced 10 pheasant chicks with an efficiency of 17.5%. Chimera offspring were also generated in zebrafish by transplanting GPC from various sources including vitrified embryoid, an aggregate of cells derived form embryonic stem cells (Kawakami et al., 2010). The male chimeras were then mated with normal females through natural spawning to produce offspring.

In conclusion, cryopreservation of embryos in the few mammalian species in which it was attempted shows some, though very limited, success. The situation is much less advanced in all other vertebrates (fish, birds, reptiles and amphibians) where noticeably less efforts have been invested and the challenges are often considerably more complex. In comparison to mammals, embryos in all these classes are usually larger in volume, with large amount of yolk and multiple membranes showing varying permeability to water and cryoprotectants. All these make embryos in these classes highly susceptible to chilling injury and, with the currently available knowledge and techniques, make their cryopreservation extremely complicated and often practically impossible. The alternative approach, at least for now, would therefore be to preserve blastodermal cells and primordial germ cells, which can be transplanted into host embryos to produce offspring.

3.3 Ovarian tissue cryopreservation

Cryopreservation of ovarian tissue has several advantages over oocyte or embryo cryopreservation, but it also comes with its unique complications. As was discussed earlier, in the section on testicular tissue cryopreservation, tissue is a complex structure and thus presenting many difficulties with respect to cryopreservation. Ovarian tissue is available at any time, season, stage in cycle, and age – from fetus to old to deceased. It contains large number of oocytes and, to overcome the problems associated with *in vitro* development and maturation, it can be implanted so that this can take place *in vivo* (Candy et al., 1995) or after partial development *in vivo*, oocytes can be retrieved and matured *in vitro* (Liu et al., 2001). Ovarian tissue also contains premeiotic germ cells, even in aged animals whose ovaries are otherwise devoid of follicles (Niikura et al., 2009). By transplanting such ovaries into recipient young adult animals can help generate new follicles. Attempts to cryopreserve ovarian tissue were reported already in 1951 (Smith & Parkes), only two years after the same group discovered the protective effect of glycerol during freezing (Polge et al., 1949). The first live birth following ovarian tissue freezing and transplantation was reported in mice, in which the tissue was frozen to -79°C (Parrott, 1960). Grafts can be transplanted to the owner of the tissue (autotransplantation), to another member of the species (allotransplantation) or to a member of a different species (xenotransplantation). All three possibilities were successfully used to support follicular development in grafted tissue. When it comes to wildlife conservation, ovarian tissue will not be used in a similar manner to the way it is used in human medicine, namely retransplanted into its donor. Rather, these cryopreserved tissues will be used to collect oocytes by isolation and maturation *in vitro* or by transplanting them to immune deficient host animals (usually mice or rats) that will support oocyte development *in vivo*. Although ovarian tissue grafting is usually done under the kidney's capsule where ample of blood vessels are found, other locations like subcutaneous grafting for easy access have also been reported (Cleary et al., 2003). Transplantation can be to either female or male recipient (Weissman et al., 1999; Snow et al., 2002) and, interestingly, in a study on human ovarian cortex transplantation to non-obese diabetic-severe combined immune deficiency (NOD-SCID) mice, more males (76.5%, 13/17) supported follicular development than females (30%, 6/20) (Weissman et al., 1999). In another study, while more xenografts were retrieved from females, the number of oocytes recovered from each xenograft was higher in those transplanted to males (Snow et al., 2002). Oocytes developed in males, however, showed reduced fertilizing ability and none of the transferred embryos resulted in implantation. The tissue, cut of its blood supply from harvesting till about 48h after transplantation, needs to rely on its surrounding for supply of oxygen and nutrients and removal of CO_2 and other wastes. If not completely lost or rejected, ischemia can thus lead to the death of more than half of the follicles in the graft (Candy et al., 1997). The surviving follicles, though may grow and develop after transplantation, often contain oocytes of suboptimal quality (Kim et al., 2005). Transplanted ovarian tissue, like any transplanted tissue, carries the risk of transmitting diseases from donor to recipients, a risk that is greatly elevated by the need to use immune-deficient recipients to reduce the risk of graft rejection. The alternative to grafting is growing the follicles to maturation *in vitro*. This, however, has been demonstrated thus far only in mice where primordial follicles (Eppig & O'Brien, 1996) or primary follicles (Lenie et al., 2004) were cultured successfully *in vitro*.

The standard cryopreservation protocol, which seems to work for many different species, is cryopreservation of ovarian cortical tissue slices with a size of 1 to 2 mm^3 in cryoprotective solution containing Me$_2$SO, ethylene glycol or 1,2-propanediol. The tissue and the cryoprotective solution are equilibrated at 0°C and then again at -5 to -7°C. Seeding to initiate extracellular freezing is performed and the sample is then cooled at a slow and constant rate of 0.3°C to 0.5°C/min till somewhere between -30°C and -80°C, before being plunged into liquid nitrogen for storage (for review see Paris et al., 2004). An alternative technique proposed a few years ago does not require expensive equipment and is suitable for work under field conditions (Cleary et al., 2003). Following this technique, equilibration is performed on ice, and the sample is then placed in a passive freezing device that is placed on dry ice. Using this device, a cooling rate of about 1°C/min can be achieved. This is faster than optimal cooling rate but still tolerable. When freezing wombat (*Vombatus ursinus*) ovarian cortical tissue slices this way, 134 ± 32 intact follicles per graft were found compared to 214 ± 55 for the controlled-rate freezing machine.

Cryopreserved ovarian tissue, which was later auto-, allo- or xenografted, has been done in a variety of species including humans (Weissman et al., 1999; Gook et al., 2001; Gook et al., 2003; Donnez et al., 2004), non-human primates - rhesus macaque (*Macaca mulatta*) (Lee et al., 2004), cynomolgus macaque (*Macaca fascicularis*) (Schnorr et al., 2002) and common marmoset (*Callytrix jacchus jacchus*) (von Schönfeldt et al., 2011), bovine (Herrera et al., 2002), sheep (Gosden et al., 1994), cats (Gosden et al., 1994; Jewgenow et al., 1997; Bosch et al., 2004; Jewgenow & Paris, 2006; Luvoni, 2006), mice (Parrott, 1960; Liu et al., 2000; Liu et al., 2001), rabbits (Almodin et al., 2004), common wombat (*Vombatus ursinus*) (Wolvekamp et al., 2001; Cleary et al., 2003), African elephant (*Loxodonta Africana*) (Gunasena et al., 1998), Amur leopard (*Panthera pardus orientalis*) and African lion (*Panthera leo*) (Jewgenow et al., 2011), tammar wallaby (*Macropus eugenii*) (Mattiske et al., 2002), and Fat-tailed dunnart

(*Sminthopsis crassicaudata*) (Shaw et al., 1996). The last two are of special interest as they demonstrate that even when xenografting between species so philogentically distant as marsupials and mice, the graft is still supported and oocytes can develop. Primordial oocytes in ovarian tissue are probably less prone to cooling and cryopreservation damages when compared to mature ones because they are smaller in size and they lack zona pellucida. Still, recovery rate is low. In cats, for example, only 10% of the follicles survived freezing, thawing and transplantation-associated ischemia (Bosch et al., 2004). To overcome this low harvesting rate, multiple grafts are required.

The alternative cryopreservation approach that has been applied to gametes and embryos, namely vitrification, has been applied to ovarian tissue as well. Naturally, to achieve good cryoprotectant penetration and proper heat transfer the sample should be thin enough, normally in the range of 1 mm or less. Several groups have experimented with this approach, cryopreserving tissue samples from humans (Isachenko et al., 2009), mice (Salehnia et al., 2002), sheep (Baudot et al., 2007), pig (Gandolfi et al., 2006), cow (Kagawa et al., 2009), goat (Santos et al., 2007), dog (Ishijima et al., 2006) and cynomolgus and rhesus macaques (Yeoman et al., 2005). The general trend in recent years is for similar outcome from slow freezing and vitrification (see recent review by Amorim et al., 2011)

3.4 Whole ovary cryopreservation

Cryopreservation of large volumes, including whole organs, involves several aspects, which make any attempt at cryopreservation a challenge (Arav & Natan, 2009). These difficulties include: 1) the need for efficient heat transfer throughout the tissue. When a thick tissue or whole organs are involved, this is very difficult to accomplish, 2) the need for efficient cryoprotectant penetration to all cells in the tissue. This is challenging because of the tissue thickness and because different cell types in it have different permeability coefficients and different sensitivities. Excessive exposure time may be damaging to some cells in the tissue due to cryoprotectant toxicity while shorter time might not provide sufficient protection to others. Thus, the optimal time slot is to be identified, 3) supercooling (cooling below the solution's freezing point without crystallization) may take place in some parts of the tissue. This may lead to damages from uncontrolled intra- and extracellular ice formation once crystallization occurs, 4) attaining homogenous cooling rate while avoiding the excessive build-up of toxic concentrations of cryoprotectants, 5) during cryopreservation, latent heat is released from the solution. This released heat can induce recrystallization and extend the isothermal stage, resulting in the development of a large temperature difference between the tissue/organ and the surrounding. This may lead to faster-than-optimal cooling once all latent heat has been released, 6) recrystallization may also occur during thawing because of inhomogeneous warming of the sample. Still, if these issues can be overcome, whole ovary presents one very important advantage over ovarian tissue when it comes to cryopreservation. One of the major problems with cryopreserving ovarian cortical tissue is the ischemia the graft goes through when transplanted. This ischemia cause both graft loss and death of large portion of the follicles within surviving grafts. Cryopreserving whole ovary, including its vascular pedicle, can ensure blood supply as soon as the organ has been transplanted (Bromer & Patrizio, 2009). For the grafted ovary to become fully functional, both ovaries of the recipient should be removed (Liu et al., 2008). Grafting the ovary can be done to its natural position or to any other location in the body that may provide easy access. Of course ovary transplanted to another location can produce oocytes that should be harvested for use *in vitro*. First whole ovary cryopreservation reported was in sheep (Revel et al., 2001; Revel et al., 2004). This report used directional freezing technique, which is claimed to provide a solution to many of the issues involved in large volume cryopreservation mentioned above (Arav & Natan, 2009). Most other cryopreservation experiments used controlled-rate freezing equipment to achieve the desired very slow (~0.1°C/min) cooling rate needed. This first report was followed by reports on cryopreserving ovaries of various other species such as rats (Wang et al., 2002; Qi et al., 2008), mice (Liu et al., 2008), bovine (Arav, 2003), pigs (Imhof et al., 2004), human (Bedaiwy et al., 2006) and another study on sheep (Onions et al., 2009) . In some of these studies, pregnancies were achieved and live young were produced. Interestingly, to date transplantation of cryopreserved whole human ovary has not been reported (Bromer & Patrizio, 2009) despite the fact that ovarian transplantation has been in practice for several years now and whole human ovary cryopreservation was attempted by several researchers.

Although vitrification is an attractive procedure for cryopreservation of whole ovaries, the current knowledge in cryobiology is insufficient to overcome the multiple problems involved in large volume vitrification (Fahy et al., 1990), primarily when tissue, rather than suspension, is involved. Keeping in mind the relationship between the three factors determining the probability of vitrification mentioned earlier (see section on semen vitrification and also Saragusty & Arav, 2011), to avoid cryoprotectant toxicity, very high cooling rates and very small sample volume are needed. Attempts at whole ovary vitrification did take place and in some cases, when the ovaries were sufficiently small, were even successful. An attempt to vitrify whole sheep ovary resulted in complete loss of all follicles (Courbiere et al., 2009). On the other hand, in studies on mice and rats, vitrification of whole ovary was successful (Migishima et al., 2003; Hoshina et al., 2009). One study showed acceptable post warming viability by *in vitro* evaluations of mice ovaries (Migishima et al., 2003). In another study follicular growth was demonstrated after autotransplantation under the kidney capsule of vitrified warmed rat ovaries (Sugimoto et al., 2000). In yet another study, live offspring were produced when the donor mice were transgenic so that their ovaries expressed anti-freeze protein type III as an additional mean of cryoprotection (Bagis et al., 2008).

With the big potential whole ovary cryopreservation holds for wildlife conservation, this procedure is yet to be reported in any animal other than laboratory or domestic species.

4. Options equally good for both males and females

Some options, as will be discussed in the following sections, are available for both sexes. These options are still largely experimental in nature, their efficiency is often low and they require well equipped laboratories with highly experienced staff so their widespread implementation in wildlife conservation is probably still years down the road. They are, however, worthy of mentioning because of the great potential they hold. These, and many of the options described in the previous sections, are not and may never become widely used techniques. They are also nowhere near the decades old slow freezing and vitrification and so, to be on the safe side one should probably opt for cryopreservation of gametes and embryos using one of the available techniques. However, by definition endangered species are species whose global population is small and declining. This means that with time the genetic diversity of such populations is dwindling. If we do not set up collections of samples (gametes, embryos, somatic cells, or anything else we can put your hands on) of the genetic diversity, and just sit and wait for some new technology to come by or for breakthrough in one of the still experimental technologies at hand, genetic diversity within species and possibly entire species will be lost for ever. We should therefore aim to create banks that will hold samples from each endangered species on earth and of as wide a diversity of genetic make up as possible in each. Cryopreservation is a more mature technology for this purpose but many other options are advancing and may one day play an important role in long-term banking for wildlife conservation. New and much better technologies may emerge with time but we cannot sit and watch species going extinct and take no action. Collections should be created with any and all possible technologies in mind.

4.1 Somatic cells cryopreservation for SCNT

To produce embryos *in vivo* or *in vitro*, conspecific spermatozoa and good quality oocytes are required, both or either of which often prove very difficult to obtain. An alternative that can circumvent this, at least in part, is preservation of somatic cells, to be later used for somatic cell nuclear transfer (SCNT, Wilmut et al., 1997). In SCNT, also known as cloning, nucleus of a somatic cell is microinjected into enucleated oocyte, which is then grown *in vitro* and can be later transferred to recipient females for development to term, with or without a cryopreservation step in between. Somatic cells from a wide variety of sources can be used for this purpose. Such diverse sources include cells from tissues preserved without cryoprotectant at -80°C for more than a decade, or cells from tissues kept at -20°C for as long as 16 years (Hoshino et al., 2009), cells isolated from mummified animals (Kato et al., 2009), freeze-dried somatic cells (Loi et al., 2008a; Ono et al., 2008; see next section), semen-derived somatic cells (Nel-Themaat et al., 2008a; Nel-Themaat et al., 2008b; Liu et al., 2010), cells collected postmortem (Oh et al., 2008), cell line (Campbell et al., 1996), and of course both fetal and adult cells are suitable for this purpose (Wilmut et al., 1997). SCNT has indeed an obvious potential for the multiplication of rare genotypes (Corley-Smith & Brandhorst, 1999; Loi et al., 2008a; Loi et al., 2008b), but its wide application is prevented by the currently low efficiency in terms of offspring outcome. To date, successful cloning was reported in sheep (Campbell et al., 1996; Wilmut et al., 1997; Loi et al., 2008a; Loi et al., 2008b), cow (Cibelli et al., 1998), mice (Wakayama & Yanagimachi, 1998), goat (Baguisi et al., 1999), pigs (Polejaeva et al., 2000), cats (Shin et al., 2002), dogs (Jang et al., 2007), rabbits (Chesne et al., 2002), ferrets (Li et al., 2006), mule (Woods et al., 2003), horse (Galli et al., 2003), gaur (*Bos gaurus*) (Lanza et al., 2000), buffalo (*Bubalus bubalis*) (Lu et al., 2005; Shi et al., 2007), mouflon (*Ovis orientalis musimon*) (Loi et al., 2001), African wild cat (*Felis silvestris libica*) (Gómez et al., 2003), wolves (*Canis lupus*) (Kim et al., 2007), mountain bongo antelope (*Tragelaphus euryceros isaaci*) (Lee et al., 2003) and eland (*Taurotragus oryx*) (Nel-Themaat et al., 2008b). When dealing with already extinct species, we can anticipate survival of nucleus DNA but not for viable oocytes. The only hope is then to use oocytes from closely related species. Interspecies SCNT (ISCNT), performed by injecting the nucleus from one species into the oocyte of another has also been carried out in a variety of species (for a recent review see Loi et al., 2011a). These include ISCNT from the endangered mouflon to a domestic sheep (*Ovis aries*) (Loi et al., 2001), from red panda (*Ailurus fulgens*) to rabbit (Tao et al., 2009), from sand cat (*Felis margarita*) to domestic cat (Gómez et al., 2008), from Canada lynx (*Lynx canadensis*) to both domestic cat and caracal (*Caracal caracal*) (Gómez et al., 2009), from water buffalo (*Bubalus bubalis*) to cow (*Bos taurus*) (Srirattana et al., 2011), and most strikingly – from a

15,000 year-old wooly mammoth (*Mammuthus primigenius*) to a mouse (Kato et al., 2009). While this technique holds much promise for the resurrection of extinct species and saving those on their way there (Loi et al., 2011b), with the exception of a few sporadic instances, all these attempts at ISCNT did not result in live offspring. Cryopreservation of reproductive tissue or any other viable body tissue or, alternatively, of *in vitro* grown cell cultures is routinely done in many places around the world and enough cells survive the process to be used in SCNT. Furthermore, obtaining tissue samples is usually much simpler than collecting gametes or embryos, so a larger and more diverse collection can be accumulated.

While SCNT has the advantage that no genetic drift takes place because recombination does not occur, when considering SCNT for wildlife species preservation, several important issues should be taken into consideration. First, as mentioned above, suitable enucleated oocytes are required. The availability of such oocytes and the ability to access them should thus be part of the program (Loi et al., 2011b). If conspecific oocytes are not available, the issues of mitochondrial inheritance and nucleus-cytoplasmic incompatibility become a problem and ways to overcome these should be sought for. When the donor and recipient are close enough, some of the donor mitochondria get transferred as well (Gómez et al., 2009; Srirattana et al., 2011). As was demonstrated for the famous sheep, Dolly, the telomere is shorter following SCNT (Shiels et al., 1999). Interestingly, it was recently shown that cloned cows with short telomeres produce normal and healthy offspring with normal telomere length following artificial insemination with sperm from normal bulls (Miyashita et al., 2011). This study suggests that cloning does not interfere with the eventual function of the germ line. Cloned offspring, however, are known to show elevated prevalence of developmental abnormalities and high mortality rate, issues that should be kept in mind when initiating a cloning program (e.g. Lanza et al., 2000). One should also keep in mind that the spermatozoa carry more than just genetic material. They come with a whole load of epigenetic factors important for proper embryonic development (Yamauchi et al., 2011). These are missing when SCNT is performed and might be one of the causes behind the relatively low efficiency of the process. As with cryopreservation of other cells and tissues, storage space and costs and environmental impact are major issue pertaining to liquid nitrogen storage so a cheaper alternative would be very attractive for long-term conservation purposes.

4.2 Somatic cell drying for SCNT

In tissue banking, as in the banking of germ cells and embryos, storage and maintenance costs are always an issue because of the properties of liquid nitrogen. Seeds of plants, having low water content are relatively easy to preserve at high subzero temperatures (-20 to -30°C). With water content of about 80%, preservation of gametes and embryos in the animal kingdom is complicated and species-specific. The use of large quantities of liquid nitrogen for cryopreservation and storage also has its toll on the environment, as the production of liquid nitrogen is energy-intensive, resulting in the release of large quantities of carbon dioxide. An alternative to cryopreservation of somatic cells, then, can be to dry them and store the dry cells at room temperature. While, as was discussed earlier, sperm drying has been achieved in a number of species, the parallel in females, namely oocyte drying, is yet to be demonstrated. Somatic cell drying is thus the way to go when long-term storage for females or of the entire genetic complement is desired. In this respect, the use of sheep freeze-dried somatic cells for SCNT was recently demonstrated (Loi et al., 2008a; Loi et al., 2008b). In their report, utilizing the directional freezing technology, freeze-dried granulosa cells, kept at room temperature for 3 years, were used to direct embryonic development following nuclear transfer into *in vitro* matured enucleated oocytes. The reconstructed oocytes initiated cleavage at similar rates to control embryos generated using fresh granulosa cells. Microsatellite DNA analysis of the cloned blastocysts matched perfectly with the lyophilized donor cells. Later, these results were confirmed by other researchers studying mouse granulosa cells (Ono et al., 2008), human hematopoietic stem and progenitor cells (Buchanan et al., 2010) or porcine fetal fibroblasts (Das et al., 2010). These studies demonstrate for the first time that dry cells maintain the development potential when injected into enucleated oocytes. Naturally, we still have a long way to go before live offspring will be generated using this technology but the potential is there.

4.3 Stem cell preservation

Embryos can be a source for primordial germ cells (PGC) which, as was shown in the zebrafish, can be vitrified, warmed and then transplanted into sterilized recipient blastulae to differentiate into males and females that produced gametes carrying the genetic material of the transplanted PGC donor (Higaki et al., 2010). Such PGC can be transplanted, along with gonadal somatic cells, and develop into normal male or female gonadal tissue with normal spermatogenesis or oogenesis. Both mouse round spermatids and GV oocytes derived from such tissues were able to direct embryonic development to term following ICSI (Matoba & Ogura, 2010). In a recent study on felids (Silva et al., 2011) it was shown that such germ line stem cells can be transplanted to the gonads of a different species and still develop normal early stage gametes. In that study, ocelot (*Leopardus pardalis*) spermatogonial stem cells were transplanted into domestic cat testis and thirteen weeks later ocelot spermatozoa were retrieved from the cat's epididymis.

Going even earlier in the development timeline, embryos can be a source for stem cells. Embryonic stem cells, being pluripotent, can differentiate *in vivo* or *in vitro* into germ cells. They can also be used for nuclear transfer. So, they, too, can be considered an optional venue. In a study on mice, transplanted embryonic stem cells were able to form testicular tissue structures and direct spermatogenesis (Toyooka et al., 2003). These cells, which can be isolated from embryos, can also be cryopreserved (Thomson et al., 1998; Toyooka et al., 2003) or vitrified (Reubinoff et al., 2001; He et al., 2008). Such stem cells can also be derived from embryos generated by nuclear transfer of freeze-dried cells (Ono et al., 2008). Embryonic stem cells can also be derived from isolated blastomeres, and blastomers can also be cryopreserved individually by inserting them into emptied zona pellucida and then vitrifying them (Escriba et al., 2010). If embryonic stem cells are not available, somatic cells can be induced to become embryonic stem cells-like (Takahashi & Yamanaka, 2006), also known as induced pluripotent stem cells or iPS cells (for recent review see: Cox & Rizzino, 2010). Being pluripotent in nature, they are also germ line competent (Okita et al., 2007) and as such can give rise to germ cells of both male and female.

The fantastic options mentioned above are theoretical and speculative in nature when it comes to wildlife preservation as currently these techniques are in their infancy and were adapted thus far only to laboratory animals, and even in these the unknown is still vast.

5. Conclusion

With the dramatically accelerated species extinction rate we see in recent decades, it is our obligation to seek any possible venue to bring this biodiversity loss to a halt and, while attempting to do so, to seek ways to safe-keep gametes, embryos and somatic cells from (ideally) all species on Earth. To our great disadvantage from a cryobiologist standpoint, species are different from each other and preservation techniques almost invariably require species-specific customization. As was discussed here, there are many options for 'putting life on hold'. Cryopreservation is by far the most advanced and widely used technique that has lead to the establishment of several genome resource banks. Within cryopreservation, slow freezing currently holds the leading role but at least for oocytes, and slowly for embryos too, vitrification is gradually replacing it to become the cryopreservation technique of choice. Due to its small size, condensed DNA and little cytoplasm, spermatozoa are relatively easy to cryopreserve and this was already done in hundreds of species. Oocytes and embryos are much more difficult to obtain in large enough numbers to develop the needed protocols and, because of their large size, more difficult to cryopreserve. It is thus not surprising that oocytes or embryos of only a handful of wildlife species have been cryopreserved. An array of other options, including gonadal tissue and whole gonads cryopreservation, freeze-drying of spermatozoa and somatic cells, SCNT and ISCNT, to name just a few, are largely in the developmental or experimental stage and, if matured and improve in efficiency, they hold great promise and will become highly attractive to wildlife conservation and other fields concerned with 'putting life on hold'. These techniques will not replace the basic and well studies equilibrium freezing and vitrification but will help in supporting them as well as in handling cases in which routine cryopreservation cannot be done. While waiting for these and future technologies to mature and improve in efficiency we should strive to preserve whatever we can – gametes, embryos, gonadal tissue, whole gonads, somatic cells, and stem cells – anything we can. Such collections should be from sufficient number of representatives of each species so that we will be well prepared in the unfortunate event that a need will arise. As Benirschke (1984) put it: 'You must collect things for reasons we don't yet understand.'

6. References

Abe, Y, Suwa, Y, Asano, T, Ueta, YY, Kobayashi, N, Ohshima, N, Shirasuna, S, Abdel-Ghani, MA, Oi, M, Kobayashi, Y, Miyoshi, M, Miyahara, K & Suzuki, H. (2011). Cryopreservation of Canine Embryos. *Biology of Reproduction* 84, 2, 363-368.

Abe, Y, Suwa, Y, Lee, DS, Kim, SK & Suzuki, H. (2007). Vitrification of canine oocytes and embryos, and pregnancy after non-surgical tranfer of vitrified embryos. *Biology of Reproduction* 77, 1 supplement, 134 (abstract).

Aller, JF, Rebuffi, GE, Cancino, AK & Alberio, RH. (2002). Successful transfer of vitrified llama (*Lama glama*) embryos. *Animal Reproduction Science* 73, 1-2, 121-127.

Almodin, CG, Minguetti-Camara, VC, Meister, H, Ferreira, JOHR, Franco, RL, Cavalcante, AA, Radaelli, MRM, Bahls, AS, Moron, AF & Murta, CGV. (2004). Recovery of fertility after grafting of cryopreserved germinative tissue in female rabbits following radiotherapy. *Human Reproduction* 19, 6, 1287-1293.

Amaral, A, Paiva, C, Baptista, M, Sousa, AP & Ramalho-Santos, J. (2011). Exogenous glucose improves long-standing human sperm motility, viability, and mitochondrial function. *Fertility and Sterility* 96, 4, 848-850.

Amorim, CA, Curaba, M, Van Langendonckt, A, Dolmans, M-M & Donnez, J. (2011). Vitrification as an alternative means of cryopreserving ovarian tissue. *Reproductive Biomedicine Online* 23, 2, 160-186.

Anel, L, Álvarez, M, Martínez-Pastor, F, Gomes, S, Nicolás, M, Mata, M, Martínez, AF, Borragán, S, Anel, E & de Paz, P. (2008). Sperm cryopreservation in brown bear (*Ursus arctos*): Preliminary aspects. *Reproduction in Domestic Animals* 43, s4, 9-17.

Arav, A; Arav, Amir, assignee. (1999). Device and methods for multigradient directional cooling and warming of biological samples. US patent 5,873,254.

Arav, A. (2003). Large tissue freezing. *Journal of Assisted Reproduction and Genetics* 20, 9, 351.

Arav, A & Natan, Y. (2009). Directional freezing: a solution to the methodological challenges to preserve large organs. *Seminars in Reproductive Medicine* 27, 6, 438-442.

Arav, A, Shehu, D & Mattioli, M. (1993). Osmotic and cytotoxic study of vitrification of immature bovine oocytes. *Journal of Reproduction and Fertility* 99, 2, 353-358.

Arav, A & Zeron, Y. (1997). Vitrification of bovine oocytes using modified minimum drop size technique (MDS) is effected by the composition and the concentration of the vitrification solution and by the cooling conditions. *Theriogenology* 47, 1, 341 (abstract).

Arav, A, Zeron, Y, Leslie, SB, Behboodi, E, Anderson, GB & Crowe, JH. (1996). Phase transition temperature and chilling sensitivity of bovine oocytes. *Cryobiology* 33, 6, 589-599.

Arav, A, Zeron, Y, Shturman, H & Gacitua, H. (2002). Successful pregnancies in cows following double freezing of a large volume of semen. *Reproduction, Nutrition, Development* 42, 6, 583-586.

Armstrong, DL, Looney, CR, Lindsey, BR, Gonseth, CL, Johnson, DL, Williams, KR, Simmons, LG & Loskutoff, NM. (1995). Transvaginal egg retrieval and *in-vitro* embryo production in gaur (*Bos gaurus*) with establishment of interspecies pregnancy. *Theriogenology* 43, 1, 162 (abstract).

Asada, M, Horii, M, Mogoe, T, Fukui, Y, Ishikawa, H & Ohsumi, S. (2000). In vitro maturation and ultrastructural observation of cryopreserved minke whale (*Balaenoptera acutorostrata*) follicular oocytes. *Biology of Reproduction* 62, 2, 253-259.

Asada, M, Wei, H, Nagayama, R, Tetsuka, M, Ishikawa, H, Ohsumi, S & Fukui, Y. (2001). An attempt at intracytoplasmic sperm injection of frozen-thawed minke whale (*Balaenoptera bonaerensis*) oocytes. *Zygote* 9, 4, 299-307.

Asher, GW, Berg, DK & Evans, G. (2000). Storage of semen and artificial insemination in deer. *Animal Reproduction Science* 62, 1-3, 195-211.

Ashwood-Smith, MJ. (1986). The cryopreservation of human embryos. *Human Reproduction* 1, 5, 319-322.

Avarbock, MR, Brinster, CJ & Brinster, RL. (1996). Reconstitution of spermatogenesis from frozen spermatogonial stem cells. *Nature Medicine* 2, 6, 693-696.

Bagis, H, Akkoc, T, Tass, A & Aktoprakligil, D. (2008). Cryogenic effect of antifreeze protein on transgenic mouse ovaries and the production of live offspring by orthotopic transplantation of cryopreserved mouse ovaries. *Molecular Reproduction and Development* 75, 4, 608-613.

Baguisi, A, Behboodi, E, Melican, DT, Pollock, JS, Destrempes, MM, Cammuso, C, Williams, JL, Nims, SD, Porter, CA, Midura, P, Palacios, MJ, Ayres, SL, Denniston, RS, Hayes, ML, Ziomek, CA, Meade, HM, Godke, RA, Gavin, WG, Overstrom, EW & Echelard, Y. (1999). Production of goats by somatic cell nuclear transfer. *Nature Biotechnology* 17, 5, 456-461.

Balmaceda, JP, Heitman, TO, Garcia, MR, Pauerstein, CJ & Pool, TB. (1986). Embryo cryopreservation in cynomolgus monkeys. *Fertility and Sterility* 45, 3, 403-406.

Balmaceda, JP, Pool, TB, Arana, JB, Heitman, TS & Asch, RH. (1984). Successful *in vitro* fertilization and embryo transfer in cynomolgus monkeys. *Fertility and Sterility* 42, 5, 791-795.

Bank, H & Maurer, RR. (1974). Survival of frozen rabbit embryos. *Experimental Cell Research* 89, 1, 188-196.

Barfield, JP, McCue, PM, Squires, EL & Seidel, GE, Jr. (2009). Effect of dehydration prior to cryopreservation of large equine embryos. *Cryobiology* 59, 1, 36-41.

Barnosky, AD, Matzke, N, Tomiya, S, Wogan, GOU, Swartz, B, Quental, TB, Marshall, C, McGuire, JL, Lindsey, EL, Maguire, KC, Mersey, B & Ferrer, EA. (2011). Has the Earth's sixth mass extinction already arrived? *Nature* 471, 7336, 51-57.

Barone, MA, Roelke, ME, Howard, J, Brown, JL, Anderson, AE & Wildt, DE. (1994). Reproductive characteristics of male Florida panthers: Comparative studies from Florida, Texas, Colorado, Latin America, and North American zoos. *Journal of Mammalogy* 75, 1, 150-162.

Bartels, P & Kotze, A. (2006). Wildlife biomaterial banking in Africa for now and the future. *Journal of Environmental Monitoring* 8, 8, 779-781.

Baudot, A, Courbiere, B, Odagescu, V, Salle, B, Mazoyer, C, Massardier, J & Lornage, J. (2007). Towards whole sheep ovary cryopreservation. *Cryobiology* 55, 3, 236-248.

Bedaiwy, MA, Hussein, MR, Biscotti, C & Falcone, T. (2006). Cryopreservation of intact human ovary with its vascular pedicle. *Human Reproduction* 21, 12, 3258-3269.

Beebe, LFS, Cameron, RDA, Blackshaw, AW & Keates, HL. (2005). Changes to porcine blastocyst vitrification methods and improved litter size after transfer. *Theriogenology* 64, 4, 879-890.

Behr, B, Rath, D, Hildebrandt, TB, Goeritz, F, Blottner, S, Portas, TJ, Bryant, BR, Sieg, B, Knieriem, A, de Graaf, SP, Maxwell, WMC & Hermes, R. (2009a). Germany/Australia index of sperm sex sortability in elephants and rhinoceros. *Reproduction in Domestic Animals* 44, 2, 273-277.

Behr, B, Rath, D, Mueller, P, Hildebrandt, TB, Goeritz, F, Braun, BC, Leahy, T, de Graaf, SP, Maxwell, WMC & Hermes, R. (2009b). Feasibility of sex-sorting sperm from the white and the black rhinoceros (*Ceratotherium simum, Diceros bicornis*). *Theriogenology* 72, 3, 353-364.

Benirschke, K. (1984). The frozen zoo concept. *Zoo Biology* 3, 4, 325-328.

Berg, DK & Asher, GW. (2003). New developments reproductive technologies in deer. *Theriogenology* 59, 1, 189-205.

Berlinguer, F, Gonzalez, R, Succu, S, del Olmo, A, Garde, JJ, Espeso, G, Gomendio, M, Ledda, S & Roldan, ER. (2008). *In vitro* oocyte maturation, fertilization and culture after ovum pick-up in an endangered gazelle (*Gazella dama mhorr*). *Theriogenology* 69, 3, 349-359.

Bilton, RJ & Moore, NW. (1976). *In vitro* culture, storage and transfer of goat embryos. *Australian Journal of Biological Sciences* 29, 1-2, 125-129.

Blondin, P, Coenen, K, Guilbault, LA & Sirard, MA. (1996). Superovulation can reduce the developmental competence of bovine embryos. *Theriogenology* 46, 7, 1191-1203.

Bosch, P, Hernandez-Fonseca, HJ, Miller, DM, Wininger, JD, Massey, JB, Lamb, SV & Brackett, BG. (2004). Development of antral follicles in cryopreserved cat ovarian tissue transplanted to immunodeficient mice. *Theriogenology* 61, 2-3, 581-594.

Boutelle, S, Lenahan, K, Krisher, R, Bauman, KL, Asa, CS & Silber, S. (2011). Vitrification of oocytes from endangered Mexican gray wolves (*Canis lupus baileyi*). *Theriogenology* 75, 4, 647-654.

Bowen, RA. (1977). Fertilization *in vitro* of feline ova by spermatozoa from the ductus deferens. *Biology of Reproduction* 17, 1, 144-147.

Bravo, PW, Skidmore, JA & Zhao, XX. (2000). Reproductive aspects and storage of semen in Camelidae. *Animal Reproduction Science* 62, 1-3, 173-193.

Breed, WG, Taggart, DA, Bradtke, V, Leigh, CM, Gameau, L & Carroll, J. (1994). Effect of cryopreservation on development and ultrastructure of preimplantation embryos from the dasyurid marsupial *Sminthopsis crassicaudata*. *Journal of Reproduction and Fertility* 100, 2, 429-438.

Brinster, RL & Zimmermann, JW. (1994). Spermatogenesis following male germ-cell transplantation. *Proceedings of the National Academy of Sciences of the United States of America* 91, 24, 11298-11302.

Bromer, JG & Patrizio, P. (2009). Fertility preservation: the rationale for cryopreservation of the whole ovary. *Seminars in Reproductive Medicine* 27, 6, 465-471.

Buchanan, SS, Pyatt, DW & Carpenter, JF. (2010). Preservation of differentiation and clonogenic potential of human hematopoietic stem and progenitor cells during lyophilization and ambient storage. *PLoS One* 5, 9, e12518.

Burrows, WH & Quinn, JP. (1937). The collection of spermatozoa from domestic fowl and turkey. *Poultry Science* 16, 1, 19-24.

Cabrita, E, Robles, V, Wallace, JC, Sarasquete, MC & Herráez, MP. (2006). Preliminary studies on the cryopreservation of gilthead seabream (*Sparus aurata*) embryos. *Aquaculture* 251, 2-4, 245-255.

Cai, XY, Chen, GA, Lian, Y, Zheng, XY & Peng, HM. (2005). Cryoloop vitrification of rabbit oocytes. *Human Reproduction* 20, 7, 1969-1974.

Callebaut, M. (1973). Correlation between germinal vesicle and oocyte development in the adult Japanese quail (*Coturnix coturnix japonica*). A cytochemical and autoradiographic study. *Journal of Embryology and Experimental Morphology* 29, 1, 145-157.

Campbell, KH, McWhir, J, Ritchie, WA & Wilmut, I. (1996). Sheep cloned by nuclear transfer from a cultured cell line. *Nature* 380, 6569, 64-66.

Campion, SN, Cappon, GD, Chapin, RE, Jamon, RT, Winton, TR & Nowland, WS. (2012). Isoflurane reduces motile sperm counts in the Sprague-Dawley rat. *Drug and Chemical Toxicology* 35, 1, 20-24.

Candy, CJ, Wood, MJ & Whittingham, DG. (1995). Ovary and ovulation: Follicular development in cryopreserved marmoset ovarian tissue after transplantation. *Human Reproduction* 10, 9, 2334-2338.

Candy, CJ, Wood, MJ & Whittingham, DG. (1997). Effect of cryoprotectants on the survival of follicles in frozen mouse ovaries. *Journal of Reproduction and Fertility* 110, 1, 11-19.

Carroll, J & Gosden, RG. (1993). Transplantation of frozen-thawed mouse primordial follicles. *Human Reproduction* 8, 8, 1163-1167.

Carroll, J, Whittingham, DG, Wood, MJ, Telfer, E & Gosden, RG. (1990). Extra-ovarian production of mature viable mouse oocytes from frozen primary follicles. *Journal of Reproduction and Fertility* 90, 1, 321-327.

Ceballos, G & Ehrlich, PR. (2002). Mammal population losses and the extinction crisis. *Science* 296, 5569, 904-907.

Chamayou, S, Bonaventura, G, Alecci, L, Tibullo, D, Di Raimondo, F, Guglielmino, A & Barcellona, ML. (2011). Consequences of metaphase II oocyte cryopreservation on mRNA content. *Cryobiology* 62, 2, 130-134.

Chapin, FSI, Zavaleta, ES, Eviner, VT, Naylor, RL, Vitousek, PM, Reynolds, HL, Hooper, DU, Lavorel, S, Sala, OE, Hobbie, SE, Mack, MC & Diaz, S. (2000). Consequences of changing biodiversity. *Nature* 405, 6783, 234-242.

Chastant-Maillard, S, Viaris de Lesegno, C, Chebrout, M, Thoumire, S, Meylheuc, T, Fontbonne, A, Chodkiewicz, M, Saint-Dizier, M & Reynaud, K. (2011). The canine oocyte: uncommon features of *in vivo* and *in vitro* maturation. *Reproduction, Fertility and Development* 23, 3, 391-402.

Chen, C. (1986). Pregnancy after human oocyte cryopreservation. *Lancet* 1, 8486, 884-886.

Chen, LM, Hou, R, Zhang, ZH, Wang, JS, An, XR, Chen, YF, Zheng, HP, Xia, GL & Zhang, MJ. (2007). Electroejaculation and semen characteristics of Asiatic black bears (*Ursus thibetanus*). *Animal Reproduction Science* 101, 3-4, 358-364.

Chen, S-U & Yang, Y-S. (2009). Slow freezing or vitrification of oocytes: Their effects on survival and meiotic spindles, and the time schedule for clinical practice. *Taiwanese Journal of Obstetrics and Gynecology* 48, 1, 15-22.

Chen, SL & Tian, YS. (2005). Cryopreservation of flounder (*Paralichthys olivaceus*) embryos by vitrification. *Theriogenology* 63, 4, 1207-1219.

Chesne, P, Adenot, PG, Viglietta, C, Baratte, M, Boulanger, L & Renard, JP. (2002). Cloned rabbits produced by nuclear transfer from adult somatic cells. *Nature Biotechnology* 20, 4, 366-369.

Cheung, H-H & Rennert, OM. (2011). Generation of fertile sperm in a culture dish: clinical implications. *Asian Journal of Andrology* 13, 4, 618-619.

Choi, YH, Hartman, DL, Bliss, SB, Hayden, SS, Blanchard, TL & Hinrichs, K. (2009). High pregnancy rates after transfer of large equine blastocysts collapsed via micromanipulation before vitrification. *Reproduction, Fertility and Development* 22, 1, 203 (abstract).

Choi, YH, Varner, DD, Love, CC, Hartman, DL & Hinrichs, K. (2011). Production of live foals via intracytoplasmic injection of lyophilized sperm and sperm extract in the horse. *Reproduction* 142, 4, 529-538.

Christensen, BW, Asa, CS, Wang, C, Vansandt, L, Bauman, K, Callahan, M, Jens, JK & Ellinwood, NM. (2011). Effect of semen collection method on sperm motility of gray wolves (*Canis lupus*) and domestic dogs (*C. l. familiaris*). *Theriogenology* 76, 5, 975-980.

Cibelli, JB, Stice, SL, Golueke, PJ, Kane, JJ, Jerry, J, Blackwell, C, Ponce de Leon, FA & Robl, JM. (1998). Cloned transgenic calves produced from nonquiescent fetal fibroblasts. *Science* 280, 5367, 1256-1258.

Ciotti, PM, Porcu, E, Notarangelo, L, Magrini, O, Bazzocchi, A & Venturoli, S. (2009). Meiotic spindle recovery is faster in vitrification of human oocytes compared to slow freezing. *Fertility and Sterility* 91, 6, 2399-2407.

Clarke, AG. (2009). The Frozen Ark Project: the role of zoos and aquariums in preserving the genetic material of threatened animals. *International Zoo Yearbook* 43, 1, 222-230.

Cleary, M, Paris, MC, Shaw, J, Jenkin, G & Trounson, A. (2003). Effect of ovariectomy and graft position on cryopreserved common wombat (*Vombatus ursinus*) ovarian tissue following xenografting to nude mice. *Reproduction Fertility and Development* 15, 6, 333-342.

Cleary, M, Snow, M, Paris, M, Shaw, J, Cox, SL & Jenkin, G. (2001). Cryopreservation of mouse ovarian tissue following prolonged exposure to an ischemic environment. *Cryobiology* 42, 2, 121-33.

Cohen, J & Garrisi, GJ. (1997). Micromanipulation of gametes and embryos: Cryopreservation of a single human spermatozoon within an isolated zona pellucida. *Human Reproduction Update* 3, 5, 453.

Cohen, J, Garrisi, GJ, Congedo-Ferrara, TA, Kieck, KA, Schimmel, TW & Scott, RT. (1997). Cryopreservation of single human spermatozoa. *Human Reproduction* 12, 5, 994-1001.

Combelles, CMH & Racowsky, C. (2005). Assessment and optimization of oocyte quality during assisted reproductive technology treatment. *Seminars in Reproductive Medicine* 23, 3, 277-284.

Comizzoli, P, Wildt, DE & Pukazhenthi, BS. (2003). Overcoming poor *in vitro* nuclear maturation and developmental competence of domestic cat oocytes during the non-breeding season. *Reproduction* 126, 6, 809-816.

Concannon, PW. (2011). Reproductive cycles of the domestic bitch. *Animal Reproduction Science* 124, 3-4, 200-210.

Contri, A, De Amicis, I, Veronesi, MC, Faustini, M, Robbe, D & Carluccio, A. (2010). Efficiency of different extenders on cooled semen collected during long and short day length seasons in Martina Franca donkey. *Animal Reproduction Science* 120, 1-4, 136-141.

Convention on Biological Diversity. Strategic Plan for the Convention on Biological Diversity. Decision VI/26 of the Conference of the Parties to the Convention on Biological Diversity. Accessed on 13 January, 2010, Available from: <http://www.cbd.int/decision/cop/?id=7200>.

Corley-Smith, GE & Brandhorst, BP. (1999). Preservation of endangered species and populations: a role for genome banking, somatic cell cloning, and androgenesis? *Molecular Reproduction and Development* 53, 3, 363-367.

Coticchio, G, Sereni, E, Serrao, L, Mazzone, S, Iadarola, I & Borini, A. (2004). What criteria for the definition of oocyte quality? *Annals of the New York Academy of Sciences* 1034, 132-144.

Courbiere, B, Caquant, L, Mazoyer, C, Franck, M, Lornage, J & Salle, B. (2009). Difficulties improving ovarian functional recovery by microvascular transplantation and whole ovary vitrification. *Fertility and Sterility* 91, 6, 2697-2706.

Cox, JL & Rizzino, A. (2010). Induced pluripotent stem cells: what lies beyond the paradigm shift. *Experimental Biology and Medicine* 235, 2, 148-158.

Crabbe, E, Verheyen, G, Tournaye, H & Van Steirteghem, A. (1999). Freezing of testicular tissue as a minced suspension preserves sperm quality better than whole-biopsy freezing when glycerol is used as cryoprotectant. *International Journal of Andrology* 22, 1, 43-48.

Cranfield, MR, Berger, NG, Kempske, S, Bavister, BD, Boatman, DE & Ialeggio, DM. (1992). Macaque monkey birth following transfer of *in vitro* fertilized, frozen-thawed embryos to a surrogate mother. *Theriogenology* 37, 1, 197 (abstract).

Criado, E, Albani, E, Novara, PV, Smeraldi, A, Cesana, A, Parini, V & Levi-Setti, PE. (2010). Human oocyte ultravitrification with a low concentration of cryoprotectants by ultrafast cooling: a new protocol. *Fertility and Sterility* 95, 3, 1101-1103.

Crichton, EG, Armstrong, DL, Vajta, G, Pope, CE & Loskutoff, NM. (2000). Developmental competence *in vitro* of embryos produced from Siberian tigers (*Panthera tigris altaica*) cryopreserved by controlled rate freezing versus vitrification. *Theriogenology* 53, 1, 328 (abstract).

Crichton, EG, Bedows, E, Miller-Lindholm, AK, Baldwin, DM, Armstrong, DL, Graham, LH, Ford, JJ, Gjorret, JO, Hyttel, P, Pope, CE, Vajta, G & Loskutoff, NM. (2003). Efficacy of porcine gonadotropins for repeated stimulation of ovarian activity for oocyte retrieval and *in vitro* embryo production and cryopreservation in Siberian tigers (*Panthera tigris altaica*). *Biology of Reproduction* 68, 1, 105-113.

Cuello, C, Gil, MA, Parrilla, I, Tornel, J, Vázquez, JM, Roca, J, Berthelot, F, Martinat-Botté, F & Martínez, EA. (2004). Vitrification of porcine embryos at various developmental stages using different ultra-rapid cooling procedures. *Theriogenology* 62, 1-2, 353-361.

Curnow, EC, Kuleshova, LL, Shaw, JM & Hayes, ES. (2002). Comparison of slow- and rapid-cooling protocols for early-cleavage-stage *Macaca fascicularis* embryos. *American Journal of Primatology* 58, 4, 169-174.

Czarny, NA, Harris, MS, De Iuliis, GN & Rodger, JC. (2009a). Acrosomal integrity, viability, and DNA damage of sperm from dasyurid marsupials after freezing or freeze drying. *Theriogenology* 72, 6, 817-825.

Czarny, NA, Harris, MS & Rodger, JC. (2009b). Dissociation and preservation of preantral follicles and immature oocytes from female dasyurid marsupials. *Reproduction, Fertility and Development* 21, 5, 640-648.

Czarny, NA & Rodger, JC. (2010). Vitrification as a method for genome resource banking oocytes from the endangered Tasmanian devil (*Sarcophilus harrisii*). *Cryobiology* 60, 3, 322-325.

Darwin, C. (1871). *The Descent of Man, and Selection in Relation to Sex*, D. Appleton and Company, New York.

Das, ZC, Gupta, MK, Uhm, SJ & Lee, HT. (2010). Lyophilized somatic cells direct embryonic development after whole cell intracytoplasmic injection into pig oocytes. *Cryobiology* 61, 2, 220-224.

de Queiroz, K. (2005). Ernst Mayr and the modern concept of species. *Proceedings of the National Academy of Sciences of the United States of America* 102, Suppl 1, 6600-6607.

Devroey, P, Liu, J, Nagy, Z, Goossens, A, Tournaye, H, Camus, M, van Steirteghem, A & Silber, S. (1995). Pregnancies after testicular sperm extraction and intracytoplasmic sperm injection in non-obstructive azoospermia. *Human Reproduction* 10, 6, 1457-1460.

Ding, FH, Xiao, ZZ & Li, J. (2007). Preliminary studies on the vitrification of red sea bream (*Pagrus major*) embryos. *Theriogenology* 68, 5, 702-708.

Dixon, TE, Hunter, JW & Beatson, NS. (1991). Pregnancies following the export of frozen red deer embryos from New Zealand to Australia. *Theriogenology* 35, 1, 193 (Abstract).

Dobrinsky, JR, Pursel, VG, Long, CR & Johnson, LA. (2000). Birth of piglets after transfer of embryos cryopreserved by cytoskeletal stabilization and vitrification. *Biology of Reproduction* 62, 3, 564-570.

Donnez, J, Dolmans, MM, Demylle, D, Jadoul, P, Pirard, C, Squifflet, J, Martinez-Madrid, B & van Langendonckt, A. (2004). Livebirth after orthotopic transplantation of cryopreserved ovarian tissue. *Lancet* 364, 9443, 1405-1410.

Donoghue, AM, Johnston, LA, Seal, US, Armstrong, DL, Tilson, RL, Wolf, P, Petrini, K, Simmons, LG, Gross, T & Wildt, DE. (1990). *In vitro* fertilization and embryo development in vitro and *in vivo* in the tiger (*Panthera tigris*). *Biology of Reproduction* 43, 5, 733-744.

Dresser, BL, Gelwicks, EJ, Wachs, KB & Keller, GL. (1988). First successful transfer of cryopreserved feline (*Felis catus*) embryos resulting in live offspring. *Journal of Experimental Zoology* 246, 2, 180-186.

Dresser, BL, Kramer, L, Dalhausen, RD, Pope, CE & Baker, RD. (1984) Cryopreservation followed by successful embryo transfer of African eland antelope. *Proceedings of the 10th International Congress on Animal Reproduction and Artificial Insemination*, University of Illinois at Urbana-Champaign, Illinois, USA, June 10-14, 1984.

Dresser, BL, Pope, CE, Kramer, L, Kuehn, G, Dahlhausen, RD, Maruska, EJ, Reece, B & Thomas, WD. (1985). Birth of bongo antelope (*Tragelaphus euryceros*) to eland antelope (*Tragelaphus oryx*) and cryopreservation of bongo embryos. *Theriogenology* 23, 1, 190 (abstract).

Durrant, BS. (1983). Reproductive studies of the Oryx. *Zoo Biology* 2, 3, 191-197.

Dziminski, MA, Roberts, JD, Beveridge, M & Simmons, LW. (2009). Sperm competitiveness in frogs: slow and steady wins the race. *Proceedings of the Royal Society B-Biological Sciences* 276, 1675, 3955-3961.

Edgar, DH & Gook, DA. (2007). How should the clinical efficiency of oocyte cryopreservation be measured? *Reproduction Biomedicine Online* 14, 4, 430-435.

Ehmcke, J & Schlatt, S. (2008). Animal models for fertility preservation in the male. *Reproduction* 136, 6, 717-723.

Ehrlich, PR & Wilson, EO. (1991). Biodiversity studies: science and policy. *Science* 253, 5021, 758-762.

Eppig, JJ & O'Brien, MJ. (1996). Development *in vitro* of mouse oocytes from primordial follicles. *Biology of Reproduction* 54, 1, 197-207.

Erwin, DH. (2001). Lessons from the past: biotic recoveries from mass extinctions. *Proceedings of the National Academy of Science of the Uunited States of America* 98, 10, 5399-5403.

Escriba, M-J, Grau, N, Escrich, L & Pellicer, A. (2010). Vitrification of isolated human blastomeres. *Fertility and Sterility* 93, 2, 669-671.

Fabbri, R, Porcu, E, Marsella, T, Primavera, MR, Rocchetta, G, Ciotti, PM, Magrini, O, Seracchioli, R, Venturoli, S & Flamigni, C. (2000). Technical aspects of oocyte cryopreservation. *Molecular and Cellular Endocrinology* 169, 1-2, 39-42.

Fahy, GM, Saur, J & Williams, RJ. (1990). Physical problems with the vitrification of large biological systems. *Cryobiology* 27, 5, 492-510.

Farstad, W. (2000a). Assisted reproductive technology in canid species. *Theriogenology* 53, 1, 175-186.

Farstad, W. (2000b). Current state in biotechnology in canine and feline reproduction. *Animal Reproduction Science* 60-61, 375-387.

Feng, L-X, Chen, Y, Dettin, L, Pera, RAR, Herr, JC, Goldberg, E & Dym, M. (2002). Generation and *in vitro* differentiation of a spermatogonial cell line. *Science* 297, 5580, 392-395.

Ferdows, M, Moore, CL & Dracy, AE. (1958). Survival of rabbit ova stored as -79°C. *Journal of Dairy Science* 41, 5, 739 (abstract).

Fisher, HS & Hoekstra, HE. (2010). Competition drives cooperation among closely related sperm of deer mice. *Nature* 463, 7282, 801-803.

Foote, RH. (2000). Fertilizing ability of epididymal sperm from dead animals. *Journal of Andrology* 21, 3, 355.

Fornari, DC, Ribeiro, RP, Streit, D, Godoy, LC, Neves, PR, de Oliveira, D & Sirol, RN. (2011). Effect of cryoprotectants on the survival of cascudo preto (*Rhinelepis aspera*) embryos stored at -8 degrees C. *Zygote* In Press, doi: 10.1017/S0967199411000517.

Fujihira, T, Kobayashi, M, Hochi, S, Hirabayashi, M, Ishikawa, H, Ohsumi, S & Fukui, Y. (2006). Developmental capacity of Antarctic minke whale (*Balaenoptera bonaerensis*) vitrified oocytes following *in vitro* maturation, and parthenogenetic activation or intracytoplasmic sperm injection. *Zygote* 14, 2, 89-95.

Galli, C, Lagutina, I, Crotti, G, Colleoni, S, Turini, P, Ponderato, N, Duchi, R & Lazzari, G. (2003). A cloned horse born to its dam twin. *Nature* 424, 635.

Gandolfi, F, Paffoni, A, Papasso Brambilla, E, Bonetti, S, Brevini, TA & Ragni, G. (2006). Efficiency of equilibrium cooling and vitrification procedures for the cryopreservation of ovarian tissue: comparative analysis between human and animal models. *Fertility and Sterility* 85 Suppl 1, 1150-1156.

Gastal, MO, Henry, M, Beker, AR, Gastal, EL & Goncalves, A. (1996). Sexual behavior of donkey jacks: Influence of ejaculatory frequency and season. *Theriogenology* 46, 4, 593-603.

Gianaroli, L, Magli, MC, Selman, HA, Colpi, G, Belgrano, E, Trombetta, C, Vitali, G & Ferraretti, AP. (1999). Diagnostic testicular biopsy and cryopreservation of testicular tissue as an alternative to repeated surgical openings in the treatment of azoospermic men. *Human Reproduction* 14, 4, 1034-1038.

Gibson, CD & Graham, EF. (1969). The relationship between fertility and post-freeze motility of bull spermatozoa (by pellet freezing) without glycerol. *Journal of Reproduction and Fertility* 20, 1, 155-157.

Gil-Salom, M, Romero, J, Rubio, C, Ruiz, A, Remohì, J & Pellicer, A. (2000). Intracytoplasmic sperm injection with cryopreserved testicular spermatozoa. *Molecular and Cellular Endocrinology* 169, 1-2, 15-19.

Gohbara, A, Katagiri, K, Sato, T, Kubota, Y, Kagechika, H, Araki, Y, Araki, Y & Ogawa, T. (2010). *In vitro* murine spermatogenesis in an organ culture system. *Biology of Reproduction* 83, 2, 261-267.

Gomendio, M, Cassinello, J & Roldan, ERS. (2000). A comparative study of ejaculate traits in three endangered ungulates with different levels of inbreeding: fluctuating asymmetry as an indicator of reproductive and genetic stress. *Proceedings of the Royal Society of London. Series B: Biological Sciences* 267, 1446, 875-882.

Gómez, MC, Jenkins, JA, Giraldo, A, Harris, RF, King, A, Dresser, BL & Pope, CE. (2003). Nuclear transfer of synchronized African wild cat somatic cells into enucleated domestic cat oocytes. *Biology of Reproduction* 69, 3, 1032-1041.

Gómez, MC, Lyons, JI, Pope, CE, Biancardi, M, Dumas, C, Galiguis, J, Wang, G & Dresser, BL. (2009). Effects of phylogenic genera of recipient cytoplasts on development and viability of Canada lynx (*Lynx canadensis*) cloned embryos. *Reproduction, Fertility and Development* 22, 1, 186 (abstract).

Gómez, MC, Pope, CE, Kutner, RH, Ricks, DM, Lyons, LA, Ruhe, M, Dumas, C, Lyons, J, Lopez, M, Dresser, BL & Reiser, J. (2008). Nuclear transfer of sand cat cells into enucleated domestic cat oocytes is affected by cryopreservation of donor cells. *Cloning and Stem Cells* 10, 4, 469-483.

Goodrowe, KL, Miller, AM & Wildt, DE. (1989). *In vitro* fertilization of gonadotropin-stimulated leopard cat (*Felis bengalensis*) follicular oocytes. *Journal of Experimental Zoology* 252, 1, 89-95.

Goodrowe, KL, Wall, RJ, O'Brien, SJ, Schmidt, PM & Wildt, DE. (1988). Developmental competence of domestic cat follicular oocytes after fertilization *in vitro*. *Biology of Reproduction* 39, 2, 355-372.

Gook, DA, Edgar, DH, Borg, J, Archer, J, Lutjen, PJ & McBain, JC. (2003). Oocyte maturation, follicle rupture and luteinization in human cryopreserved ovarian tissue following xenografting. *Human Reproduction* 18, 9, 1772-1781.

Gook, DA, McCully, BA, Edgar, DH & McBain, JC. (2001). Development of antral follicles in human cryopreserved ovarian tissue following xenografting. *Human Reproduction* 16, 3, 417-422.

Gook, DA, Osborn, SM & Johnston, WIH. (1993). Cryopreservation of mouse and human oocytes using 1, 2-propanediol and the configuration of the meiotic spindle. *Human Reproduction* 8, 7, 1101-1109.

Gosden, RG, Boulton, MI, Grant, K & Webb, R. (1994). Follicular development from ovarian xenografts in SCID mice. *Journal of Reproduction and Fertility* 101, 3, 619-623.

Guenther, JF, Seki, S, Kleinhans, FW, Edashige, K, Roberts, DM & Mazur, P. (2006). Extra- and intra-cellular ice formation in Stage I and II *Xenopus laevis* oocytes. *Cryobiology* 52, 3, 401-416.

Gunasena, KT, Lakey, JRT, Villines, PM, Bush, M, Raath, C, Critser, ES, McGann, LE & Critser, JK. (1998). Antral follicles develop in xenografted cryopreserved African elephant (*Loxodonta africana*) ovarian tissue. *Animal Reproduction Science* 53, 1-4, 265-275.

Gupta, MK, Uhm, SJ & Lee, HT. (2010). Effect of vitrification and beta-mercaptoethanol on reactive oxygen species activity and *in vitro* development of oocytes vitrified before or after *in vitro* fertilization. *Fertility and Sterility* 93, 8, 2602-2607.

Gvakharia, M & Adamson, GD. (2001). A method of successful cryopreservation of small numbers of human spermatozoa. *Fertility and Sterility* 76, 3, Supplement 1, S101 (Abstract).

Hagedorn, M, Hsu, E, Kleinhans, FW & Wildt, DE. (1997a). New approaches for studying the permeability of fish embryos: Toward successful cryopreservation. *Cryobiology* 34, 4, 335-347.

Hagedorn, M, Hsu, EW, Pilatus, U, Wildt, DE, Rall, WR & Blackband, SJ. (1996). Magnetic resonance microscopy and spectroscopy reveal kinetics of cryoprotectant permeation in a multicompartmental biological system. *Proceedings of the National Academy of Science of the Uunited States of America* 93, 15, 7454-7459.

Hagedorn, M, Kleinhans, FW, Freitas, R, Liu, J, Hsu, EW, Wildt, DE & Rall, WF. (1997b). Water distribution and permeability of zebrafish embryos, *Brachydanio rerio*. *Journal of Experimental Zoology* 278, 6, 356-371.

Hagedorn, M, Kleinhans, FW, Wildt, DE & Rall, WF. (1997c). Chill sensitivity and cryoprotectant permeability of dechorionated zebrafish embryos, *Brachydanio rerio*. *Cryobiology* 34, 3, 251-263.

Hagedorn, M, Lance, SL, Fonseca, DM, Kleinhans, FW, Artimov, D, Fleischer, R, Hoque, AT, Hamilton, MB & Pukazhenthi, BS. (2002). Altering fish embryos with aquaporin-3: an essential step toward successful cryopreservation. *Biology of Reproduction* 67, 3, 961-966.

Harvey, B. (1983). Cooling of embryonic cells, isolated blastoderms, and intact embryos of the zebra fish *Brachydanio rerio* to -196°C. *Cryobiology* 20, 4, 440-447.

Hayashi, S, Kobayashi, K, Mizuno, J, Saitoh, K & Hirano, S. (1989). Birth of piglets from frozen embryos. *The Veterinary Record* 125, 2, 43-44.

He, X, Park, EYH, Fowler, A, Yarmush, ML & Toner, M. (2008). Vitrification by ultra-fast cooling at a low concentration of cryoprotectants in a quartz micro-capillary: A study using murine embryonic stem cells. *Cryobiology* 56, 3, 223-232.

Hearn, JP & Summers, PM. (1986). Experimental manipulation of embryo implantation in the marmoset monkey and exotic equids. *Theriogenology* 25, 1, 3-11.

Hermes, R, Behr, B, Hildebrandt, TB, Blottner, S, Sieg, B, Frenzel, A, Knieriem, A, Saragusty, J & Rath, D. (2009a). Sperm sex-sorting in the Asian elephant (*Elephas maximus*). *Animal Reproduction Science* 112, 3-4, 390-396.

Hermes, R, Göritz, F, Saragusty, J, Sos, E, Molnar, V, Reid, CE, Schwarzenberger, F & Hildebrandt, TB. (2009b). First successful artificial insemination with frozen-thawed semen in rhinoceros. *Theriogenology* 71, 3, 393-399.

Herrera, C, Conde, P, Donaldson, M, Quintans, C, Cortvrindt, R & de Matos, DG. (2002). Bovine follicular development up to antral stages after frozen-thawed ovarian tissue transplantation into nude mice. *Theriogenology* 57, 1, 608 (anstract).

Herrick, JR, Campbell, M, Levens, G, Moore, T, Benson, K, D'Agostino, J, West, G, Okeson, DM, Coke, R, Portacio, SC, Leiske, K, Kreider, C, Polumbo, PJ & Swanson, WF. (2010). *In vitro* fertilization and sperm cryopreservation in the black-footed cat (*Felis nigripes*) and sand cat (*Felis margarita*). *Biology of Reproduction* 82, 3, 552-562.

Herrler, A, Eisner, S, Bach, V, Weissenborn, U & Beier, HM. (2006). Cryopreservation of spermatozoa in alginic acid capsules. *Fertility and Sterility* 85, 1, 208-213.

Hewitson, L, Martinovich, C, Simerly, C, Takahashi, D & Schatten, G. (2002). Rhesus offspring produced by intracytoplasmic injection of testicular sperm and elongated spermatids. *Fertility and Sterility* 77, 4, 794-801.

Higaki, S, Eto, Y, Kawakami, Y, Yamaha, E, Kagawa, N, Kuwayama, M, Nagano, M, Katagiri, S & Takahashi, Y. (2010). Production of fertile zebrafish (*Danio rerio*) possessing germ cells (gametes) originated from primordial germ cells recovered from vitrified embryos. *Reproduction* 139, 4, 733-740.

Hildebrandt, TB, Hermes, R, Jewgenow, K & Göritz, F. (2000). Ultrasonography as an important tool for the development and application of reproductive technologies in non-domestic species. *Theriogenology* 53, 1, 73-84.

Hildebrandt, TB, Roellig, K, Goeritz, F, Fassbender, M, Krieg, R, Blottner, S, Behr, B & Hermes, R. (2009). Artificial insemination of captive European brown hares (*Lepus europaeus* PALLAS, 1778) with fresh and cryopreserved semen derived from free-ranging males. *Theriogenology* 72, 8, 1065-1072.

Hirabayashi, M, Kato, M & Hochi, S. (2008). Factors affecting full-term development of rat oocytes microinjected with fresh or cryopreserved round spermatids. *Experimental Animals* 57, 4, 401-405.

Hirabayashi, M, Kato, M, Ito, J & Hochi, S. (2005). Viable rat offspring derived from oocytes intracytoplasmically injected with freeze-dried sperm heads. *Zygote* 13, 1, 79-85.

Hishinuma, M, Suzuki, K & Sekine, J. (2003). Recovery and cryopreservation of Sika deer (*Cervus nippon*) spermatozoa from epididymides stored at 4°C. *Theriogenology* 59, 3-4, 813-820.

Hochi, S, Watanabe, K, Kato, M & Hirabayashi, M. (2008). Live rats resulting from injection of oocytes with spermatozoa freeze-dried and stored for one year. *Molecular Reproduction and Development* 75, 5, 890-894.

Hollinshead, FK, O'Brien, JK, Maxwell, WM & Evans, G. (2004). Assessment of *in vitro* sperm characteristics after flow cytometric sorting of frozen-thawed bull spermatozoa. *Theriogenology* 62, 5, 958-968.

Holt, WV. (2001). Germplasm Cryopreservation in Elephants and Wild Ungulates. In: *Cryobanking the Genetic Resource: Wildlife Conservation for the Future*, Watson, PF, Holt, WV, editors, p 317-348, Taylor & Francis, New York.

Holt, WV. (2011). Mechanisms of sperm storage in the female reproductive tract: an interspecies comparison. *Reproduction in Domestic Animals* 46, Suppl. S2, 68-74.

Holt, WV & Lloyd, RE. (2010). Sperm storage in the vertebrate female reproductive tract: How does it work so well? *Theriogenology* 73, 6, 713-722.

Honaramooz, A, Behboodi, E, Blash, S, Megee, SO & Dobrinski, I. (2003a). Germ cell transplantation in goats. *Molecular Reproduction and Development* 64, 4, 422-428.

Honaramooz, A, Behboodi, E, Megee, SO, Overton, SA, Galantino-Homer, H, Echelard, Y & Dobrinski, I. (2003b). Fertility and germline transmission of donor haplotype following germ cell transplantation in immunocompetent goats. *Biology of Reproduction* 69, 4, 1260-1264.

Honaramooz, A, Li, MW, Penedo, MC, Meyers, S & Dobrinski, I. (2004). Accelerated maturation of primate testis by xenografting into mice. *Biology of Reproduction* 70, 5, 1500-1503.

Honaramooz, A, Megee, SO & Dobrinski, I. (2002). Germ cell transplantation in pigs. *Biology of Reproduction* 66, 1, 21-28.

Hong, Y, Liu, T, Zhao, H, Xu, H, Wang, W, Liu, R, Chen, T, Deng, J & Gui, J. (2004). Establishment of a normal medakafish spermatogonial cell line capable of sperm production *in vitro*. *Proceedings of the National Academy of Science of the Uunited States of America* 101, 21, 8011-8016.

Hoshina, M, Furugaichi, A, Kuji, N, Ito, J & Kashiwazaki, N. (2009). Vitrification of whole ovaries in young rats. *Reproduction, Fertility and Development* 22, 1, 206 (abstract).

Hoshino, Y, Hayashi, N, Taniguchi, S, Kobayashi, N, Sakai, K, Otani, T, Iritani, A & Saeki, K. (2009). Resurrection of a bull by cloning from organs frozen without cryoprotectant in a -80 degrees c freezer for a decade. *PLoS One* 4, 1, e4142.

Hovatta, O, Foudila, T, Siegberg, R, Johansson, K, von Smitten, K & Reima, I. (1996). Pregnancy resulting from intracytoplasmic injection of spermatozoa from a frozen-thawed testicular biopsy specimen. *Human Reproduction* 11, 11, 2472-2473.

Howell-Stephens, JA, Brown, J, Santymire, R, Bernier, D & Mulkerin, D. (2009) Using giving-up densities and adrenocortical activity to determine the state of Southern three-banded armadillos (*Tolypeutes matacus*) housed in North America. *10th International Mammalogical Congress*, Mendoza, Argentina, 9-14.8.2009.

Hunter, P. (2011). To protect and save. *EMBO Reports* 12, 3, 205-207.

Ikeda, H, Kikuchi, K, Noguchi, J, Takeda, H, Shimada, A, Mizokami, T & Kaneko, H. (2002). Effect of preincubation of cryopreserved porcine epididymal sperm. *Theriogenology* 57, 4, 1309-1318.

Imhof, M, Hofstetter, G, Bergmeister, H, Rudas, M, Kain, R, Lipovac, M & Huber, J. (2004). Cryopreservation of a whole ovary as a strategy for restoring ovarian function. *Journal of Assisted Reproduction and Genetics* 21, 12, 459-465.

Isachenko, E, Isachenko, V, Katkov, II, Dessole, S & Nawroth, F. (2003). Vitrification of mammalian spermatozoa in the absence of cryoprotectants: from past practical difficulties to present success. *Reproductive Biomedicine Online* 6, 2, 191-200.

Isachenko, V, Alabart, JL, Nawroth, F, Isachenko, E, Vajta, G & Folch, J. (2001). The open pulled straw vitrification of ovine GV-oocytes: positive effect of rapid cooling or rapid thawing or both? *Cryo Letters* 22, 3, 157-162.

Isachenko, V, Isachenko, E, Katkov, II, Montag, M, Dessole, S, Nawroth, F & van der Ven, H. (2004). Cryoprotectant-free cryopreservation of human spermatozoa by vitrification and freezing in vapor: Effect on motility, DNA integrity, and fertilization ability. *Biology of Reproduction* 71, 4, 1167-1173.

Isachenko, V, Lapidus, I, Isachenko, E, Krivokharchenko, A, Kreienberg, R, Woriedh, M, Bader, M & Weiss, JM. (2009). Human ovarian tissue vitrification versus conventional freezing: morphological, endocrinological, and molecular biological evaluation. *Reproduction* 138, 2, 319-327.

Isaev, DA, Zaletov, SY, Zaeva, VV, Zakharova, EE, Shafei, RA & Krivokharchenko, IS. (2007). Artificial microcontainers for cryopreservation of solitary spermatozoa. *Human Reproduction* 22, suppl 1, i154-i155 (abstract).

Ishijima, T, Kobayashi, Y, Lee, D-S, Ueta, YY, Matsui, M, Lee, J-Y, Suwa, Y, Miyahara, K & Suzuki, H. (2006). Cryopreservation of canine ovaries by vitrification. *The Journal of Reproduction and Development* 52, 2, 293-299.

IUCN. IUCN Red List of Threatened Species. Accessed on August 6, 2007, Available from: <http://www.iucnredlist.org>.

IUCN. IUCN Red List of Threatened Species. Version 2010.4. Accessed on 31 December, 2010, Available from: <http://www.iucnredlist.org/>.

Iwayama, H, Hochi, S, Kato, M, Hirabayashi, M, Kuwayama, M, Ishikawa, H, Ohsumi, S & Fukui, Y. (2005). Effects of cryodevice type and donors' sexual maturity on vitrification of minke whale (*Balaenoptera bonaerensis*) oocytes at germinal vesicle stage. *Zygote* 12, 4, 333-338.

Izadyar, F, Den Ouden, K, Stout, TA, Stout, J, Coret, J, Lankveld, DP, Spoormakers, TJ, Colenbrander, B, Oldenbroek, JK, Van der Ploeg, KD, Woelders, H, Kal, HB & De Rooij, DG. (2003). Autologous and homologous transplantation of bovine spermatogonial stem cells. *Reproduction* 126, 6, 765-774.

Jackowski, S, Leibo, SP & Mazur, P. (1980). Glycerol permeabilities of fertilized and unfertilized mouse ova. *Journal of Experimental Zoology* 212, 3, 329-341.

Jang, G, Kim, MK, Oh, HJ, Hossein, MS, Fibrianto, YH, Hong, SG, Park, JE, Kim, JJ, Kim, HJ, Kang, SK, Kim, DY & Lee, BC. (2007). Birth of viable female dogs produced by somatic cell nuclear transfer. *Theriogenology* 67, 5, 941-947.

Janik, M, Kleinhans, FW & Hagedorn, M. (2000). Overcoming a permeability barrier by microinjecting cryoprotectants into zebrafish embryos (*Brachydanio rerio*). *Cryobiology* 41, 1, 25-34.

Jewgenow, K, Blottner, S, Goritz, F, Hildebrandt, T, Hingst, O, Lengwinat, T, Lucker, H & Schneider, HE. (1994). Application of assisted reproductive technology in puma (*Felis concolor*). *Erkrankungen der Zootiere* 36, 59-65.

Jewgenow, K, Blottner, S, Lengwinat, T & Meyer, HH. (1997). New methods for gamete rescue from gonads of nondomestic felids. *Journal of Reproduction and Fertility, supplement* 51, 33-39.

Jewgenow, K & Paris, MC. (2006). Preservation of female germ cells from ovaries of cat species. *Theriogenology* 66, 1, 93-100.

Jewgenow, K, Penfold, LM, Meyer, HHD & Wildt, DE. (1998). Viability of small preantral ovarian follicles from domestic cats after cryoprotectant exposure and cryopreservation. *Journal of Reproduction and Fertility* 112, 1, 39-47.

Jewgenow, K, Wiedemann, C, Bertelsen, MF & Ringleb, J. (2011). Cryopreservation of mammalian ovaries and oocytes. *International Zoo Yearbook* 45, 124-132.

Jin, B, Kawai, Y, Hara, T, Takeda, S, Seki, S, Nakata, Y-i, Matsukawa, K, Koshimoto, C, Kasai, M & Edashige, K. (2011). Pathway for the movement of water and cryoprotectants in bovine oocytes and embryos. *Biology of Reproduction* 85, 4, 834-847.

Johnson, LA, Flook, JP & Look, MV. (1987a). Flow cytometry of X and Y chromosome-bearing sperm for DNA using an improved preparation method and staining with Hoechst 33342. *Gamete Research* 17, 3, 203-212.

Johnson, LA, Flook, JP, Look, MV & Pinkel, D. (1987b). Flow sorting of X and Y chromosome-bearing spermatozoa into two populations. *Gamete Research* 16, 1, 1-9.

Johnson, MH, Pickering, SJ & George, MA. (1988). The influence of cooling on the properties of the zona pellucida of the mouse oocyte. *Human Reproduction* 3, 3, 383-387.

Johnston, LA, Donoghue, AM, O'Brien, SJ & Wildt, DE. (1991). Rescue and maturation *in vitro* of follicular oocytes collected from nondomestic felid species. *Biology of Reproduction* 45, 6, 898-906.

Johnston, SD, McGowan, MR, Carrick, FN & Tribe, A. (1993). Preliminary investigations into the feasibility of freezing koala (*Phascolarctos cinereus*) semen. *Australian Veterinary Journal* 70, 11, 424-425.

Just, A, Gruber, I, Wöber, M, Lahodny, J, Obruca, A & Strohmer, H. (2004). Novel method for the cryopreservation of testicular sperm and ejaculated spermatozoa from patients with severe oligospermia: A pilot study. *Fertility and Sterility* 82, 2, 445-447.

Kagawa, N, Silber, S & Kuwayama, M. (2009). Successful vitrification of bovine and human ovarian tissue. *Reproductive Biomedicine Online* 18, 4, 568-577.

Kalicharan, D, Jongebloed, WL, Rawson, DM & Zhang, T. (1998). Variations in fixation techniques for field emission SEM and TEM of zebrafish (*Branchydanio rerio*) embryo inner and outer membranes. *Journal of Electron Microscopy (Tokyo)* 47, 6, 645-658.

Kaneko, T & Nakagata, N. (2005). Relation between storage temperature and fertilizing ability of freeze-dried mouse spermatozoa. *Comparative Medicine* 55, 2, 140-144.

Kaneko, T, Whittingham, DG & Yanagimachi, R. (2003). Effect of pH value of freeze-drying solution on the chromosome integrity and developmental ability of mouse spermatozoa. *Biology of Reproduction* 68, 1, 136-139.

Kang, SJ, Choi, JW, Kim, SY, Park, KJ, Kim, TM, Lee, YM, Kim, H, Lim, JM & Han, JY. (2008). Reproduction of wild birds via interspecies germ cell transplantation. *Biology of Reproduction* 79, 5, 931-937.

Kanno, H, Saito, K, Ogawa, T, Takeda, M, Iwasaki, A & Kinoshita, Y. (1998). Viability and function of human sperm in electrolyte-free cold preservation. *Fertility and Sterility* 69, 1, 127-131.

Kasai, M, Iritani, A & Chang, MC. (1979). Fertilization *in vitro* of rat ovarian oocytes after freezing and thawing. *Biology of Reproduction* 21, 4, 839-844.

Kasiraj, R, Misra, AK, Mutha Rao, M, Jaiswal, RS & Rangareddi, NS. (1993). Successful culmination of pregnancy and live birth following the transfer of frozen-thawed buffalo embryos. *Theriogenology* 39, 5, 1187-1192.

Katayose, H, Matsuda, J & Yanagimachi, R. (1992). The ability of dehydrated hamster and human sperm nuclei to develop into pronuclei. *Biology of Reproduction* 47, 2, 277-284.

Katkov, II, Isachenko, V, Isachenko, E, Kim, MS, Lulat, AGMI, Mackay, AM & Levine, F. (2006). Low- and high-temperature vitrification as a new approach to biostabilization of reproductive and progenitor cells. *International Journal of Refrigeration* 29, 3, 346-357.

Kato, H, Anzai, M, Mitani, T, Morita, M, Nishiyama, Y, Nakao, A, Kondo, K, Lazarev, PA, Ohtani, T, Shibata, Y & Iritani, A. (2009). Recovery of cell nuclei from 15,000-year-old mammoth tissue and injection into mouse enucleated matured oocytes. *Reproduction, Fertility and Development* 22, 1, 189 (abstract).

Kawakami, Y, Goto-Kazeto, R, Saito, T, Fujimoto, T, Higaki, S, Takahashi, Y, Arai, K & Yamaha, E. (2010). Generation of germ-line chimera zebrafish using primordial germ cells isolated from cultured blastomeres and cryopreserved embryoids. *International Journal of Developmental Biology* 54, 10, 1493-1501.

Keeley, T, McGreevy, PD & O'Brien, JK. (2011). Characterization and short-term storage of Tasmanian devil sperm collected post-mortem. *Theriogenology* 76, 4, 705-714.

Keskintepe, L, Pacholczyk, G, Machnicka, A, Norris, K, Curuk, MA, Khan, I & Brackett, BG. (2002). Bovine blastocyst development from oocytes injected with freeze-dried spermatozoa. *Biology of Reproduction* 67, 2, 409-415.

Kikuchi, K, Nagai, T, Kashiwazaki, N, Ikeda, H, Noguchi, J, Shimada, A, Soloy, E & Kaneko, H. (1998). Cryopreservation and ensuing *in vitro* fertilization ability of boar spermatozoa from epididymides stored at 4ºC. *Theriogenology* 50, 4, 615-623.

Kim, MK, Jang, G, Oh, HJ, Yuda, F, Kim, HJ, Hwang, WS, Hossein, MS, Kim, JJ, Shin, NS, Kang, SK & Lee, BC. (2007). Endangered wolves cloned from adult somatic cells. *Cloning and Stem Cells* 9, 1, 130-137.

Kim, SS, Kang, HG, Kim, NH, Lee, HC & Lee, HH. (2005). Assessment of the integrity of human oocytes retrieved from cryopreserved ovarian tissue after xenotransplantation. *Human Reproduction* 20, 9, 2502-2508.

Kimura, Y & Yanagimachi, R. (1995). Mouse oocytes injected with testicular spermatozoa or round spermatids can develop into normal offspring. *Development* 121, 8, 2397-2405.

Kleinhans, FW, Guenther, JF, Roberts, DM & Mazur, P. (2006). Analysis of intracellular ice nucleation in *Xenopus* oocytes by differential scanning calorimetry. *Cryobiology* 52, 1, 128-138.

Klooster, KL, Burruel, VR & Meyers, SA. (2011). Loss of fertilization potential of desiccated rhesus macaque spermatozoa following prolonged storage. *Cryobiology* 62, 3, 161-166.

Koh, LP, Dunn, RR, Sodhi, NS, Colwell, RK, Proctor, HC & Smith, VS. (2004). Species coextinctions and the biodiversity crisis. *Science* 305, 5690, 1632-1634.

Kohara, Y, Kanai, Y & Tajima, A. (2008). Cryopreservation of gonadal germ cells (GGCs) from the domestic chicken using vitrification. *The Journal of Poultry Science* 45, 1, 57-61.

Kono, T, Suzuki, O & Tsunoda, Y. (1988). Cryopreservation of rat blastocysts by vitrification. *Cryobiology* 25, 2, 170-173.

Koscinski, I, Wittemer, C, Lefebvre-Khalil, V, Marcelli, F, Defossez, A & Rigot, JM. (2007). Optimal management of extreme oligozoospermia by an appropriate cryopreservation programme. *Human Reproduction* 22, 10, 2679-2684.

Kramer, L, Dresser, BL, Pope, CE, Dalhausen, RD & Baker, RD. (1983) The non-surgical transfer of frozen-thawed eland (*Tragelaphus oryx*) embryos. *Annual Proceedings of the American Association of Zoo Veterinarians*, Tampa, Florida, USA, October 24-27, 1983.

Krisher, RL. (2004). The effect of oocyte quality on development. *Journal of Animal Science* 82, 13_suppl, E14-E23.

Krzywinski, A. (1981). Freezing of post-mortem collected semen from moose and red deer. *Acta Theriologica* 26, 424-426.

Kusakabe, H & Tateno, H. (2011). Characterization of chromosomal damage accumulated in freeze-dried mouse spermatozoa preserved under ambient and heat stress conditions. *Mutagenesis* 26, 3, 447-453.

Kusakabe, H, Yanagimachi, R & Kamiguchi, Y. (2008). Mouse and human spermatozoa can be freeze-dried without damaging their chromosomes. *Human Reproduction* 23, 2, 233-239.

Kwon, IK, Park, KE & Niwa, K. (2004). Activation, pronuclear formation, and development *in vitro* of pig oocytes following intracytoplasmic injection of freeze-dried spermatozoa. *Biology of Reproduction* 71, 5, 1430-1436.

Kyriakopoulou-Sklavounou, P & Loumbourdis, N. (1990). Annual ovarian cycle in the frog, *Rana ridibunda*, in northern Greece. *Journal of Herpetology* 24, 2, 185-191.

Landa, CA & Almquist, JO. (1979). Effect of freezing large numbers of straws of bovine spermatozoa in an automatic freezer on post-thaw motility and acrosomal retention. *Journal of Animal Science* 49, 5, 1190-1194.

Lane, M, Forest, KT, Lyons, EA & Bavister, BD. (1999). Live births following vitrification of hamster embryos using a novel containerless technique. *Theriogenology* 51, 1, 167 (abstract).

Lanza, RP, Cibelli, JB, Diaz, F, Moraes, CT, Farin, PW, Farin, CE, Hammer, CJ, West, MD & Damiani, P. (2000). Cloning of an endangered species (*Bos gaurus*) using interspecies nuclear transfer. *Cloning* 2, 2, 79-90.

Lanzendorf, SE, Zelinski-Wooten, MB, Stouffer, RL & Wolf, DP. (1990). Maturity at collection and the developmental potential of rhesus monkey oocytes. *Biology of Reproduction* 42, 4, 703-711.

Lanzendorf SE, Holmgren WJ, Schaffer N, Hatasaka H, Wentz AC, Jeyendran RS. (1992) *In vitro* fertilization and gamete micromanipulation in the lowland gorilla. *Journal of Assisted Reproduction and Genetics* 9, 4, 358-364.

Lattanzi, M, Santos, C, Chaves, G, Miragaya, M, Capdevielle, E, Judith, E, Agüero, A & Baranao, L. (2002). Cryopreservation of llama (*Lama glama*) embryos by slow freezing and vitrification. *Theriogenology* 57, 1, 585 (abstract).

Lee, B, Wirtu, GG, Damiani, P, Pope, E, Dresser, BL, Hwang, W & Bavister, BD. (2003). Blastocyst development after intergeneric nuclear transfer of mountain bongo antelope somatic cells into bovine oocytes. *Cloning and Stem Cells* 5, 1, 25-33.

Lee, DM, Yeoman, RR, Battaglia, DE, Stouffer, RL, Zelinski-Wooten, MB, Fanton, JW & Wolf, DP. (2004). Live birth after ovarian tissue transplant. *Nature* 428, 6979, 137-138.

Lee, DR, Yang, YH, Eum, JH, Seo, JS, Ko, JJ, Chung, HM & Yoon, TK. (2007). Effect of using slush nitrogen (SN2) on development of microsurgically manipulated vitrified/warmed mouse embryos. *Human Reproduction* 22, 9, 2509-2514.

Lee, H-J, Elmoazzen, H, Wright, D, Biggers, J, Rueda, BR, Heo, YS, Toner, M & Toth, TL. (2010). Ultra-rapid vitrification of mouse oocytes in low cryoprotectant concentrations. *Reproductive Biomedicine Online* 20, 2, 201-208.

Leibo, SP. (1980). Water permeability and its activation energy of fertilized and unfertilized mouse ova. *Journal of Membrane Biology* 53, 3, 179-188.

Leibo, SP & Oda, K. (1993). High survival of mouse zygotes and embryos cooled rapidly or slowly in ehtylene glycol plus polyvinylpyrrolidone. *Cryo Letters* 14, 133-144.

Lenie, S, Cortvrindt, R, Adriaenssens, T & Smitz, J. (2004). A reproducible two-step culture system for isolated primary mouse ovarian follicles as single functional units. *Biology of Reproduction* 71, 5, 1730-1738.

Li, MW, Biggers, JD, Elmoazzen, HY, Toner, M, McGinnis, L & Lloyd, KC. (2007). Long-term storage of mouse spermatozoa after evaporative drying. *Reproduction* 133, 5, 919-929.

Li, Z, Deng, H, Liu, C, Song, Y, Sha, J, Wang, N & Wei, H. (2002). Production of duck-chicken chimeras by transferring early blastodermal cells. *Poultry Science* 81, 9, 1360-1364.

Li, Z, Sun, X, Chen, J, Liu, X, Wisely, SM, Zhou, Q, Renard, JP, Leno, GH & Engelhardt, JF. (2006). Cloned ferrets produced by somatic cell nuclear transfer. *Developmental Biology* 293, 2, 439-448.

Lin, T-T, Pitt, RE & Steponkus, PL. (1989). Osmometric behavior of *Drosophila melanogaster* embryos. *Cryobiology* 26, 5, 453-471.

Lindeberg, H, Aalto, J, Amstislavsky, S, Piltti, K, Järvinen, M & Valtonen, M. (2003). Surgical recovery and successful surgical transfer of conventionally frozen-thawed embryos in the farmed European polecat (*Mustela putorius*). *Theriogenology* 60, 8, 1515-1525.

Liu, J, Van der Elst, J, Van den Broecke, R & Dhont, M. (2001). Live offspring by *in vitro* fertilization of oocytes from cryopreserved primordial mouse follicles after sequential in vivo transplantation and *in vitro* maturation. *Biology of Reproduction* 64, 1, 171-178.

Liu, J, Van Der Elst, J, Van Den Broecke, R, Dumortier, F & Dhont, M. (2000). Maturation of mouse primordial follicles by combination of grafting and *in vitro* culture. *Biology of Reproduction* 62, 5, 1218-1223.

Liu, J, Westhusin, M, Long, C, Johnson, G, Burghardt, R & Kraemer, D. (2010). Embryo production and possible species preservation by nuclear transfer of somatic cells isolated from bovine semen. *Theriogenology* 74, 9, 1629-1635.

Liu, JL, Kusakabe, H, Chang, CC, Suzuki, H, Schmidt, DW, Julian, M, Pfeffer, R, Bormann, CL, Tian, XC, Yanagimachi, R & Yang, X. (2004). Freeze-dried sperm fertilization leads to full-term development in rabbits. *Biology of Reproduction* 70, 6, 1776-1781.

Liu, L, Wood, GA, Morikawa, L, Ayearst, R, Fleming, C & McKerlie, C. (2008). Restoration of fertility by orthotopic transplantation of frozen adult mouse ovaries. *Human Reproduction* 23, 1, 122-128.

Living Planet Report. (2008). Gland, Switzerland: World Wildlife Fund (WWF).

Locatelli, Y, Vallet, J-C, Baril, G, Touzé, J-L, Hendricks, A, Legendre, X, Verdier, M & Mermillod, P. (2008). Successful interspecific pregnancy after transfer of *in vitro* produced sika deer (*Cervus nippon nippon*) embryo in red deer (*Cervus elaphus hippelaphus*) surrogate hind. *Reproduction, Fertility and Development* 20, 1, 160-161 (abstract).

Loi, P, Matsukawa, K, Ptak, G, Clinton, M, Fulka Jr., J, Natan, Y & Arav, A. (2008a). Freeze-dried somatic cells direct embryonic development after nuclear transfer. *PLoS One* 3, 8, e2978.

Loi, P, Matsukawa, K, Ptak, G, Nathan, Y, Fulka, J, Jr. & Arav, A. (2008b). Nuclear transfer of freeze dried somatic cells into enucleated sheep oocytes. *Reproduction in Domestic Animals* 43, Suppl. 2, 417-422.

Loi, P, Modlinski, JA & Ptak, G. (2011a). Interspecies somatic cell nuclear transfer: a salvage tool seeking first aid. *Theriogenology* 76, 2, 217-228.

Loi, P, Ptak, G, Barboni, B, Fulka, J, Jr., Cappai, P & Clinton, M. (2001). Genetic rescue of an endangered mammal by cross-species nuclear transfer using post-mortem somatic cells. *Nature Biotechnology* 19, 10, 962-964.

Loi, P, Wakayama, T, Saragusty, J, Fulka, J, Jr. & Ptak, G. (2011b). Biological time machines: A realistic approach for cloning an extinct mammal. *Endangered Species Research* 14, 3, 227-233.

Loomis, PR & Graham, JK. (2008). Commercial semen freezing: Individual male variation in cryosurvival and the response of stallion sperm to customized freezing protocols. *Animal Reproduction Science* 105, 1-2, 119-128.

Loskutoff, NM, Bartels, P, Meintjes, M, Godke, RA & Schiewe, MC. (1995). Assisted reproductive technology in nondomestic ungulates: A model approach to preserving and managing genetic diversity. *Theriogenology* 43, 1, 3-12.

Lu, F, Shi, D, Wei, J, Yang, S & Wei, Y. (2005). Development of embryos reconstructed by interspecies nuclear transfer of adult fibroblasts between buffalo (*Bubalus bubalis*) and cattle (*Bos indicus*). *Theriogenology* 64, 6, 1309-1319.

Luvoni, GC. (2006). Gamete cryopreservation in the domestic cat. *Theriogenology* 66, 1, 101-111.

Luvoni, GC & Pellizzari, P. (2000). Embryo development *in vitro* of cat oocytes cryopreserved at different maturation stages. *Theriogenology* 53, 8, 1529-1540.

Luvoni, GC, Pellizzari, P & Battocchio, M. (1997). Effects of slow and ultrarapid freezing on morphology and resumption of meiosis in immature cat oocytes. *Journal of Reproduction and Fertility Supplement* 51, 93-98.

Luyet, B. (1937). The vitrification of organic colloids and protoplasm. *Biodynamica* 1, 29, 1-14.

Mace, G, Masundire, H, Baillie, J, Ricketts, T, Brooks, T, Hoffman, M, Stuart, S, Andrew, B, Purvis, A, Reyers, B, Wang, J, Revenga, C, Kennedy, E, Naeem, S, Alkemade, R, Allnutt, T, Bakarr, M, Bond, W, Chanson, J, Cox, N, Fonseca, G, Hilton-Taylor, C, Loucks, C, Rodrigues, A, Sechrest, W, Stattersfield, A, van Rensburg, BJ, Whiteman, C, Abell, R, Cokeliss, Z, Lamoreux, J, Pereira, HM, Thonell, J & Williams, P. (2005). Biodiversity. In: *Ecosystem and Human Well-being: Current State and Trends*, Hassan, R, Scholes, R, Ash, N, editors, p 77-122, Island Press, Washington.

Manno III, FAM. (2010). Cryopreservation of mouse embryos by vitrification: A meta-analysis. *Theriogenology* 74, 2, 165-172.

Martinez-Pastor, F, Garcia-Macias, V, Alvarez, M, Chamorro, C, Herraez, P, Paz, Pd & Anel, L. (2006). Comparison of two methods for obtaining spermatozoa from the cauda epididymis of Iberian red deer. *Theriogenology* 65, 3, 471-485.

Martino, A, Songsasen, N & Leibo, SP. (1996). Development into blastocysts of bovine oocytes cryopreserved by ultra-rapid cooling. *Biology of Reproduction* 54, 5, 1059-1069.

Martins, CF, Bao, SN, Dode, MN, Correa, GA & Rumpf, R. (2007). Effects of freeze-drying on cytology, ultrastructure, DNA fragmentation, and fertilizing ability of bovine sperm. *Theriogenology* 67, 8, 1307-1315.

Martins, CF, Driessen, K, Melo Costa, P, Carvalho-Neto, JO, de Sousa, RV, Rumpf, R & Dode, MN. (2009). Recovery, cryopreservation and fertilization potential of bovine spermatozoa obtained from epididymides stored at 5°C by different periods of time. *Animal Reproduction Science* 116, 1-2, 50-57.

Mastromonaco, GF & King, WA. (2007). Cloning in companion animal, non-domestic and endangered species: can the technology become a practical reality? *Reproduction, Fertility and Development* 19, 6, 748-761.

Matoba, S & Ogura, A. (2010). Generation of functional oocytes and spermatids from fetal primordial germ cells after ectopic transplantation in adult mice. *Biology of Reproduction* 84, 4, 631-638.

Matsumoto, H, Jiang, JY, Tanaka, T, Sasada, H & Sato, E. (2001). Vitrification of large quantities of immature bovine oocytes using nylon mesh. *Cryobiology* 42, 2, 139-144.

Mattiske, D, Shaw, G & Shaw, JM. (2002). Influence of donor age on development of gonadal tissue from pouch young of the tammar wallaby, *Macropus eugenii*, after cryopreservation and xenografting into mice. *Reproduction* 123, 1, 143-153.

Maxwell, WMC, Parrilla, I, Caballero, I, Garcia, E, Roca, J, Martinez, EA, Vazquez, JM & Rath, D. (2007). Retained functional integrity of bull spermatozoa after double freezing and thawing using PureSperm(R) density gradient centrifugation. *Reproduction in Domestic Animals* 42, 5, 489-494.

Mazur, P, Leibo, SP & Seidel, GE, Jr. (2008). Cryopreservation of the germplasm of animals used in biological and medical research: importance, impact, status, and future directions. *Biology of Reproduction* 78, 1, 2-12.

McDonnell, SM. (2001). Oral imipramine and intravenous xylazine for pharmacologically-induced ex copula ejaculation in stallions. *Animal Reproduction Science* 68, 3-4, 153-159.

Melville, DF, Crichton, EG, Paterson-Wimberley, T & Johnston, SD. (2008). Collection of semen by manual stimulation and ejaculate characteristics of the black flying-fox (*Pteropus alecto*). *Zoo Biology* 27, 2, 159-164.

Merino, O, Sánchez, R, Risopatrón, J, Isachenko, E, Katkov, II, Figueroa, E, Valdebenito, I, Mallmann, P & Isachenko, V. (2011). Cryoprotectant-free vitrification of fish (*Oncorhynchus mykiss*) spermatozoa: first report. *Andrologia* 124, 1-2, 125-131.

Meryman, HT & Kafig, E. (1959). Survival of spermatozoa following drying. *Nature* 184, 4684, 470-471.

Meyers, SA. (2006). Dry storage of sperm: applications in primates and domestic animals. *Reproduction, Fertility and Development* 18, 1-2, 1-5.

Migishima, F, Suzuki-Migishima, R, Song, SY, Kuramochi, T, Azuma, S, Nishijima, M & Yokoyama, M. (2003). Successful cryopreservation of mouse ovaries by vitrification. *Biology of Reproduction* 68, 3, 881-887.

Miller, DL, Waldhalm, SJ, Leopold, BD & Estill, C. (2002). Embryo transfer and embryonic capsules in the bobcat (*Lynx rufus*). *Anatomia, Histologia, Embryologia* 31, 2, 119-125.

Miyashita, N, Kubo, Y, Yonai, M, Kaneyama, K, Saito, N, Sawai, K, Minamihashi, A, Suzuki, T, Kojima, T & Nagai, T. (2011). Cloned cows with short telomeres deliver healthy offspring with normal-length telomeres. *The Journal of Reproduction and Development* 57, 5, 636-642.

Mocè, ML, Blasco, A & Santacreu, MA. (2010). *In vivo* development of vitrified rabbit embryos: Effects on prenatal survival and placental development. *Theriogenology* 73, 5, 704-710.

Mochida, K, Wakayama, T, Takano, K, Noguchi, Y, Yamamoto, Y, Suzuki, O, Matsuda, J & Ogura, A. (2005). Birth of offspring after transfer of Mongolian gerbil (*Meriones unguiculatus*) embryos cryopreserved by vitrification. *Molecular Reproduction and Development* 70, 4, 464-470.

Mochida, K, Wakayama, T, Takano, K, Noguchi, Y, Yamamoto, Y, Suzuki, O, Ogura, A & Matsuda, J. (1999). Successful cryopreservation of mongolian gerbil embryos by vitrification. *Theriogenology* 51, 1, 171 (abstract).

Mochida, K, Yamamoto, Y, Noguchi, Y, Takano, K, Matsuda, J & Ogura, A. (2000). Survival and subsequent *in vitro* development of hamster embryos after exposure to cryoprotectant solutions. *Journal of Assisted Reproduction and Genetics* 17, 3, 182-185.

Moghadam, KK, Nett, R, Robins, JC, Thomas, MA, Awadalla, SG, Scheiber, MD & Williams, DB. (2005). The motility of epididymal or testicular spermatozoa does not directly affect IVF/ICSI pregnancy outcomes. *Journal of Andrology* 26, 5, 619-623.

Moisan, AE, Leibo, SP, Lynn, JW, Gómez, MC, Pope, CE, Dresser, BL & Godke, RA. (2005). Embryonic development of felid oocytes injected with freeze-dried or air-dried spermatozoa. *Cryobiology* 51, 373 (abstract).

Montano, GA, Kraemer, DC, Love, CC, Robeck, TR & O'Brien, JK. (2010). *In vitro* function of frozen-thawed bottlenose dolphin (*Tursiops truncatus*) sperm undergoing sorting and recryopreservation. *Reproduction, Fertility and Development* 23, 1, 240 (abstract).

Moor, RM, Dai, Y, Lee, C & Fulka, J, Jr. (1998). Oocyte maturation and embryonic failure. *Human Reproduction Update* 4, 3, 223-226.

Moore, HD & Taggart, DA. (1995). Sperm pairing in the opossum increases the efficiency of sperm movement in a viscous environment. *Biology of Reproduction* 52, 4, 947-953.

Morato, RG, Conforti, VA, Azevedo, FC, Jacomo, AT, Silveira, L, Sana, D, Nunes, AL, Guimaraes, MA & Barnabe, RC. (2001). Comparative analyses of semen and endocrine characteristics of free-living versus captive jaguars (*Panthera onca*). *Reproduction* 122, 5, 745-751.

Morrow, CJ, Asher, GW, Berg, DK, Tervit, HR, Pugh, PA, McMillan, WH, Beaumont, S, Hall, DRH & Bell, ACS. (1994). Embryo transfer in fallow deer (*Dama dama*): Superovulation, embryo recovery and laparoscopic transfer of fresh and cryopreserved embryos. *Theriogenology* 42, 4, 579-590.

Murakami, M, Otoi, T, Karja, NW, Wongsrikeao, P, Agung, B & Suzuki, T. (2004). Blastocysts derived from *in vitro*-fertilized cat oocytes after vitrification and dilution with sucrose. *Cryobiology* 48, 3, 341-348.

Nagano, M, Avarbock, MR, Leonida, EB, Brinster, CJ & Brinster, RL. (1998). Culture of mouse spermatogonial stem cells. *Tissue and Cell* 30, 4, 389-397.

Nagashima, H, Kashiwazaki, N, Ashman, RJ, Grupen, CG & Nottle, MB. (1995). Cryopreservation of porcine embryos. *Nature* 374, 6521, 416.

Nagy, ZP, Chang, CC, Shapiro, DB, Bernal, DP, Kort, HI & Vajta, G. (2009). The efficacy and safety of human oocyte vitrification. *Seminars in Reproductive Medicine* 27, 6, 450-455.

Naitana, S, Bogliolo, L, Ledda, S, Leoni, G, Madau, L, Falchi, S & Muzzeddu, M. (2000). Survival of vitrified mouflon (*Ovis g. musimon*) blastocysts. *Theriogenology* 53, 1, 340 (abstract).

Naitana, S, Ledda, S, Loi, P, Leoni, G, Bogliolo, L, Dattena, M & Cappai, P. (1997). Polyvinyl alcohol as a defined substitute for serum in vitrification and warming solutions to cryopreserve ovine embryos at different stages of development. *Animal Reproduction Science* 48, 2-4, 247-256.

Naito, M, Nirasawa, K & Oishi, T. (1992). Preservation of quail blastoderm cells in liquid nitrogen. *British Poultry Science* 33, 2, 449-453.

Naito, M, Tajima, A, Tagami, T, Yasuda, Y & Kuwana, T. (1994). Preservation of chick primordial germ cells in liquid nitrogen and subsequent production of viable offspring. *Journal of Reproduction and Fertility* 102, 2, 321-325.

Nawroth, F, Isachenko, V, Dessole, S, Rahimi, G, Farina, M, Vargiu, N, Mallmann, P, Dattena, M, Capobianco, G, Peters, D, Orth, I & Isachenko, E. (2002). Vitrification of human spermatozoa without cryoprotectants. *Cryo Letters* 23, 2, 93-102.

Nayudu, P, Wu, J & Michelmann, H. (2003). *In vitro* development of marmoset monkey oocytes by pre-antral follicle culture. *Reproduction in Domestic Animals* 38, 2, 90-96.

Nel-Themaat, L, Gomez, MC, Pope, CE, Lopez, M, Wirtu, G, Cole, A, Dresser, BL, Lyons, LA, Bondioli, KR & Godke, RA. (2008a). Cloned embryos from semen. Part 1: *In vitro* proliferation of epithelial cells on embryonic fibroblasts after isolation from semen by gradient centrifugation. *Cloning and Stem Cells* 10, 1, 143-160.

Nel-Themaat, L, Gomez, MC, Pope, CE, Lopez, M, Wirtu, G, Jenkins, JA, Cole, A, Dresser, BL, Bondioli, KR & Godke, RA. (2008b). Cloned embryos from semen. Part 2: Intergeneric nuclear transfer of semen-derived eland (*Taurotragus oryx*) epithelial cells into bovine oocytes. *Cloning and Stem Cells* 10, 1, 161-172.

Niikura, Y, Niikura, T & Tilly, JL. (2009). Aged mouse ovaries possess rare premeiotic germ cells that can generate oocytes following transplantation into a young host environment. *Aging* 1, 12, 971-978.

Nowshari, MA, Ali, SA & Saleem, S. (2005). Offspring resulting from transfer of cryopreserved embryos in camel (*Camelus dromedarius*). *Theriogenology* 63, 9, 2513-2522.

Noyes, N, Boldt, J & Nagy, ZP. (2010). Oocyte cryopreservation: is it time to remove its experimental label? *Journal of Assisted Reproduction and Genetics* 27, 2-3, 69-74.

O'Brien, JK, Hollinshead, FK, Evans, KM, Evans, G & Maxwell, WM. (2004). Flow cytometric sorting of frozen-thawed spermatozoa in sheep and non-human primates. *Reproduction, Fertility and Development* 15, 7, 367-375.

O'Brien, JK & Robeck, TR. (2006). Development of sperm sexing and associated assisted reproductive technology for sex preselection of captive bottlenose dolphins (*Tursiops truncatus*). *Reproduction, Fertility and Development* 18, 3, 319-329.

O'Brien, JK & Robeck, TR. (2010). The value of *ex situ* Cetacean populations in understanding reproductive physiology and developing assisted reproductive technology for *ex situ* species management and conservation efforts. *International Journal of Comparative Psychology* 23, 3, 227-248.

O'Brien, JK & Roth, TL. (2000). Post-coital sperm recovery and cryopreservation in the Sumatran rhinoceros (*Dicerorhinus sumatrensis*) and application to gamete rescue in the African black rhinoceros (*Diceros bicornis*). *Journal of Reproduction and Fertility* 118, 2, 263-271.

Ogonuki, N, Mochida, K, Miki, H, Inoue, K, Fray, M, Iwaki, T, Moriwaki, K, Obata, Y, Morozumi, K, Yanagimachi, R & Ogura, A. (2006). Spermatozoa and spermatids retrieved from frozen reproductive organs or frozen whole bodies of male mice can produce normal offspring. *Proceedings of the National Academy of Sciences of the United States of America* 103, 35, 13098-13103.

Oh, HJ, Kim, MK, Jang, G, Kim, HJ, Hong, SG, Park, JE, Park, K, Park, C, Sohn, SH, Kim, DY, Shin, NS & Lee, BC. (2008). Cloning endangered gray wolves (*Canis lupus*) from somatic cells collected postmortem. *Theriogenology* 70, 4, 638-647.

Okita, K, Ichisaka, T & Yamanaka, S. (2007). Generation of germline-competent induced pluripotent stem cells. *Nature* 448, 7151, 313-317.

Oliveira, KG, Miranda, SA, Leão, DL, Brito, AB, Santos, RR & Domingues, SFS. (2010). Semen coagulum liquefaction; sperm activation and cryopreservation of capuchin monkey (*Cebus apella*) semen in coconut water solution (CWS) and TES-TRIS. *Animal Reproduction Science* 123, 1-2, 75-80.

Onions, VJ, Webb, R, McNeilly, AS & Campbell, BK. (2009). Ovarian endocrine profile and long-term vascular patency following heterotopic autotransplantation of cryopreserved whole ovine ovaries. *Human Reproduction* 24, 11, 2845-2855.

Ono, T, Mizutani, E, Li, C & Wakayama, T. (2008). Nuclear transfer preserves the nuclear genome of freeze-dried mouse cells. *Journal of Reproduction and Development* 54, 6, 486-491.

Pan, G, Chen, Z, Liu, X, Li, D, Xie, Q, Ling, F & Fang, L. (2001). Isolation and purification of the ovulation-inducing factor from seminal plasma in the bactrian camel (*Camelus bactrianus*). *Theriogenology* 55, 9, 1863-1879.

Papis, K, Korwin-Kossakowski, M & Wenta-Muchalska, E. (2009). Comparison of traditional and modified (VitMaster) methods of rabbit embryo vitrification. *Acta Veterinaria Hungarica* 57, 3, 411-416.

Paris, M, Millar, R, Colenbrander, B & Schwarzenberger, F. (2008). Non-invasive assessment of female reproductive physiology in the pygmy hippopotamus (*Choeropsis liberiensis*). *Reproduction in Domestic Animals* 43, s3, 21 (abstract).

Paris, MCJ, Snow, M, Cox, S-L & Shaw, JM. (2004). Xenotransplantation: a tool for reproductive biology and animal conservation? *Theriogenology* 61, 2-3, 277-291.

Parkening, TA, Tsunoda, Y & Chang, MC. (1976). Effects of various low temperatures, cryoprotective agents and cooling rates on the survival, fertilizability and development of frozen-thawed mouse eggs. *Journal of Experimental Zoology* 197, 3, 369-374.

Parrott, DMV. (1960). The fertility of mice with orthotopic ovarian grafts derived from frozen tissue. *Journal of Reproduction and Fertility* 1, 3, 230-241.

Patakine Varkonyi, E, Vegi, B, Varadi, E, Liptoi, K & Barna, J. (2007). Preliminary results of cryopreservation of early embryonic cells of goose. 5th Vietnamese-Hungarian International Conference on „Animal Production and Aquaculture for Sustainable Farming". Can Tho, Vietnam: Can Tho University. p 82-85.

Pickering, SJ & Johnson, MH. (1987). The influence of cooling on the oranization of meiotic spindle of the mouse oocyte. *Human Reproduction* 2, 3, 207-216.

Piltti, K, Lindeberg, H, Aalto, J & Korhonen, H. (2004). Live cubs born after transfer of OPS vitrified-warmed embryos in the farmed European polecat (*Mustela putorius*). *Theriogenology* 61, 5, 811-820.

Pimm, SL, Russell, GJ, Gittleman, JL & Brooks, TM. (1995). The future of biodiversity. *Science* 269, 5222, 347-350.

Polejaeva, IA, Chen, SH, Vaught, TD, Page, RL, Mullins, J, Ball, S, Dai, Y, Boone, J, Walker, S, Ayares, DL, Colman, A & Campbell, KH. (2000). Cloned pigs produced by nuclear transfer from adult somatic cells. *Nature* 407, 6800, 86-90.

Poleo, GA, Godke, RR & Tiersch, TR. (2005). Intracytoplasmic sperm injection using cryopreserved, fixed, and freeze-dried sperm in eggs of Nile tilapia. *Marine Biotechnology (New York, NY)* 7, 2, 104-111.

Polge, C, Smith, AU & Parkes, AS. (1949). Revival of spermatozoa after vitrification and dehydration at low temperatures. *Nature* 164, 4172, 666.

Pope, CE. (2000). Embryo technology in conservation efforts for endangered felids. *Theriogenology* 53, 1, 163-174.

Pope, CE, Dresser, BL, Chin, NW, Liu, JH, Loskutoff, NM, Behnke, EJ, Brown, C, McRae, MA, Sinoway, CE, Campbell, MK, Cameron, KN, Owens, ODM, Johnson, CA, Evans, RR & Cedars, MI. (1997a). Birth of a Western Lowland gorilla (*Gorilla gorilla gorilla*) following *in vitro* fertilization and embryo transfer. *American Journal of Primatology* 41, 3, 247-260.

Pope, CE, Gómez, MC, Cole, A, Dumas, C & Dresser, BL. (2005). Oocyte recovery, *in vitro* fertilization and embryo transfer in the serval (*Leptailurus serval*). *Reproduction, Fertility and Development* 18, 2, 223 (abstract).

Pope, CE, Gomez, MC & Dresser, BL. (2006). *In vitro* embryo production and embryo transfer in domestic and non-domestic cats. *Theriogenology* 66, 6-7, 1518-1524.

Pope, CE, Gomez, MC & Dresser, BL. (2009). *In vitro* embryo production in the clouded leopard (*Neofelis nebulosa*). *Reproduction, Fertility and Development* 22, 1, 258 (abstract).

Pope, CE, Gomez, MC, Harris, RF & Dresser, BL. (2002). Development of *in vitro* matured, *in vitro* fertilized cat embryos following cryopreservation, culture and transfer. *Theriogenology* 57, 1, 464 (abstract).

Pope, CE, Gómez, MC, Mikota, SK & Dresser, BL. (2000). Development of *in vitro* produced African wild cat (*Felis silvestris*) embryos after cryopreservation and transfer into domestic cat recipients. *Biology of Reproduction* 62 (Suppl 1), 321 (abstract).

Pope, CE, McRae, MA, Plair, BL, Keller, GL & Dresser, BL. (1994). Successful *in vitro* and *in vivo* development of *in vitro* fertilized two- to four-cell cat embryos following cryopreservation, culture and transfer. *Theriogenology* 42, 3, 513-525.

Pope, CE, McRae, MA, Plair, BL, Keller, GL & Dresser, BL. (1997b). *In vitro* and *in vivo* development of embryos produced by *in vitro* maturation and *in vitro* fertilization of cat oocytes. *Journal of Reproduction and Fertility Supplement* 51, 69-82.

Pope, CE, Pope, VZ & Beck, LR. (1984). Live birth following cryopreservation and transfer of a baboon embryo. *Fertility and Sterility* 42, 1, 143-145.

Popelkova, M, Turanova, Z, Koprdova, L, Ostro, A, Toporcerova, S, Makarevich, AV & Chrenek, P. (2009). Effect of vitrification technique and assisted hatching on rabbit embryo developmental rate. *Zygote* 17, 1, 57-61.

Pribenszky, C, Molnar, M, Horvath, A, Harnos, A & Szenci, O. (2006). Hydrostatic pressure induced increase in post-thaw motility of frozen boar spermatozoa. *Reproduction, Fertility and Development* 18, 2, 162-163 (abstract).

Pribenszky, C & Vajta, G. (2010). Cells under pressure: how sublethal hydrostatic pressure stress treatment increases gametes' and embryos' performance? *Reproduction, Fertility and Development* 23, 1, 48-55.

Ptak, G, Clinton, M, Barboni, B, Muzzeddu, M, Cappai, P, Tischner, M & Loi, P. (2002). Preservation of the wild European mouflon: the first example of genetic management using a complete program of reproductive biotechnologies. *Biology of Reproduction* 66, 3, 796-801.

Qi, S, Ma, A, Xu, D, Daloze, P & Chen, H. (2008). Cryopreservation of vascularized ovary: an evaluation of histology and function in rats. *Microsurgery* 28, 5, 380-386.

Quinn, PJ & White, IG. (1967). Phospholipid and cholesterol content of epididymal and ejaculated ram spermatozoa and seminal plasma in relation to cold shock. *Australian Journal of Biological Science* 20, 1205-1215.

Quintans, CJ, Donaldson, MJ, Bertolino, MV & Pasqualini, RS. (2002). Birth of two babies using oocytes that were cryopreserved in a choline-based freezing medium. *Human Reproduction* 17, 12, 3149-3152.

Rabon, DRJ & Waddell, W. (2010). Effects of inbreeding on reproductive success, performance, litter size, and survival in captive red wolves (*Canis rufus*). *Zoo Biology* 29, 1, 36-49.

Rall, WF. (2001). Cryopreservation of mammalian embryos, gametes, and ovarian tissue. In: *Assisted Fertilization and Nuclear Transfer in Mammals*, Wolf, DP, Zelinski-Wooten, MB, editors, p 173-187, Humana Press Inc., Totowa, NJ.

Rao, BS, Mahesh, YU, Suman, K, Charan, KV, Lakshmikantan, U, Gibence, HRW & Shivaji, S. (2011). Meiotic maturation of vitrified immature chousingha (*Tetracerus quadricorni*) oocytes recovered postmortem. *Cryobiology* 62, 1, 47-52.

Rawson, DM, Zhang, T, Kalicharan, D & Jongebloed, WL. (2000). Field emission scanning electron microscopy and transmission electron microscopy studies of the chorion, plasma membrane and syncytial layers of the gastrula-stage embryo of the zebrafish *Brachydanio rerio*: a consideration of the structural and functional relationships with respect to cryoprotectant penetration. *Aquaculture Research* 31, 3, 325-336.

Reid, CE, Hermes, R, Blottner, S, Goeritz, F, Wibbelt, G, Walzer, C, Bryant, BR, Portas, TJ, Streich, WJ & Hildebrandt, TB. (2009). Split-sample comparison of directional and liquid nitrogen vapour freezing method on post-thaw semen quality in white rhinoceroses (*Ceratotherium simum simum* and *Ceratotherium simum cottoni*). *Theriogenology* 71, 2, 275-291.

Reid, WV & Miller, KR. (1989). Keeping Options Alive: The scientific basis for the conservation of biodiversity. Washington, DC: World Resources Institute.

Reubinoff, BE, Pera, MF, Vajta, G & Trounson, AO. (2001). Effective cryopreservation of human embryonic stem cells by the open pulled straw vitrification method. *Human Reproduction* 16, 10, 2187-2194.

Revel, A, A, E, Bor, A, Yavin, S, Natan, Y & Arav, A. (2001). Intact sheep ovary cryopreservation and transplantation. *Fertility and Sterility* 76, 3, Suppl. 1, S42-S43 (abstract).

Revel, A, Elami, A, Bor, A, Yavin, S, Natan, Y & Arav, A. (2004). Whole sheep ovary cryopreservation and transplantation. *Fertility and Sterility* 82, 6, 1714-1715.

Reynaud, K, Fontbonne, A, Marseloo, N, Thoumire, S, Chebrout, M, de Lesegno, CV & Chastant-Maillard, S. (2005). *In vivo* meiotic resumption, fertilization and early embryonic development in the bitch. *Reproduction* 130, 2, 193-201.

Ridha, MT & Dukelow, WR. (1985). The developmental potential of frozen-thawed hamster preimplantation embryos following embryo transfer: Viability of slowly frozen embryos following slow and rapid thawing. *Animal Reproduction Science* 9, 3, 253-259.

Riel, JM, Huang, TT & Ward, MA. (2007). Freezing-free preservation of human spermatozoa--a pilot study. *Archives of Andrology* 53, 5, 275-284.

Riel, JM, Yamauchi, Y, Huang, TTF, Grove, J & Ward, MA. (2011). Short-term storage of human spermatozoa in electrolyte-free medium without freezing maintains sperm chromatin integrity better than cryopreservation. *Biology of Reproduction* 85, 3, 536-547.

Ringleb, J, Waurich, R, Wibbelt, G, Streich, WJ & Jewgenow, K. (2011). Prolonged storage of epididymal spermatozoa does not affect their capacity to fertilise *in vitro*-matured domestic cat (*Felis catus*) oocytes when using ICSI. *Reproduction, Fertility and Development* 23, 6, 818-825.

Risco, R, Elmoazzen, H, Doughty, M, He, X & Toner, M. (2007). Thermal performance of quartz capillaries for vitrification. *Cryobiology* 55, 3, 222-229.

Rizos, D, Ward, F, Duffy, P, Boland, MP & Lonergan, P. (2002). Consequences of bovine oocyte maturation, fertilization or early embryo development *in vitro* versus *in vivo*:

Implications for blastocyst yield and blastocyst quality. *Molecular Reproduction and Development* 61, 2, 234-248.

Robeck, TR, Gearhart, SA, Steinman, KJ, Katsumata, E, Loureiro, JD & O'Brien, JK. (2011). *In vitro* sperm characterization and development of a sperm cryopreservation method using directional solidification in the killer whale (*Orcinus orca*). *Theriogenology* 76, 2, 267-279.

Robeck, TR & O'Brien, JK. (2004). Effect of cryopreservation methods and precryopreservation storage on bottlenose dolphin (*Tursiops truncatus*) Spermatozoa. *Biology of Reproduction* 70, 5, 1340-1348.

Robles, V, Cabrita, E & Herraez, MP. (2009). Germplasm cryobanking in zebrafish and other aquarium model species. *Zebrafish* 6, 3, 281-293.

Robles, V, Cabrita, E, Real, M, Álvarez, R & Herráez, MP. (2003). Vitrification of turbot embryos: preliminary assays. *Cryobiology* 47, 1, 30-39.

Rockstrom, J, Steffen, W, Noone, K, Persson, A, Chapin, FSr, Lambin, EF, Lenton, TM, Scheffer, M, Folke, C, Schellnhuber, HJ, Nykvist, B, de Wit, CA, Hughes, T, van der Leeuw, S, Rodhe, H, Sorlin, S, Snyder, PK, Costanza, R, Svedin, U, Falkenmark, M, Karlberg, L, Corell, RW, Fabry, VJ, Hansen, J, Walker, B, Liverman, D, Richardson, K, Crutzen, P & Foley, JA. (2009). A safe operating space for humanity. *Nature* 461, 7263, 472-475.

Rodger, JC, Giles, I & Mate, KE. (1992). Unexpected oocyte growth after follicular antrum formation in four marsupial species. *Journal of Reproduction and Fertility* 96, 2, 755-763.

Rodrigues, BA & Rodrigues, JL. (2006). Responses of canine oocytes to *in vitro* maturation and *in vitro* fertilization outcome. *Theriogenology* 66, 6-7, 1667-1672.

Rodriguez-Sosa, JR, Foster, RA & Hahnel, A. (2010). Development of strips of ovine testes after xenografting under the skin of mice and co-transplantation of exogenous spermatogonia with grafts. *Reproduction* 139, 1, 227-235.

Roldan, ERS, Cassinello, J, Abaigar, T & Gomendio, M. (1998). Inbreeding, fluctuating asymmetry, and ejaculate quality in an endangered ungulate. *Proceedings of the Royal Society of London. Series B: Biological Sciences* 265, 1392, 243-248.

Roth, TL, Bush, LM, Wildt, DE & Weiss, RB. (1999). Scimitar-horned oryx (*Oryx dammah*) spermatozoa are functionally competent in a heterologous bovine *in vitro* fertilization system after cryopreservation on dry ice, in a dry shipper, or over liquid nitrogen vapor. *Biology of Reproduction* 60, 2, 493-8.

Roth, TL, Stoops, MA, Atkinson, MW, Blumer, ES, Campbell, MK, Cameron, KN, Citino, SB & Maas, AK. (2005). Semen collection in rhinoceroses (*Rhinoceros unicornis, Diceros bicornis, Ceratotherium simum*) by electroejaculation with a uniquely designed probe. *Journal of Zoo and Wildlife Medicine* 36, 4, 617-627.

Roth, TL, Stoops, MA, Robeck, TR, Ball, RL, Wolf, BA, Finnegan, MV & O'Brien, JK. (2010). Alkaline phosphatase as an indicator of true ejaculation in the rhinoceros. *Theriogenology* 74, 9, 1701-1706.

Roth, TL, Swanson, WF, Collins, D, Burton, M, Garell, DM & Wildt, DE. (1996). Snow leopard (*Panthera uncia*) spermatozoa are sensitive to alkaline pH, but motility in vitro is not influenced by protein or energy supplements. *Journal of Andrology* 17, 5, 558-566.

Roussel, JD, Kellgren, HC & Patrick, TE. (1964). Bovine semen frozen in liquid nitrogen vapor. *Journal of Dairy Science* 47, 12, 1403-1406.

Roy, VK & Krishna, A. (2011). Sperm storage in the female reproductive tract of *Scotophilus heathii*: Role of androgen. *Molecular Reproduction and Development* 78, 7, 477-487.

Ruffing, NA, Steponkus, PL, Pitt, RE & Parks, JE. (1993). Osmometric behavior, hydraulic conductivity, and incidence of intracellular ice formation in bovine oocytes at different developmental stages. *Cryobiology* 30, 6, 562-580.

Ruiz-Lopez, MJ, Evenson, DP, Espeso, G, Gomendio, M & Roldan, ER. (2010). High levels of DNA fragmentation in spermatozoa are associated with inbreeding and poor sperm quality in endangered ungulates. *Biology of Reproduction* 83, 3, 332-338.

Ruttimann, J. (2006). Doomsday food store takes pole position. *Nature* 441, 7096, 912-913.

Salehnia, M, Moghadam, EA & Velojerdi, MR. (2002). Ultrastructure of follicles after vitrification of mouse ovarian tissue. *Fertility and Sterility* 78, 3, 644-645.

Sanchez, R, Isachenko, V, Petrunkina, AM, Risopatron, J, Schulz, M & Isachenko, E. (2011). Live birth after intrauterine insemination with spermatozoa from an oligo-asthenozoospermic patient vitrified without permeable cryoprotectants. *Journal of Andrology*. In Press.

Sanchez-Partida, LG, Simerly, CR & Ramalho-Santos, J. (2008). Freeze-dried primate sperm retains early reproductive potential after intracytoplasmic sperm injection. *Fertility and Sterility* 89, 3, 742-745.

Santiago-Moreno, J, Toledano-Díaz, A, Sookhthezary, A, Gómez-Guillamón, F, de la Vega, RS, Pulido-Pastor, A & López-Sebastián, A. (2010). Effects of anesthetic protocols on electroejaculation variables of Iberian ibex (*Capra pyrenaica*). *Research in Veterinary Science* 90, 1, 150-155.

Santos, R, Tharasanit, T, Van Haeften, T, Figueiredo, J, Silva, J & Van den Hurk, R. (2007). Vitrification of goat preantral follicles enclosed in ovarian tissue by using conventional and solid-surface vitrification methods. *Cell and Tissue Research* 327, 1, 167-176.

Santos, RMd, Barreta, MH, Frajblat, M, Cucco, DC, Mezzalira, JC, Bunn, S, Cruz, FB, Vieira, AD & Mezzalira, A. (2006). Vacuum-cooled liquid nitrogen increases the developmental ability of vitrified-warmed bovine oocytes. *Ciencia Rural* 36, 5, 1501-1506.

Saragusty, J & Arav, A. (2011). Current progress in oocyte and embryo cryopreservation by slow freezing and vitrification. *Reproduction* 141, 1, 1-19.

Saragusty, J, Gacitua, H, King, R & Arav, A. (2006). Post-mortem semen cryopreservation and characterization in two different endangered gazelle species (*Gazella gazella and Gazella dorcas*) and one subspecies (*Gazella gazelle acaiae*). *Theriogenology* 66, 4, 775-784.

Saragusty, J, Gacitua, H, Pettit, MT & Arav, A. (2007). Directional freezing of equine semen in large volumes. *Reproduction in Domestic Animals* 42, 6, 610-615.

Saragusty, J, Gacitua, H, Rozenboim, I & Arav, A. (2009a). Do physical forces contribute to cryodamage? *Biotechnology and Bioengineering* 104, 4, 719-728.

Saragusty, J, Gacitua, H, Rozenboim, I & Arav, A. (2009b). Protective effects of iodixanol during bovine sperm cryopreservation. *Theriogenology* 71, 9, 1425-1432.

Saragusty, J, Gacitua, H, Zeron, Y, Rozenboim, I & Arav, A. (2009c). Double freezing of bovine semen. *Animal Reproduction Science* 115, 1-4, 10-17.

Saragusty, J, Hermes, R, Göritz, F, Schmitt, DL & Hildebrandt, TB. (2009d). Skewed birth sex ratio and premature mortality in elephants. *Animal Reproduction Science* 115, 1-4, 247-254.

Saragusty, J, Hildebrandt, TB, Behr, B, Knieriem, A, Kruse, J & Hermes, R. (2009e). Successful cryopreservation of Asian elephant (*Elephas maximus*) spermatozoa. *Animal Reproduction Science* 115, 1-4, 255-266.

Saragusty, J, Hildebrandt, TB, Bouts, T, Göritz, F & Hermes, R. (2010a). Collection and preservation of pygmy hippopotamus (*Choeropsis liberiensis*) semen. *Theriogenology* 74, 4, 652-657.

Saragusty, J, Hildebrandt, TB, Natan, Y, Hermes, R, Yavin, S, Göritz, F & Arav, A. (2005). Effect of egg-phosphatidylcholine on the chilling sensitivity and lipid phase transition of Asian elephant (*Elephas maximus*) spermatozoa. *Zoo Biology* 24, 3, 233-245.

Saragusty, J, Walzer, C, Petit, T, Stalder, G, Horowitz, I & Hermes, R. (2010b). Cooling and freezing of epididymal sperm in the common hippopotamus (*Hippopotamus amphibius*). *Theriogenology* 74, 7, 1256-1263.

Saroff, J & Mixner, JP. (1955). The relationship of egg yolk and glycerol content of diluters and glycerol equilibration time to survival of bull spermatozoa after low temperature freezing. *Journal of Dairy Science* 38, 3, 292-297.

Sathananthan, AH, Ng, SC, Trounson, AO, Bongso, A, Ratnam, SS, Ho, J, Mok, H & Lee, MN. (1988). The effects of ultrarapid freezing on meiotic and mitotic spindles of mouse oocytes and embryos. *Gamete Research* 21, 4, 385-401.

Sato, T, Katagiri, K, Gohbara, A, Inoue, K, Ogonuki, N, Ogura, A, Kubota, Y & Ogawa, T. (2011). *In vitro* production of functional sperm in cultured neonatal mouse testes. *Nature* 471, 7339, 504-507.

Schiewe, MC. (1991). The science and significance of embryo cryopreservation. *Journal of Zoo and Wildlife Medicine* 22, 1, 6-22.

Schiewe, MC, Bush, M, Phillips, LG, Citino, S & Wildt, DE. (1991a). Comparative aspects of estrus synchronization, ovulation induction, and embryo cryopreservation in the scimitar-horned oryx, bongo, eland, and greater kudu. *Journal of Experimental Zoology* 258, 1, 75-88.

Schlatt, S, Foppiani, L, Rolf, C, Weinbauer, GF & Nieschlag, E. (2002a). Germ cell transplantation into X-irradiated monkey testes. *Human Reproduction* 17, 1, 55-62.

Schlatt, S, Kim, SS & Gosden, R. (2002b). Spermatogenesis and steroidogenesis in mouse, hamster and monkey testicular tissue after cryopreservation and heterotopic grafting to castrated hosts. *Reproduction* 124, 3, 339-346.

Schmitt, DL & Hildebrandt, TB. (1998). Manual collection and characterization of semen from Asian elephants (*Elephas maximus*). *Animal Reproduction Science* 53, 1-4, 309-314.

Schmitt, DL & Hildebrandt, TB. (2000). Corrigendum to "Manual collection and characterization of semen from asian elephants" [Anim. Reprod. Sci. 53 (1998) 309-314]. *Animal Reproduction Science* 59, 1-2, 119.

Schneider, W. (1992). Lipoprotein receptors in oocyte growth. *Journal of Molecular Medicine* 70, 5, 385-390.

Schneiders, A, Sonksen, J & Hodges, JK. (2004). Penile vibratory stimulation in the marmoset monkey: a practical alternative to electro-ejaculation, yielding ejaculates of enhanced quality. *Journal of Medical Primatology* 33, 2, 98-104.

Schnorr, J, Oehninger, S, Toner, J, Hsiu, J, Lanzendorf, S, Williams, R & Hodgen, G. (2002). Functional studies of subcutaneous ovarian transplants in non-human primates: steroidogenesis, endometrial development, ovulation, menstrual patterns and gamete morphology. *Human Reproduction* 17, 3, 612-619.

Schoysman, R, Vanderzwalmen, P, Nijs, M, Segal, L, Segal-Bertin, G, Geerts, L, van Roosendaal, E & Schoysman, D. (1993). Pregnancy after fertilisation with human testicular spermatozoa. *The Lancet* 342, 8881, 1237.

Selman, K, Wallace, RA, Sarka, A & Qi, X. (1993). Stages of oocyte development in the zebrafish, *Brachydanio rerio*. *Journal of Morphology* 218, 2, 203-224.

Shaw, J, Temple-Smith, P, Trounson, A & Lamden, K. (1996). Fresh and frozen marsupial (*Sminthopsis crassicaudata*) ovaries develop after grafting to SCID mice. *Cryobiology* 33, 6, 631 (abstract).

Sherman, JK. (1954). Freezing and freeze-drying of human spermatozoa. *Fertility and Sterility* 5, 4, 357-371.

Sherman, JK. (1957). Freezing and freeze-drying of bull spermatozoa. *American Journal of Physiology* 190, 2, 281-286.

Sherman, JK. (1963). Improved methods of preservation of human spermatozoa by freezing and freeze-drying. *Fertility and Sterility* 14, 49-64.

Shi, D, Lu, F, Wei, Y, Cui, K, Yang, S, Wei, J & Liu, Q. (2007). Buffalos (*Bubalus bubalis*) cloned by nuclear transfer of somatic cells. *Biology of Reproduction* 77, 2, 285-291.

Shiels, PG, Kind, AJ, Campbell, KHS, Wilmut, I, Waddington, D, Colman, A & Schnieke, AE. (1999). Analysis of telomere length in Dolly, a sheep derived by nuclear transfer. *Cloning* 1, 2, 119-125.

Shin, T, Kraemer, D, Pryor, J, Liu, L, Rugila, J, Howe, L, Buck, S, Murphy, K, Lyons, L & Westhusin, M. (2002). A cat cloned by nuclear transplantation. *Nature* 415, 6874, 859.

Shinohara, T, Inoue, K, Ogonuki, N, Kanatsu-Shinohara, M, Miki, H, Nakata, K, Kurome, M, Nagashima, H, Toyokuni, S, Kogishi, K, Honjo, T & Ogura, A. (2002). Birth of offspring following transplantation of cryopreserved immature testicular pieces and *in-vitro* microinsemination. *Human Reproduction* 17, 12, 3039-3045.

Shinohara, T, Kato, M, Takehashi, M, Lee, J, Chuma, S, Nakatsuji, N, Kanatsu-Shinohara, M & Hirabayashi, M. (2006). Rats produced by interspecies spermatogonial transplantation in mice and *in vitro* microinsemination. *Proceedings of the National Academy of Sciences of the United States of America* 103, 37, 13624-13628.

Si, W, Hildebrandt, TB, Reid, C, Krieg, R, Ji, W, Fassbender, M & Hermes, R. (2006). The successful double cryopreservation of rabbit (*Oryctolagus cuniculus*) semen in large volume using the directional freezing technique with reduced concentration of cryoprotectant. *Theriogenology* 65, 4, 788-798.

Silva, RC, Costa, GMJ, Lacerda, SMSN, Batlouni, SR, Soares, JM, Avelar, GF, Bottger, KB, Silva, SF, Jr., Noqueira, MS, Andrade, LM & Franca, LR. (2011). Germ cell transplantation in Felids: A potential approach to preserving endangered species. *Journal of Andrology* 76, 6, 1084-1089.

Skidmore, JA & Loskutoff, NM. (1999). Developmental competence *in vitro* and *in vivo* of cryopreserved expanding blastocysts from the dromedary camel (*Camelus dromedarius*). *Theriogenology* 51, 1, 293 (abstract).

Slade, NP, Takeda, T, Squires, EL, Elsden, RP & Seidel, GE, Jr. (1985). A new procedure for the cryopreservation of equine embryos. *Theriogenology* 24, 1, 45-58.

Smith, AU & Parkes, AS. (1951). Preservation of ovarian tissue at low temperatures. *The Lancet* 258, 6683, 570-572.

Snedaker, AK, Honaramooz, A & Dobrinski, I. (2004). A game of cat and mouse: Xenografting of testis tissue from domestic kittens results in complete cat spermatogenesis in a mouse host. *Journal of Andrology* 25, 6, 926-930.

Snow, M, Cox, SL, Jenkin, G, Trounson, A & Shaw, J. (2002). Generation of live young from xenografted mouse ovaries. *Science* 297, 5590, 2227.

Soler, JP, Mucci, N, Kaiser, GG, Aller, J, Hunter, JW, Dixon, TE & Alberio, RH. (2007). Multiple ovulation and embryo transfer with fresh, frozen and vitrified red deer (*Cervus elaphus*) embryos in Argentina. *Animal Reproduction Science* 102, 3-4, 322-327.

Spindler, RE, Pukazhenthi, BS & Wildt, DE. (2000). Oocyte metabolism predicts the development of cat embryos to blastocyst *in vitro*. *Molecular Reproduction and Development* 56, 2, 163-171.

Srirattana, K, Matsukawa, K, Akagi, S, Tasai, M, Tagami, T, Nirasawa, K, Nagai, T, Kanai, Y, Parnpai, R & Takeda, K. (2011). Constant transmission of mitochondrial DNA in intergeneric cloned embryos reconstructed from swamp buffalo fibroblasts and bovine ooplasm. *Animal Science Journal* 82, 2, 236-243.

Stachecki, JJ, Cohen, J & Willadsen, S. (1998a). Detrimental effects of sodium during mouse oocyte cryopreservation. *Biology of Reproduction* 59, 2, 395-400.

Stachecki, JJ, Cohen, J & Willadsen, SM. (1998b). Cryopreservation of unfertilized mouse oocytes: the effect of replacing sodium with choline in the freezing medium. *Cryobiology* 37, 4, 346-354.

Steponkus, PL, Myers, SP, Lynch, DV, Gardner, L, Bronshteyn, V, Leibo, SP, Rall, WF, Pitt, RE, Lin, TT & MacIntyre, RJ. (1990). Cryopreservation of *Drosophila melanogaster* embryos. *Nature* 345, 6271, 170-172.

Stover, J & Evans, J. (1984) Interspecies embryos transfer from gaur (*Bos gaurus*) to domestic Holstein cattle (*Bos taurus*) at the New York Zoological Park. *Proceedings of the 10th International Congress on Animal Reproduction and Artificial Insemination,* University of Illinois at Urbana-Champaign, Illinois, USA, June 10-14, 1984.

Stukenborg, J-B, Schlatt, S, Simoni, M, Yeung, C-H, Elhija, MA, Luetjens, CM, Huleihel, M & Wistuba, J. (2009). New horizons for *in vitro* spermatogenesis? An update on novel three-dimensional culture systems as tools for meiotic and post-meiotic differentiation of testicular germ cells. *Molecular Human Reproduction* 15, 9, 521-529.

Sugimoto, M, Maeda, S, Manabe, N & Miyamoto, H. (2000). Development of infantile rat ovaries autotransplanted after cryopreservation by vitrification. *Theriogenology* 53, 5, 1093-1103.

Summers, PM, Shephard, AM, Taylor, CT & Hearn, JP. (1987). The effects of cryopreservation and transfer on embryonic development in the common marmoset monkey, *Callithrix jacchus*. *Journal of Reproduction and Fertility* 79, 1, 241-250.

Sun, X, Li, Z, Yi, Y, Chen, J, Leno, GH & Engelhardt, JF. (2008). Efficient term development of vitrified ferret embryos using a novel pipette chamber technique. *Biology of Reproduction* 79, 5, 832-840.

Suzuki, H, Asano, T, Suwa, Y & Abe, Y. (2009). Successful delivery of pups from cryopreserved canine embryos. *Biology of Reproduction* 81, 1 supplement, 619 (abstract).

Swanson, WF. (2001) Reproductive biotechnology and conservation of the forgotten felids - the small cats. *The 1st International Symposium on Assisted Reproductive Technology for the Conservation and Genetic Management of Wildlife*, Omaha, 17-18 Januar, 2001.

Swanson, WF. (2003). Research in nondomestic species: Experiences in reproductive physiology research for conservation of endangered felids. *ILAR Journal* 44, 4, 307-316.

Swanson, WF. (2006). Application of assisted reproduction for population management in felids: The potential and reality for conservation of small cats. *Theriogenology* 66, 1, 49-58.

Swanson, WF & Brown, JL. (2004). International training programs in reproductive sciences for conservation of Latin American felids. *Animal Reproduction Science* 82-83, 21-34.

Swanson, WF, Paz, RCR, Morais, RN, Gomes, MLF, Moraes, W & Adania, CH. (2002). Influence of species and diet on efficiency of *in vitro* fertilization in two endangered Brazilian felids - the ocelot (*Leopardus pardalis*) and tigrina (*Leopardus tigrinus*). *Theriogenology* 57, 1, 593 (abstract).

Takagi, M, Kim, IH, Izadyar, F, Hyttel, P, Bevers, MM, Dieleman, SJ, Hendriksen, PJ & Vos, PL. (2001). Impaired final follicular maturation in heifers after superovulation with recombinant human FSH. *Reproduction* 121, 6, 941-951.

Takahashi, K & Yamanaka, S. (2006). Induction of pluripotent stem cells from mouse embryonic and adult fibroblast cultures by defined factors. *Cell* 126, 4, 663-676.

Tamayo-Canul, J, Alvarez, M, LÛpez-UrueÒa, E, Nicolas, M, Martinez-Pastor, F, Anel, E, Anel, L & de Paz, P. (2011). Undiluted or extended storage of ram epididymal spermatozoa as alternatives to refrigerating the whole epididymes. *Animal Reproduction Science* 126, 1-2, 76-82.

Tao, Y, Liu, J, Zhang, Y, Zhang, M, Fang, J, Han, W, Zhang, Z, Liu, Y, Ding, J & Zhang, X. (2009). Fibroblast cell line establishment, cryopreservation and interspecies embryos reconstruction in red panda (*Ailurus fulgens*). *Molecular Ecology* 17, 2, 117-124.

Tharasanit, T & Techakumphu, M. (2010). The effect of chemical delipidation on cryopreservability of cat embryos. *Reproduction, Fertility and Development* 23, 1, 153 (abstract).

Thibier, M. (2009). Data Retrieval Committee statistics of embryo transfer - year 2008. The worldwide statistics of embryo transfers in farm animals. *International Embryo Transfer Society Newsletter* 27, 4, 13-19.

Thomas, CD, Cameron, A, Green, RE, Bakkenes, M, Beaumont, LJ, Collingham, YC, Erasmus, BFN, de Siqueira, MF, Grainger, A, Hannah, L, Hughes, L, Huntley, B, van Jaarsveld, AS, Midgley, GF, Miles, L, Ortega-Huerta, MA, Townsend Peterson, A, Phillips, OL & Williams, SE. (2004). Extinction risk from climate change. *Nature* 427, 6970, 145-148.

Thomson, JA, Itskovitz-Eldor, J, Shapiro, SS, Waknitz, MA, Swiergiel, JJ, Marshall, VS & Jones, JM. (1998). Embryonic stem cell lines derived from human blastocysts. *Science* 282, 5391, 1145-1147.

Thundathil, J, Whiteside, D, Shea, B, Ludbrook, D, Elkin, B & Nishi, J. (2007). Preliminary assessment of reproductive technologies in wood bison (*Bison bison athabascae*): Implications for preserving genetic diversity. *Theriogenology* 68, 1, 93-99.

Thurston, LM, Siggins, K, Mileham, AJ, Watson, PF & Holt, WV. (2002). Identification of amplified restriction fragment length polymorphism markers linked to genes controlling boar sperm viability following cryopreservation. *Biology of Reproduction* 66, 3, 545-554.

Tilly, JL, Niikura, Y & Rueda, BR. (2009). The current status of evidence for and against postnatal oogenesis in mammals: A case of ovarian optimism versus pessimism? *Biology of Reproduction* 80, 1, 2-12.

Tourmente, M, Gomendio, M & Roldan, E. (2011). Sperm competition and the evolution of sperm design in mammals. *BMC Evolutionary Biology* 11, 1, 12.

Toyooka, Y, Tsunekawa, N, Akasu, R & Noce, T. (2003). Embryonic stem cells can form germ cells *in vitro*. *Proceedings of the National Academy of Science of the Uunited States of America* 100, 20, 11457-11462.

Trimeche, A, Renard, P & Tainturier, D. (1998). A procedure for Poitou jackass sperm cryopreservation. *Theriogenology* 50, 5, 793-806.

Trounson, A & Kirby, C. (1989). Problems in the cryopreservation of unfertilized eggs by slow cooling in dimethyl sulfoxide. *Fertility and Sterility* 52, 5, 778-786.

Trounson, A & Mohr, L. (1983). Human pregnancy following cryopreservation, thawing and transfer of an eight-cell embryo. *Nature* 305, 5936, 707-709.

Tryon, BW & Murphy, JB. (1982). Miscellaneous notes on the reproductive biology of reptiles. 5. Thirteen varieties of the genus *Lampropeltis*, species *mexicana*, *triangulum* and *zonata*. *Transactions of the Kansas Academy of Science* 85, 2, 96-119.

Uribe, MCA & Guillette, LJ. (2000). Oogenesis and ovarian histology of the American alligator *Alligator mississippiensis*. *Journal of Morphology* 245, 3, 225-240.

Van Blerkom, J. (1989). Maturation at high frequency of germinal-vesicle-stage mouse oocytes after cryopreservation: alterations in cytoplasmic, nuclear, nucleolar and chromosomal structure and organization associated with vitrification. *Human Reproduction* 4, 8, 883-898.

Van Thuan, N, Wakayama, S, Kishigami, S & Wakayama, T. (2005). New preservation method for mouse spermatozoa without freezing. *Biology of Reproduction* 72, 2, 444-450.

Vendramini, OM, Bruyas, JF, Fieni, F, Battut, I & Tainturier, D. (1997). Embryo transfer in Poitou donkeys, preliminary results. *Theriogenology* 47, 1, 409 (abstract).

Veprintsev, BN & Rott, NN. (1979). Conserving genetic resources of animal species. *Nature* 280, 5724, 633-634.

von Baer, A, Del Campo, MR, Donoso, X, Toro, F, von Baer, L, Montecinos, S, Rodriguez-Martinez, H & Palasz, T. (2002). Vitrification and cold storage of llama (*Lama glama*) hatched blastocysts. *Theriogenology* 57, 1, 489 (abstract).

von Schönfeldt, V, Chandolia, R, Kiesel, L, Nieschlag, E, Schlatt, S & Sonntag, B. (2011). Assessment of follicular development in cryopreserved primate ovarian tissue by xenografting: prepubertal tissues are less sensitive to the choice of cryoprotectant. *Reproduction* 141, 4, 481-490.

Wakayama, T & Yanagimachi, R. (1998). Development of normal mice from oocytes injected with freeze-dried spermatozoa. *Nature Biotechnology* 16, 7, 639-641.

Wake, DB & Vredenburg, VT. (2008). Colloquium paper: are we in the midst of the sixth mass extinction? A view from the world of amphibians. *Proceedings of the National Academy of Science of the Uunited States of America* 105 Suppl 1, 11466-11473.

Walmsley, R, Cohen, J, Ferrara-Congedo, T, Reing, A & Garrisi, J. (1998). The first births and ongoing pregnancies associated with sperm cryopreservation within evacuated egg zonae. *Human Reproduction* 13, suppl 4, 61-70.

Walpole, M, Almond, REA, Besancon, C, Butchart, SHM, Campbell-Lendrum, D, Carr, GM, Collen, B, Collette, L, Davidson, NC, Dulloo, E, Fazel, AM, Galloway, JN, Gill, M, Goverse, T, Hockings, M, Leaman, DJ, Morgan, DHW, Revenga, C, Rickwood, CJ, Schutyser, F, Simons, S, Stattersfield, AJ, Tyrrell, TD, Vie, J-C & Zimsky, M. (2009). Tracking progress toward the 2010 biodiversity target and beyond. *Science* 325, 5947, 1503-1504.

Wang, X, Chen, H, Yin, H, Kim, SS, Lin Tan, S & Gosden, RG. (2002). Fertility after intact ovary transplantation. *Nature* 415, 6870, 385.

Ward, MA, Kaneko, T, Kusakabe, H, Biggers, JD, Whittingham, DG & Yanagimachi, R. (2003). Long-term preservation of mouse spermatozoa after freeze-drying and freezing without cryoprotection. *Biology of Reproduction* 69, 6, 2100-2108.

Weissman, A, Gotlieb, L, Colgan, T, Jurisicova, A, Greenblatt, EM & Casper, RF. (1999). Preliminary experience with subcutaneous human ovarian cortex transplantation in the NOD-SCID mouse. *Biology of Reproduction* 60, 6, 1462-1467.

Whittingham, DG. (1971). Survival of mouse embryos after freezing and thawing. *Nature* 233, 5315, 125-126.

Whittingham, DG. (1975). Survival of rat embryos after freezing and thawing. *Journal of Reproduction and Fertility* 43, 3, 575-578.

Whittingham, DG. (1977). Fertilization *in vitro* and development to term of unfertilized mouse oocytes previously stored at -196°C. *Journal of Reproduction and Fertility* 49, 1, 89-94.

Whittingham, DG & Adams, CE. (1974). Low temperature preservation of rabbit embryos. *Cryobiology* 11, 6, 560-561 (abstract).

Whittingham, DG & Adams, CE. (1976). Low temperature preservation of rabbit embryos. *Journal of Reproduction and Fertility* 47, 2, 269-274.

Whittingham, DG, Leibo, SP & Mazur, P. (1972). Survival of mouse embryos frozen to -196° and -269°C. *Science* 178, 4059, 411-414.

Whittington, K & Ford. (1999). Relative contribution of leukocytes and of spermatozoa to reactive oxygen species production in human sperm suspensions. *International Journal of Andrology* 22, 4, 229-235.

Wildt, DE. (1992). Genetic resource banks for conserving wildlife species: justification, examples and becoming organized on a global basis. *Animal Reproduction Science* 28, 1-4, 247-257.

Wildt, DE, Schiewe, MC, Schmidt, PM, Goodrowe, KL, Howard, JG, Phillips, LG, O'Brien, SJ & Bush, M. (1986). Developing animal model systems for embryo technologies in rare and endangered wildlife. *Theriogenology* 25, 1, 33-51.

Willadsen, S, Polge, C & Rowson, LEA. (1978). The viability of deep-frozen cow embryos. *Journal of Reproduction and Fertility* 52, 2, 391-393.

Willadsen, SM, Polge, C, Rowson, LEA & Moor, RM. (1974). Preservation of sheep embryos in liquid nitrogen. *Cryobiology* 11, 6, 560 (abstract).

Willadsen, SM, Polge, C, Rowson, LEA & Moor, RM. (1976). Deep freezing of sheep embryos. *Journal of Reproduction and Fertility* 46, 1, 151-154.

Wilmut, I. (1972). The effect of cooling rate, warming rate, cryoprotective agent and stage of development of survival of mouse embryos during freezing and thawing. *Life Sciences* 11, 22, Part 2, 1071-1079.

Wilmut, I & Rowson, LE. (1973). Experiments on the low-temperature preservation of cow embryos. *The Veterinary Record* 92, 26, 686-690.

Wilmut, I, Schnieke, AE, McWhir, J, Kind, AJ & Campbell, KH. (1997). Viable offspring derived from fetal and adult mammalian cells. *Nature* 385, 6619, 810-813.

Wolf, DP, Vandevoort, CA, Meyer-Haas, GR, Zelinski-Wooten, MB, Hess, DL, Baughman, WL & Stouffer, RL. (1989). *In vitro* fertilization and embryo transfer in the rhesus monkey. *Biology of Reproduction* 41, 2, 335-346.

Wolfe, BA & Wildt, DE. (1996). Development to blastocysts of domestic cat oocytes matured and fertilized *in vitro* after prolonged cold storage. *Journal of Reproduction and Fertility* 106, 1, 135-141.

Wolvekamp, MC, Cleary, ML, Cox, SL, Shaw, JM, Jenkin, G & Trounson, AO. (2001). Follicular development in cryopreserved common wombat ovarian tissue xenografted to nude rats. *Animal Reproduction Science* 65, 1-2, 135-147.

Wood, TC, Montali, RJ & Wildt, DE. (1997). Follicle-oocyte atresia and temporal taphonomy in cold-stored domestic cat ovaries. *Molecular Reproduction and Development* 46, 2, 190-200.

Wood, TC & Wildt, DE. (1997). Effect of the quality of the cumulus-oocyte complex in the domestic cat on the ability of oocytes to mature, fertilize and develop into blastocysts *in vitro*. *Journal of Reproduction and Fertility* 110, 2, 355-360.

Woods, EJ, Newton, L & Critser, JK. (2010). A novel closed system vial with sentinel test segment for sperm cryopreservation. *Fertility and Sterility* 94, 4, Supplement 1, S239-S239.

Woods, GL, White, KL, Vanderwall, DK, Li, GP, Aston, KI, Bunch, TD, Meerdo, LN & Pate, BJ. (2003). A mule cloned from fetal cells by nuclear transfer. *Science* 301, 5636, 1063.

Xu, Y-N, Cui, X-S, Sun, S-C, Jin, Y-X & Kim, N-H. (2011). Cross species fertilization and development investigated by cat sperm injection into mouse oocytes. *Journal of Experimental Zoology Part A: Ecological Genetics and Physiology* 315A, 6, 349-357.

Yamamoto, Y, Oguri, N, Tsutsumi, Y & Hachinohe, Y. (1982). Experiments in the freezing and storage of equine embryos. *Journal of Reproduction and Fertility Supplement* 32, 399-403.

Yamauchi, Y, Shaman, JA & Ward, WS. (2011). Non-genetic contributions of the sperm nucleus to embryonic development. *Asian Journal of Andrology* 13, 1, 31-35.

Yang, C-R, Miao, D-Q, Zhang, Q-H, Guo, L, Tong, J-S, Wei, Y, Huang, X, Hou, Y, Schatten, H, Liu, Z & Sun, Q-Y. (2010). Short-term preservation of porcine oocytes in ambient temperature: Novel approaches. *PLoS One* 5, 12, e14242.

Yavin, S, Aroyo, A, Roth, Z & Arav, A. (2009). Embryo cryopreservation in the presence of low concentration of vitrification solution with sealed pulled straws in liquid nitrogen slush. *Human Reproduction* 24, 4, 797-804.

Yeoman, RR, Gerami-Naini, B, Mitalipov, S, Nusser, KD, Widmann-Browning, AA & Wolf, DP. (2001). Cryoloop vitrification yields superior survival of Rhesus monkey blastocysts. *Human Reproduction* 16, 9, 1965-1969.

Yeoman, RR, Wolf, DP & Lee, DM. (2005). Coculture of monkey ovarian tissue increases survival after vitrification and slow-rate freezing. *Fertility and Sterility* 83, 4, Supplement 1, 1248-1254.

Yin, H, Wang, X, Kim, SS, Chen, H, Tan, SL & Gosden, RG. (2003). Transplantation of intact rat gonads using vascular anastomosis: effects of cryopreservation, ischaemia and genotype. *Human Reproduction* 18, 6, 1165-1172.

Yu, I & Leibo, SP. (2002). Recovery of motile, membrane-intact spermatozoa from canine epididymides stored for 8 days at 4°C. *Theriogenology* 57, 3, 1179-1190.

Yushchenko, NP. (1957). Proof of the possibility of preserving mammalian spermatozoa in a dried state. *Proceedings of the Lenin Academy of Agricultural Sciences of the USSR* 22, 37-40.

Zambelli, D, Prati, F, Cunto, M, Iacono, E & Merlo, B. (2008). Quality and *in vitro* fertilizing ability of cryopreserved cat spermatozoa obtained by urethral catheterization after medetomidine administration. *Theriogenology* 69, 4, 485-490.

Zeilmaker, GH, Alberda, AT, van Gent, I, Rijkmans, CM & Drogendijk, AC. (1984). Two pregnancies following transfer of intact frozen-thawed embryos. *Fertility and Sterility* 42, 2, 293-296.

Zeng, W, Avelar, GF, Rathi, R, Franca, LR & Dobrinski, I. (2006). The length of the spermatogenic cycle is conserved in porcine and ovine testis xenografts. *Journal of Andrology* 27, 4, 527-33.

Zeron, Y, Pearl, M, Borochov, A & Arav, A. (1999). Kinetic and temporal factors influence chilling injury to germinal vesicle and mature bovine oocytes. *Cryobiology* 38, 1, 35-42.

Zeron, Y, Tomczak, M, Crowe, J & Arav, A. (2002). The effect of liposomes on thermotropic membrane phase transitions of bovine spermatozoa and oocytes: implications for reducing chilling sensitivity. *Cryobiology* 45, 2, 143-152.

Zhang, T & Rawson, DM. (1995). Studies on chilling sensitivity of zebrafish (*Brachydanio rerio*) embryos. *Cryobiology* 32, 3, 239-246.

Zhang, T & Rawson, DM. (1996). Feasibility studies on vitrification of intact zebrafish (*Brachydanio rerio*) embryos. *Cryobiology* 33, 1, 1-13.

Zuckerman, S. (1951). The number of oocytes in the mature ovary. *Recent Progress in Hormon Research* 6, 1, 63-109.

Part 2

Cryopreservation of Aquatic Species

3

Effect of Cryopreservation on Bio-Chemical Parameters, DNA Integrity, Protein Profile and Phosphorylation State of Proteins of Seawater Fish Spermatozoa

Loredana Zilli and Sebastiano Vilella
Laboratory of Comparative Physiology, Department of Biological and Environmental Sciences and Technologies,
University of Salento, Via Provinciale Lecce-Monteroni, Lecce
Italy

1. Introduction

Fish sperm cryopreservation is considered as a valuable technique for artificial reproduction and genetic improvement (Chao & Liao, 2001; Kopeika et al., 2007; Rana, 1995; Suquet et al., 2000). Semen quality must be monitored when attempts are made to increase the efficiency of artificial fertilization, to cryopreserve only sperm of high quality, and to evaluate frozen-thawed sperm. Cryopreserved sperm usually shows, with respect to fresh sperm, a lower quality, since the freezing–thawing procedure affects DNA and protein integrity (Labbe et al., 2001; Zilli et al., 2003, 2005), membrane lipids (Maldjian et al., 2005; Müller et al., 2008), sperm motility (Linhart et al., 2000; Ritar, 1999; Rodina et al., 2007; Zilli et al., 2005), fertilization ability (Gwo & Arnold, 1992; Rana, 1995), and also larval survival (Suquet et al., 1998). Spermatozoa genome alteration due to cryopreservation may affect only late embryonic development and larval survival (Kopeika et al., 2003a, 2003b, 2004; Suquet et al., 1998), but not the early events in embryonic development, because these are controlled by maternally inherited information (Braude et al., 1988). On the contrary, defects in sperm proteins (degradation and/or change of the phosphorylation state) may compromise sperm motility, fertilization ability, and the early events after fertilization (Cao et al., 2003; Huang et al., 1999; Lessard et al., 2000).

The most common parameters used to evaluate sperm quality are fertilization ability, motility (rate and duration) and cellular (chemical and/or biochemical) parameters. Fertilizing capacity is the most conclusive test of sperm quality but the use of this marker is laborious and requires the availability of eggs (McNiven et al., 1992). Motility is normally evaluated as percentage and duration, but some authors also use velocity, flagellum beat frequency, or other parameters measured by computer-assisted sperm analysis (Ciereszko et al., 1996; Cosson et al., 2000; Rurangwa et al., 2001). Cellular bio markers has been used to evaluate spermatozoa quality of different fish species such as Atlantic salmon (Aas et al., 1991; Hwang & Idler, 1969), rainbow trout (Ciereszko & Dabrowski, 1994; Lahnsteiner et al.,1996a, 1998) and sea bass (Zilli et al., 2004). All these parameters have been also used to evaluate the effect of cryopreservation on spermatozoa quality.

Here we reviewed data obtained by our group, on the effect of freezing-thawing procedures on sea bass and sea bream sperm. In particular, data concerning the effect of cryopreservation on bio-chemical parameters, DNA integrity, protein profile and phosphorylation state, are reported.

2. Effect of freezing-thawing procedure on sea bass spermatozoa bio-chemical parameters. Use of intracellular ATP concentration and seminal plasma ß-D-glucuronidase activity as quality marker of fresh and frozen-thawed semen

2.1 Effect of cryopreservation on sea bass semen

The cryopreservation of spermatozoa is known to result in considerable damage to cellular structures such as plasma membrane, nucleus, mitochondria, and flagellum (Conget et al., 1996; Drokin et al., 1998; Lahnsteiner et al., 1992; 1996b; Zhang et al., 2003). Spermatozoa plasma membrane is the cellular structure most susceptible to damage during cryopreservation (Baynes & Scott, 1987). It is well known that the activity of intracellular enzymes, the seminal plasma protein concentrations and the seminal plasma enzyme activities can be used to evaluate spermatozoan plasma membrane integrity (Babiak & Glogowski, 1997; Lahnsteiner et al., 1998; McNiven et al., 1992). Our group has evaluated (Zilli et al., 2004) the effect of cryopreservation on sea bass semen quality by measuring chemical and biochemical parameters (reported in Table 1), before and after sperm cryopreservation. Results obtained demonstrated that, the used cryopreservation protocol did not cause significant injuries to spermatozoan plasma membranes, since the activity of the intracellular enzymes as well as the seminal plasma protein concentrations and ß-D-glucuronidase activity were not affected by the freezing-thawing procedure. The absence of injuries at the plasma membrane level was also supported by the observation that eosin uptake was similar in fresh and frozen-thawed spermatozoa (75-80%). Only malate dehydrogenase activity and intracellular ATP concentration resulted significantly higher in cryopreserved than in fresh samples, while the pH of seminal plasma resulted significantly lower after freezing

Since respiration rate, malate dehydrogenase activity, and intracellular ATP concentration did not decrease in frozen/thawed spermatozoa we concluded that spermatozoan mitochondria were intact and active after cryopreservation procedure. The increases in both intracellular ATP concentration and malate dehydrogenase activity following the freezing-thawing procedure has been also reported in *Silurus glanis* (Ogier de Baulnyet al., 1999). The increase of intracellular ATP concentration occurs during early freezing of sperm (from +20°C to -10°C) (Baynes & Scott, 1987) and is most probably attributable to dimethyl sulfoxide, used as cryoprotectant, which interferes with cellular metabolism (McConnell et al., 1999). No data is available concerning the kinetics and structure of mitochondrial malate dehydrogenase in fish. In other vertebrates, mitochondrial malate dehydrogenase shows a complex dependence on the ionic environment, which influences both kinetics and structure (Birktoft et al., 1989; Bleile et al., 1977; Harada & Wolf, 1968;

Effect of Cryopreservation on Bio-Chemical Parameters, DNA Integrity, Protein Profile and
Phosphorylation State of Proteins of Seawater Fish Spermatozoa

103

Ruggia et al., 2001; Wood et al., 1981). The increase of malate dehydrogenase activity after cryopreservation could be a consequence of the oxidative stress that occurs during the freezing phase, as previously suggested (Lahnsteiner et al., 1998), or could be due to the presence of anions that increase the activity of the enzyme by stabilizing the dimeric form (Ruggia et al., 2001).

Parameters	Fresh sperm	Cryopreserved sperm
Respiration rate (μg O_2/min $\times$ ml seminal fluid)	6.71±0.76a (N=22)	7.09±1.22a (N=18)
Aspartate aminotransferase (U/mg protein)	0.010±0.008a (N=62)	0.012±0.009a (N=45)
Malate dehydrogenase (U/mg protein)	0.054±0.03a (N=65)	0.079±0.046b (N=61)
Isocitrate dehydrogenase (U/mg protein)	0.11±0.05a (N=45)	0.18±0.10a (N=40)
Intracellular ATP (μmoli/protein)	1.22±0.65a (N=72)	1.92±1.11b (N=66)
Intracellular triglycerides (μmoli/protein)	0.33±0.24a (N=48)	0.25±0.14a (N=38)
Intracellular glycerol (μmoli/protein)	0.28±0.18a (N=45)	0.21±0.15a (N=27)
Seminal plasma osmolality (mOsm/Kg)	352.1±19.7a (N=22)	354.9±20.7a (N=18)
Seminal plasma pH	8.21±0.45a (N=62)	7.65±0.61b (N=45)
Seminal plasma protein (mg/l)	815.3±174.5a (N=45)	837.1±180.0a (N=40)
Seminal plasma triglycerides (μmoli/l)	226.2±107.2a (N=72)	181.8±103.0a (N=66)
Seminal plasma ß-D-glucuronidase (U/l)	0.0083±0.0066a (N=48)	0.0093±0.0049a (N=38)

Table 1. Chemical and biochemical parameters measured in sea bass spermatozoa and seminal plasma before and after cryopreservation. Values (± SD) in a row with the same letter are not significantly different (P>0.01). N=Number of sperm samples from different males. (This table was originally published in Zilli et al., Biol. Reprod 2004)

2.2 Relationship of sperm and seminal plasma parameters and fertilization rate in fresh and cryopreserved semen samples

The most common parameters used to evaluate sperm quality are fertilization ability, motility (rate and duration) and cellular (chemical and/or biochemical) parameters. In sea bass we identified simple and cost-effective markers of sperm quality that would replace conventional motility and fertility evaluation assays, using both fresh and frozen-thawed sperm. Parameters of sperm metabolism and seminal plasma were tested by evaluating correlations with the fertilization rate using simple regression analysis and square relationship analysis.

In fresh sperm, among the measured cellular metabolites and enzymes, only ATP concentration and aspartate aminotransferase activity showed significant linear correlations (P<0.0001) with fertilization rate (Fig. 1) and the calculation of the partial correlation coefficient revealed that these two parameters were not correlated (Pr=-0.323). Malate dehydrogenase activity and sperm triglyceride concentration had a quadratic relation with fertilization rate: R^2=0.31, P<0.001 for malate dehydrogenase; R^2=0.28, P<0.01 for triglyceride concentration (see Zilli et al. 2004 for details).

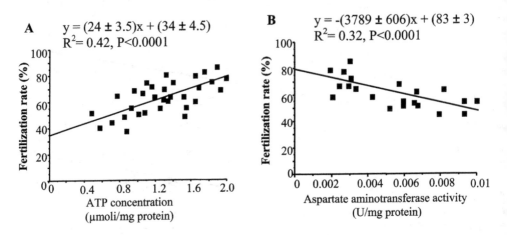

Fig. 1. Relationship between fertilization rate and ATP concentration (A, N=32) or aspartate aminotransferase activity (B, N=21) using sea bass fresh sperm samples. Samples obtained from different males were used to measure the ATP concentration and the aspartate aminotransferase activity and to perform the fertilization trials. N=Number of sperm samples from different males. (This figure was originally published in Zilli et al., Biol. Reprod 2004).

ATP concentrations of >1.8 mmol/mg protein characterized sperm with fertilization rates ≥75%. The relationship between ATP concentration and fertilization rate is due to the fact that the flagellar beat frequency of spermatozoa depends on ATP concentration and dynein ATPase activity (Christen et al., 1987; Lahnsteiner et al., 1998). Thus, intracellular ATP concentration could be used instead of sperm motility as a predictor of fertilization ability. Determination of ATP concentration has some advantages over motility assessment: it is not subjective as is motility determination based on microscopic observation (McNiven et al., 1992) and it is faster and less expensive with respect to the Computer Assisted Sperm Analysis system. Aspartate aminotransferase activity activities of 0.3 mU/100 mg protein characterized sperm with fertilization rates of 75%. A correlation between the activity of this enzyme and fertilization rate was also found in lake whitefish (*Coregonus clupeaformis*) (Ciereszko & Dabrowski, 1994) and rainbow trout (Lahnsteiner et al., 1998). The physiological meaning of this relationship is uncertain.

Among the seminal plasma (tested) parameters, only β-D-glucuronidase activity and potassium concentration had a significant linear relation (P<0.01) with fertilization rate (Fig.

2) and also in this case the calculation of the partial correlation coefficient revealed that these two parameters were not correlated.

β-D-Glucuronidase activity is negatively correlated with fertilization rate. This enzyme is involved in hydrolysis of β-glucuronides to glucuronic acid and is located most frequently in lysosomes (Rawn, 1983). It is located in the spermatic duct epithelium, usually in areas where lytic processes occur, and is also secreted into the seminal fluid (Lahnsteiner et al., 1994). A correlation between this activity of this enzyme and fertilization rate was also found in rainbow trout (Lahnsteiner et al., 1998). An increase of seminal plasma β-D-glucuronidase activity indicates degeneration or aging processes in the semen (Lahnsteiner et al., 1998). Seminal plasma potassium concentration concentrations of 17 mM characterized sperm with fertilization rates of 75%. Quadratic functions were used to described the relationship between fertilization rate and potassium concentration in other fish species, i.e., bleak (*Alburnus alburnus*), Atlantic salmon, and rainbow trout (Aas et al., 1991; Lahnsteiner et al., 1998).

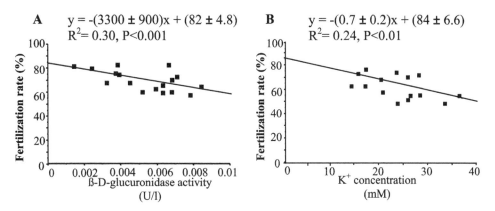

Fig. 2. Relationship between fertilization rate and seminal plasma β-D-glucuronidase activity (A, N=17) or seminal plasma potassium concentration (B, N=15) using fresh sea bass sperm samples. Samples obtained from different males were used to measure β-D-glucuronidase activity and seminal plasma potassium concentration and to perform the fertilization trials. N=Number of sperm samples from different males. (This figure was originally published in Zilli et al, Biol Reprod 2004)

Sperm and seminal plasma parameters of fresh semen that showed linear correlation with the fertilization rate have been also used in multiple regression models to predict the fertilization ability. Three models were tested: the first model included sperm ATP concentration and aspartate aminotransferase activity, the second model included the seminal plasma β-D-glucuronidase activity and potassium concentration while the third model included ATP, aspartate aminotransferase and β-D-glucuronidase (potassium concentration was exclude due to its linear relathionshep with ATP). Results indicated that sea bass fresh semen fertilization rate was well predicted by the first multiple regression model, which included cellular parameters (see Zilli et al 2004 for details).

Because sperm ATP concentration and seminal plasma β-D-glucuronidase activity among the tested parameters produced the highest correlation coefficients, we also investigated their relationship with fertilization rate in frozen-thawed samples. These parameters showed a linear relationship with fertilization rate also after the freezing-thawing procedure (Fig. 3) similar to what happens for fresh semen.

For pratical application the measuraments of ATP concentration and seminal plasma β-D-glucuronidase activity represents an alternative simple and cost-effective tests for evaluating sea bass sperm fertilization ability before and after cryopreservation.

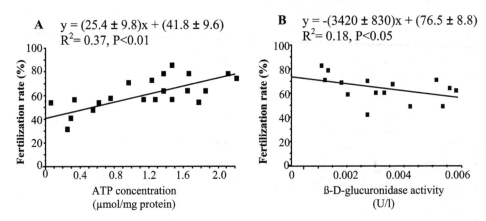

Fig. 3. Relationship between fertilization rate and ATP concentration (A, N=21) or seminal plasma β-D-glucuronidase activity (B, N=15) using cryopreserved sea bass sperm samples. Samples obtained from different males were used to measure the ATP concentration and β-D-glucuronidase activity and to perform the fertilization trials. N=Number of sperm samples from different males. (This figure was originally published in Zilli et al., Biol Reprod 2004).

3. Effect of cryopreservation on DNA integrity on sea bass spermatozoa

Sperm DNA fragmentation could be a consequence of the freezing-thawing process and the resulting genome alterations could affect late embryonic development and survival of larvae (Suquet et al., 1998). There are different methods to determine the DNA fragmentation, among these an effective tool is single-cell gel electrophoresis (SCGE). Introduced by Ostling & Johanson in the 1984 it has become a recognized method for detecting DNA damage in a variety of vertebrate cell types, including sperm (Fairbairn et al., 1995; Hughes et al., 1997; Steele et al., 2000). In this assay, the fragmented DNA migrates toward the anode, giving the appearance of a "comet tail " while the undamaged DNA appears as intact comet heads (lacking tail). These comets can be easily visualized when stained with DAPI. By using this technique we have demonstrated (Zilli et al, 2003) that the cryopreservation protocol (Fauvel et al 1998) used to cryopreserve the sea bass sperm cause significantly damage at DNA level (Figure 4). Results, expressed in terms of the "percent tail DNA" (% DNA_T) and "tail moment" (MT) (Ashby et al., 1995; Helma & Uhl, 2000; Johnson & Ferris, 2002; Piperakis et al., 1999) were reported in table 2.

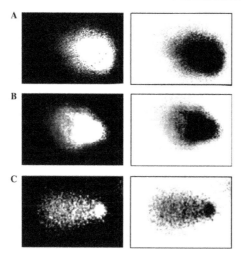

Fig. 4. The appearance of fresh (A), frozen-thawed (B) and unprotected frozen-thawed (C)
sea bass sperm following preparation by the SCGE assay. On the right are shown the
negative images of the same preparation used to perform the analysis. (This figure was
originally published in Zilli et al., Cryobiology 2003)

Parameter analysed	Fresh sperm	Frozen-thawed sperm	Unprotected frozen-thawed sperm
Motile sperm (%)	75±15a	67±18a	n.d.
Fertilization rate (%)	74±15a	70±12a	n.d.
Motility duration (sec)	129±46a	28±8b	n.d.
Percent Tail DNA	32.7±11.1a	38.2±11.2b	65.2±10.2c
Tail Moment	375.2±190.7a	498.9±166.4b	2345.1±725.2c

Table 2. Effect of cryopreservation on DNA integrity, sperm motility and fertilizing ability
determined on fresh and frozen-thawed in the presence or absence of cryoprotectant
(Me$_2$SO). Values in a row with the same letter are not significantly different (P>0.01). n.d.:
not detectable. (This table was originally published in Zilli et al., Cryobiology 2003).

The results obtained indicate that the cryopreservation protocol used for sea bass sperm
(Fauvel et al 1998): (1) is without effect on both sperm rate motility and fertilizing ability; (2)
significantly reduced the duration of motility, (3) is associated with DNA damage that,
although significant, is of low magnitude and (4) demonstrated the fundamental role
played by cryoprotectant (Me$_2$SO) in reducing fish sperm DNA fragmentation. The role
played by Me$_2$SO was also demonstrated by using DNA laddering (Fig. 5A). When the
analysis was performed on fresh semen samples no smearing was detectable (lanes 4 and 5).
In some frozen-thawed semen samples (lanes 7 and 11) but not in all (see lanes 6, 8, 9, and
10) a small degree of laddering seems to be present. On the contrary, in unprotected frozen-
thawed semen DNA laddering was clearly evident (lane 13).

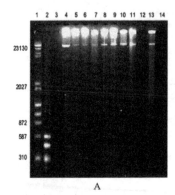

Sample	% DNA$_T$	M$_T$
Sample 4	32.8±9.9 a	371.7±189.8 c
Sample 5	32.3±11.8 a	391.2±180.1 c
Sample 6	38.3±11.2 b	486.7±154.2 d
Sample 7	39.9±6.6 b	508.2±146.6 d
Sample 8	38.6±11.1 b	498.1±184.1 d
Sample 9	38.4±12.3 b	491.3±165.3 d
Sample 10	38.8±10.7 b	496.9±161.2 d
Sample 11	40.1±7.7 b	514.3±178.0 d

A   B

Fig. 5. DNA fragmentation of sea bass sperm samples. (A) Agarose gel electrophoresis of DNA isolated from sea bass sperm. Lanes 1 and 2: DNA molecular weight markers (pb); lanes 4 and 5: 2 lg of DNA isolated from fresh sperm; lanes 6, 7, 8, 9, 10, and 11: 2 lg of DNA isolated from frozen-thawed sperm; lane 13: 2 lg of DNA isolated from unprotected frozen-thawed sperm; (B) SCGE assay in fresh samples (4 and 5) and frozen-thawed samples (6-11) of sea bass sperm. Values are given as mean ± SD. Values within a column followed by the same letter are not significantly different (P>0.01). (Modified from Zilli et al., Cryobiology 2003)

Since in some frozen-thawed semen samples (lanes 7 and 11 of Fig. 5A) analyzed by DNA laddering analysis, but not in all (lanes 6, 8, 9, and 10), a small degree of laddering seems to occur, we analyzed the same samples with the SCGE method. The results reported in Fig 5B confirmed the presence of DNA fragmentation in the samples 7 and 11; in addition it revealed a significant degree of DNA fragmentation in the samples 6, 8, 9, and 10 with respect to fresh samples (4 and 5). In any case it must be underlined that within the frozen/thawed samples (6, 7, 8, 9, 10, and 11) no statistically significant differences in the DNA fragmentation was revealed by the SCGE method (Fig 5B).

DNA laddering has been used in many studies to obtain a qualitative analysis of DNA fragmentation (Duke & Cohen, 1986; Homma-Takeda et al., 2001; Sun et al., 1999). It is a very simple method, but the most critical problem with DNA electrophoretical analysis are its inability to provide quantitative measurement and its low sensitivity. In fact, random double-stranded or rare single-stranded DNA fragmentation in cells, cannot be detected by this technique. On the contrary, the SCGE or Comet assay has been recognized as one of the most sensitive techniques for measuring DNA strand breaks (Collins et al., 1997). For human sperm, comet assay has been shown to have a significant relationship both to the SCSA (Sperm Chromatin Structure Assay) (Larson et al., 2001) and the TUNEL assay (Terminal Deoxynucleotidyl Transferasemediated Nick End Labelling), another technique for detecting the incidence of DNA fragmentation (Sakkas et al., 1999). The use of the Comet assay in alkaline conditions is a usefull tool to carry out a quantitative analysis of DNA fragmentation. Previous works (Collins et al., 1997; McKelvey-Martin et al., 1993) have reported that the assay resolves break frequencies up to a few hundred per cell, definitely well beyond the range of fragment size for which conventional electrophoresis is suitable. Since introduction of the Comet assay protocol (Ostling & Johanson, 1984), there have been modifications of it for use with various cell types, including sperm (Fairbairn et al., 1995; Hughes et al., 1997; Steele et al., 2000). We have adapted to fish sperm the method developed by Steele et al. (2000) and we have evaluated the effect of cryopreservation on sea bass sperm DNA.

A small but significant effect of cryopreservation on DNA integrity has been demonstrated in studies carried out by Labbe et al. (2001) using sperm trout. They tested how the sperm cryopreservation affected the nuclear DNA stability and whether the progeny development was modified when eggs were fertilised with cryopreserved spermatozoa. They concluded that cryopreservation of trout sperm only slightly affected sperm DNA stability and that the use of cryopreserved sperm did not impair offspring survival and quality. Analogous studies carried out on sperm of other fish species have not revealed DNA damage after cryopreservation. The freeze-thaw process did not cause genome alterations in turbot sperm since the fertilisation rate, the hatching rate, the larval survival rate (up to ten days) and the larval weight, were similar with both fresh and frozen-thawed sperm (Suquet et al., 1998). Similarly, no effect of the freeze-thaw process on the nucleus of Atlantic croaker spermatozoa was reported (Gwo et al., 2003). Moreover, the growth of tilapias (up to 800 g) and channel catfish (up to 130 g) were not altered using thawed spermatozoa (Chao et al., 1987; Tiersch et al., 1994). The DNA damage that we observed in the cryopreserved sea bass sperm did not affect fertilization capacity and motility. Different authors have reported that the DNA fragmentation is associated with a decrease of fertilization ability, abnormal embryo cleavage and decreased embryo survival (Gwo et al., 2003; Kopeika et al., 2003a, 2003b, 2004; Sun et al., 2000). Fauvel et al. (1998b) found a lower hatching rate for eggs inseminated with frozen-thawed sea bass sperm (69%) when compared with those obtained with fresh sperm (81%), although the fertilisation rates were similar. The presence of the significant degree of DNA fragmentation that we measured after cryopreservation of sea bass sperm could explain, at least partially, this observation. Since the establishment of fish sperm cryobanks could play a crucial role in the genetic management and conservation of aquatic resources the advancement of cryopreservation protocols that avoid DNA fragmentation/aberration are necessary and the SCGE technique is a useful tool to rich this goal.

4. Effect of cryopreservation on sea bass protein profile

Defects in sperm proteins may compromise sperm motility, fertilization ability, and the early events after fertilization (Cao et al., 2003; Huanget al., 1999; Lessard et al., 2000). Protein screening has become an excellent approach with which to evaluate changes in expression due to different stresses. Using this method it has been demonstrated that the reduction in motility observed in boar and human spermatozoa following cryopreservation was associated with a decrease in heat shock protein 90 during cooling (Cao et al., 2003; Huanget al., 1999). Similarly, the loss of P25b (a protein associated with the plasma membrane covering the acrosome) may be responsible, at least in part, for the decrease in fertility following the freezing/thawing procedure of bull semen (Lessard et al., 2000). Cryoinjuries due to cryopreservation have been reported for thawed spermatozoa of many freshwater (Rana, 1995) and marine fish species (Gwo et al., 1992; Lahnsteiner et al., 2000). Shrinkage of the plasma membrane of the midpiece, breakage of mitochondria, and coiling of the axoneme have been observed.Cryopreserved sea bass sperm showed similar fertilization rates and class motility compared with fresh sperm, but also showed a decline in motility duration (Fauvel et al., 1998a), changes in metabolism (Zilli et al., 2004), and lower hatching rates (Fauvel et al., 1998b). For these reason we used (Zilli et al., 2005) the 2-DE to verify whether the cryopreservation procedure, applied to sea bass milt, affected the expression of proteins involved in the control of sperm functions and, in addition, matrix-associated laser desorption/ionization time-of-flight (MALDI-TOF) mass spectrometry to identify some of these proteins.

4.1 Proteins expression in two-dimensional electrophoresis gels: Differences between fresh and frozen-thawed sea bass sperm samples

To perform two-dimensional analysis sperm samples with similar fertilization rates (70%–90%) and percentage of motility (80%–100%), before and after cryopreservation, were used to extract proteins. All the sperm samples used showed lower motility duration after the cryopreservation procedure. 163 spots were detected in all gels prepared from fresh samples (with molecular masses ranging between 190 and 10 kDa and isoelectric points between 3.5 and 8.0) and were used for comparative analysis. Results of a typical experiment performed on sperm samples before and after cryopreservation are showed in figure 6 (A and B).

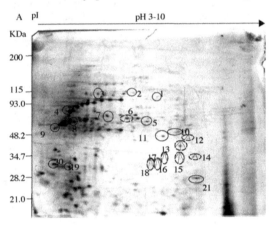

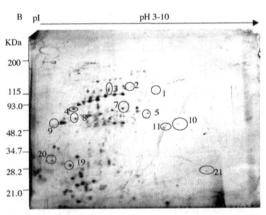

Fig. 6. Two-dimensional Electrophoresis (2-DE) maps of fresh (A) and cryopreserved (B) sea bass sperm proteins. 2-DE was performed on an immobilized pH 3–10 strip, followed by the second-dimensional separation on 12.5% polyacrylamide gels. The separated proteins were stained with silver staining. Spots that are less expressed after cryopreservation are highlighted with a continuous line; spots that are entirely absent after cryopreservation are marked with a dotted line. (This figure was originally published in Zilli et al., Biol Reprod 2005).

Differences were observed (by visual inspection and by using image analysis software) in the protein profiles of fresh and cryopreserved sperm samples. In fact, in the cryopreserved sperm samples, among the 163 spots considered, 13 were significantly ($P<0.05$) less expressed, and 8 completely disappeared. These 21 spots are highlighted in Figure 6 and the normalized spot volumes listed in Table 3. A decrease in protein abundance or spot disappearance in sperm after the cryopreservation procedure may be due to either leakage of proteins from spermatozoa to the extracellular medium or to degradation following freezing-thawing stress. The leakage of proteins is ruled out because we have previously demonstrated that the intracellular protein concentration and the seminal plasma protein concentrations do not change after cryopreservation (Zilli et al., 2004). Consequently, protein degradation seems to be responsible for the reduction in spot abundance (and disappearance). Similar results have been also reported in human and boar semen (Cao et al., 2003; Huanget al., 1999) and bull sperm (Lessard et al., 2000).

Spot Number	Normalised spot volume in fresh sperm (N=6)	Normalised spot volume in cryopreserved sperm (N=6)
SPOT 1	17.8±2.1	7.0±1.1
SPOT 2	10.7±1.1	5.9±1.5
SPOT 3	48.0±9.1	20.2±5.1
SPOT 4	40.8±5.2	29.3±4.1
SPOT 5	54.9±11.3	20.6±3.0
SPOT 6	42.9±2.2	-
SPOT 7	36.8±3.3	25.5±5.2
SPOT 8	110.0±23.0	33.4±6.3
SPOT 9	52.7±12.0	38.7±9.0
SPOT 10	31.9±8.0	4.7±2.0
SPOT 11	30.6±7.0	7.4±5.0
SPOT 12	34.9±15.2	-
SPOT 13	44.3±13.	-
SPOT 14	34.3±11.3	-
SPOT 15	11.4±3.2	-
SPOT 16	8.4±2.8	-
SPOT 17	6.2±3.5	-
SPOT 18	11.8±4.2	-
SPOT 19	30.6±7.2	10.6±4.2
SPOT 20	102.8±3.6	42.9±3.3
SPOT 21	46.3±6.6	29.2±7.3

Table 3. Differences in abundance of spots in fresh and cryopreserved sea bass sperm. Spot adundance is expressed as mean ± SD of normalized spot volume measured in six different gels. (This table was originally published in Zilli et al., Biol Reprod 2005).

4.2 Identification of protein spots by MALDI-TOF

Five of the protein spots shown in Table 3 were analyzed by MALDI-TOF for protein identification. Three were selected among the spots that significantly decreased after cryopreservation (5, 8, and 20) and two (6 and 13) were taken from among those that were absent in the gel obtained with frozen-thawed sperm (Fig. 6 and Table 3). Protein identification was performed by three search programs (PeptIdent, Mascot, and MS-Fit).

Three out of five sea bass proteins processed were found to have homologies with existing sequences in the databases used (Table 4). These proteins were identified from protein sequences already described in other teleost species and amphibians. In particular, two were from *Brachidanio rerio* (spots 5 and 20) and one was from *Xenopus laevis* (spot 13). Table 4 summarizes the data of the bio-informatics analysis for these proteins. For spot 5, the search engine PeptIdent found a homology with a protein of *Brachidanio rerio* (similar to SKB1 of human and mouse). This is a highly conserved cytoplasmic protein with methyltransferase activity that interacts with the members of the Janus family tyrosine kinases (JAK) (Pollack et al., 1999). Genome activation is one of the first critical events in the life of a new organism. Both the timing of genome activation and the array of genes activated must be controlled correctly, and these events depend on changes in chromatin structure and availability of transcription factors (Latham & Schultz, 2001).

In sea bass the observed reduction in SBK1 proteins in cryopreserved sperm could be responsible for abnormal early embryo development, which in turn, could determine the lower hatching rate observed (personal observation). The spot protein 13 matched in Mascot and MS-Fit with a G1/S-specific cyclin E2 protein, which is essential in the control of the cell cycle at the G1/S (start) transition (Moore et al., 2002). Cyclin E is involved in the activation of cyclin-dependent kinase 2 (cdk2). Recently, it has been demonstrated that cdk2 phosphorylates the protein phosphatase, PP1gamma2, a key enzyme in the development and regulation of sperm motility (Huang &Vijayaraghavan, 2004). The observed reduction in sea bass sperm motility duration in frozen-thawed spermatozoa could be a consequence of the cyclin E degradation. The protein spot 20 matched, in MS-Fit, with the hypothetical protein DKFZp566A1524 of unknown function.

Reference Spot	EWM (kDa)	EIP	Identified protein	SWISSPR-OT accession no.	Species ident-ified	TWM (kDa)	TIP	Homology	
								Matched peptides	Coverage %
SPOT 5	80	6.5	Novel protein similar to SKB1 human and mouse (PEPTIDENT)	Q7ZZ07	*Brachidanio rerio*	71.8	5.98	8	22.0
SPOT 6	110	6.0	–	–	–	–	–	–	–
SPOT 8	100	5.2	–	–	–	–	–	–	–
SPOT 13	40	6.8	G1/S-specific cyclin E2 (MASCOT, MS-FIT)	Q91780	*Xenopus laevis*	47.78	6.3	6	20.0
SPOT 20	30	4.5	Similar to hypothetical protein DKFZp566A1 524 (MS-FIT)	Q96AZ5	*Brachidanio rerio*	37.13	5.6	4	21.0

Table 4. Results from peptide mass fingerprinting of protein spots excised from 2D gels. EWM: Experimental Weight Mass; EIP: Experimental Isoelectric Point; TWM: Theoretical Weight Mass; TIP: Theoretical Isoelectric Point. (This table was originally published in Zilli et al., Biol Reprod 2005).

Effect of Cryopreservation on Bio-Chemical Parameters, DNA Integrity, Protein Profile and
Phosphorylation State of Proteins of Seawater Fish Spermatozoa

113

The results reported in figure 6 and tables 3 and 4 show that in sea bass spermatozoa the used cryopreservation procedure causes the degradation of 21 sperm proteins, and among these, 2 could be at least partially responsible for the observed decrease in sperm motility duration and the lower hatching rate of eggs fertilized with cryopreserved sperm. In addition, these observations suggest that two-dimensional electrophoresis coupled with MALDI-TOF analysis could be used as a tool to improve cryopreservation procedures.

5. Effect of cryopreservation on proteins phosphorylation state of sea bream sperm

5.1 Molecular mechanisms determining sperm motility initiation in sea bream *sparus aurata*

Most fish spermatozoa are quiescent in the testes, because the osmolality and composition of seminal plasma usually prevent motility in sperm ducts (Billard, 1986). During natural reproduction, fish sperm become motile after discharge into the aqueous environment (in oviparous species) or the female genital tract (in viviparous and ovoviviparous species) (Billard, 1986; Billard & Cosson, 1983; Stoss, 1983). Changes in the ionic and osmotic environment of the sperm cells have been identified as being critical external factors that may be responsible for initiating motility in fish spermatozoa (Morisawa, 1994). Several extracellular factors controlling sperm motility have been reported. In marine (Gwo et al, 1993; Krasznai et al, 2003a, 2003b; Morisawa & Suzuki, 1980; Oda & Morisawa, 1993) and freshwater (Billard, 1986; Morisawa et al., 1983; Stoss, 1983) teleosts, sperm motility is initiated by osmotic shock when sperm are ejaculated. In these species, spermatozoa are quiescent at the osmolality of seminal plasma (referred to as isotonic condition). In freshwater teleost sperm, flagellar motility is initiated by the hypo-osmotic shock, whereas in marine teleost sperm, flagellar motility is initiated by hyperosmotic shock. Furthermore, in medaka (Inoue & Takei, 2003) and tilapia (Linhart et al., 1999), motility regulatory mechanisms of sperm flagella are modulated to suit the spawning environment when they are in freshwater or acclimated to seawater. In herring sperm, motility initiation requires trypsin inhibitor-like sperm-activating peptide from the eggs (HSAPS), and the sperm exhibits chemotaxis when they are close to eggs (Oda et al., 1998; Yanagimachi et al., 1992).The extracellular factors controlling sperm motility (osmolality, ions, sperm-activating peptides, and chemoattractants) act on the flagellar motile apparatus, the axoneme, through signal transduction across the plasma membrane. Second messengers, such as cAMP and Ca, play key roles in the initiation of sperm motility in many animal groups, such as mammals (Lindemann, 1978; Okamura et al., 1985; Tash & Means, 1983), salmonid fish (Morisawa & Okuno, 1982), sea urchin (Cook et al., 1994), mussel (Stephens & Prior, 1992), and tunicate (Opresko & Brokaw, 1983). A cAMP-independent initiation of flagellar motility in sperm was observed in puffer fish (Morisawa, 1994) and striped bass (Shuyang et al., 2004). Second messengers (cAMP and Ca) determine the sperm motility initiation modifying dynein-mediated sliding of the axonemal outer-doublet microtubules through protein phosphorylation/dephosphorylation in different species, such as mammals (Lindemann & Kanous, 1989), rainbow trout, chum salmon, sea urchin (Inaba et al., 1999), and tunicate (Nomura et al., 2000).

In *Sparus aurata* osmolality is the key signal in sperm motility activation and motility initiation depends on a cAMP-dependent protein phosphorylation (Zilli et al., 2008). To elucidate which proteins are involved (phosphorilated/dephosphorilated) in the initiation of sea bream spermatozoa motility, proteins extracted from spermatozoa before and after motility activation were separated on SDS PAGE, blotted on nitrocellulose membrane, and treated with anti-phosphotyrosine, anti-phosphothreonine, or anti-phosphoserine antibodies.

After motility activation we observed that: 1) two protein bands (76 kDa and 57 kDa) were dephosphorylated and an unspecified number of proteins corresponding to a large band of 9-15 kDa were phosphorylated at tyrosine residues (Fig. 7A); 2) two protein bands (174 kDa and 147 kDa) resulted phosphorilated and an unspecified number of proteins with molecular weights ranging between 15 and 9 kDa were dephophorilated at threonine residues (Fig. 7B); 3) three protein bands (174 kDa, 138 kDa and 70 kDa) and an unspecified number of proteins from 9 to 12 kDa were phosphorylated and only one protein band of 33 kDa was dephosphorylated at serine residues (Fig. 7 C).

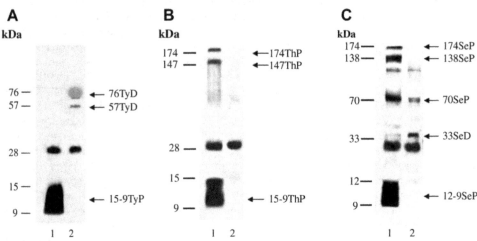

Fig. 7. Motility-dependent phosphorylation/dephosphorylation at tyrosine residues (A), threonine residues, (B) and serine residues (C) in fresh sperm of gilthead sea bream before and after motility activation. Sperm were either activated in seawater (lane 1) or maintained immotile by dilution in non-activating medium (lane 2). Sperm proteins were subjected to Western blotting (30 μg/lane) with anti-phosphotyrosine, anti-phosphothreonine and anti-phosphoserine antibodies. Number on the left indicates the molecular mass of bands. On the right, the names of proteins of interest are indicated. (This figure was originally published in Zilli et al., Cryobiology 2008).

We characterized some of these proteins by using two-dimensional gel electrophoresis (2DE) and the antibody against phosphothreonine. This antibody revealed (figure 8) that: 1) the protein band of 174 kDa (named that 1ThP in figure 7 and identified as 1THPSa in figure 8) was not a single protein but, rather, a cluster of proteins with the same molecular weight (174 kDa) but different pI (5.9–6.29); 2) the protein band of 147 kDa (named 2Thp in figure 7 and identified as 2THPSa in figure 8) was a protein with a pI of 8.7; and 3) the cluster of proteins of 9-15 kDa (named 3ThP in figure 7 and identified as 3THPSa in figure 8) consisted of 10 proteins with pI between 6.1 and 7.6 and molecular weights between 9 and 15 kDa.

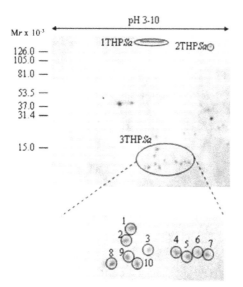

Spot	MW (kDa)	pI
1THP*Sa*	174	5.9-6.2
2THP*Sa*	147	8.7
3THP*Sa*		
1	15	6.1
2	13	6.1
3	11	6.2
4	10.5	6.6
5	10	6.8
6	10.5	7.4
7	10	7.6
8	9	6.0
9	10	6.1
10	9	6.1

Notes: 1THP*Sa*, 2THP*Sa*, 3THP*Sa*: spots
1, 2 and 3 Threonine-Phosphorylated in
Sparus aurata.

Fig. 8. Western blot analysis with antiphosphothreonine antibody of gilthead sea bream
(*Sparus aurata*) sperm proteins separated by 2DE. The 2DE was performed on an
immobilized pH 3–10 NL strip, followed by the second-dimensional separation on 13%
polyacrylamide gels. The separated proteins were then blotted on nitrocellulose and
incubated with antibody. Molecular mass and isoelectric point of proteins of interest are
listed. (This figure was originally published in Zilli et al., Biol. Reprod. 2008).

Some of these proteins have been identified by mass spectrometry and results are listed in
Table 5. In particular, spots 1 and 2 (belonging to 3THPSa) were identified as acetyl-
coenzyme A (CoA) synthetase, spot 5 as A kinase anchor protein (AKAP), and spot 7 as an
unnamed protein of *Tetraodon nigroviridis*, which have 70% identity with a novel protein
similar to phosphatase and actin regulator 3 of *Danio rerio*. Acetyl-CoA synthetase is well
known as an enzyme whose activity is central to the metabolism of prokaryotic and
eukaryotic cells. In particular, acetyl-CoA synthetase activates acetate to acetyl-CoA, and it
provides the cell with the two-carbon metabolite used in many anabolic and energy
generation processes. Therefore, we suppose that this enzyme was activated in motile sperm
to increase the level of ATP, which is necessary for flagellar movement. PKA localizes to
specific cellular structures and organelles by binding to AKAP molecules via interaction
with the regulatory subunits (RI and RII). Therefore, cAMP levels temporally regulate PKA,
whereas the spatial regulation within the cell occurs through compartmentalization by
binding to AKAP, thus assuring specificity of PKA function. The important role of AKAP as
a key regulator of sperm motility is already established (Vijayaraghavan et al., 1997a). In
addition, a recent study demonstrated that phosphorylation of AKAP in human sperm
results in tail recruitment of PKA and increase of sperm motility, providing evidence for a
functional role of phosphorylation of AKAP (Luconi et al., 2004). Regarding the role of
phosphatases and kinases in the initiation of sperm motility, many studies have
demonstrated that the development and maintenance of motility is regulated by a complex

balance between kinase and phosphatase activities (Tash & Brach, 1994; Vijayaraghavan et al., 1997b).

Reference Spot	EWM (kDa)	EIP	Identified protein	SWISSPROT accession no.	Species identified	TWM (kDa)	TIP	Homology Score	Coverage %
3THPSa									
Spot1/ Spot2	15-13	6.1	LOC568763 similar to Acetyl-coenzyme A synthetase	Q0VG88	*Danio rerio*	15.5	5.3	0.97a	27%
Spot 5	10	6.8	novel protein similar to human A kinase (PRKA) anchor protein 7 (AKAP7)	CAI11962	*Danio rerio*	8.1	6.0	0.79a	26%
Spot 7	10	7.6	Chromosome 11 SCAF14528, whole genome shotgun sequence - Unnamed protein product that have a 70% of identity with novel protein similar to phosphatase and actin regulator 3 (PHACTR3, zgc:109967) [*Danio rerio*]	Q4SQZ9	*Tetraodon nigroviridis*	8.0	7.7	0.28b	21%

Table 5. Results from peptide mass fingerprinting of protein spots excised from 2-D gels of gilthead sea bream sperm.

EWM: Experimental Weight Mass; EIP: Experimental Isoelectric Point; TWM: Theoretical
Weight Mass; TIP: Theoretical Isoelectric Point. a: Z score of ProFound; b: Normalized score
of Aldente. (Modified from Zilli et al., Biol. Reprod. 2008).

5.2 Effect of cryopreservation on phosphorylation state of proteins involved in sperm motility initiation in sea bream *Sparus aurata*

The quality of gilthead sea bream semen was decreased by cryopreservation procedure.
Even though the viability (82 ± 5%) of spermatozoa following the freezing–thawing
procedure was only slightly (but significant) decreased with respect to that measured in
fresh samples (93 ± 4), only the 50% of the thawed spermatozoa could be activated, and
showed a motility duration which was one third of that measured in fresh samples. The
reduction of sperm motility (percent and duration) is attributable (at least partially) to the
effect that the freezing-thawing procedure has on the phosphorylation state of proteins
involved in motility initiation (Zilli et al., 2008b).

Phosphorylation/dephosphorylation of tyrosine residues:

Two protein bands (76TyD and 57TyD of figure 7A) which in fresh sperm were completely
dephosphorylated after motility initiation, in frozen– thawed remained phosphorylated
(Fig. 9A), while the cluster of proteins of 15-9 TyP (Fig. 7A), that were phosphorylated when
fresh sperm shifted from the immotile to the motile phase, were much less phosphorylated
in frozen–thawed activated sperm (Fig. 9A).

Phosphorylation/dephosphorylation of threonine residues:

Among the proteins that were phosphorylated following motility activation in fresh sperm
(Fig. 7B), two bands (named 15-9ThP and 147ThP) were phosphorylated after activation in
frozen–thawed spermatozoa (Fig. 9B). However, it must be underlined that within the
proteins belonging to the 15-9 ThP band, only one (11 kDa) previously identified as acetyl-
coenzyme A synthetase (Zilli et al, 2008a) was phosphorylated after motility activation in
cryopreserved sperm (Fig. 9B).

Phosphorylation/dephosphorylation of serine residues:

Among the five previously identified protein bands (Zilli et al, 2008a) that changed their
phosphorylation state after motility activation in fresh sperm (Fig. 7C), only two (70SeP and
12-9SeP) were phosphorylated in frozen–thawed sperm after activation (Fig. 9C). The other
bands, named 174SeP, 138SeP and 33SeD, did not change their phosphorylation state after
activation (Fig. 9C), unlike what happens in fresh sperm (Fig. 7C).

Some proteins (76TyD, 57TyD and 33SeD) that were dephosphorylated after motility
activaton in fresh sperm (7A and 7C) but not in cryopreserved spermatozoa (9A and 9C)
could not play a key role in sperm motility initiation but could be involved in sperm
motility duration and motility characteristics, since the kinematic parameters were
significantly reduced by the freezing–thawing procedure.

Our studies also demonstrated that in gilthead sea bream spermatozoa the freezing-
thawing procedure increased, independently from the motility activation procedure, protein
phosphorylation (mainly at threonine residues), since more phosphorylated proteins were
present in non-activated cryopreserved sperm with respect to the fresh sperm. This could be

due, as previously proposed by Perez-Pe et al., (2002), to a membrane modifications that determine conformational changes of these proteins or facilitate calcium influx into the cell (Bailey & Buhr, 1994; McLaughlin & Ford, 1994). This ion could stimulate adenylyl cyclase to initiate cAMP-mediated phosphorylation of sperm protein. Alternatively (or in addition), the cryopreservation procedure could also determine the activation of protein kinases different from PKA (Pommer et al., 2003).

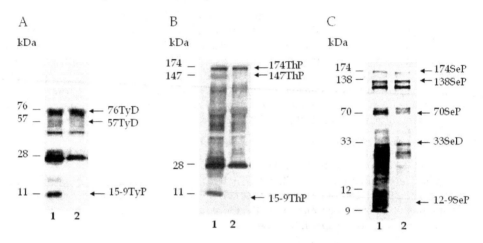

Fig. 9. Motility-dependent phosphorylation/dephosphorylation at tyrosine residues (A), threonine residues (B), and serine residues (C) in frozen–thawed sperm of gilthead sea bream before and after motility activation. Sperm were either activated in seawater (lane 1) or maintained immotile by dilution in non-activating medium (lane 2). Sperm proteins were subjected to Western blotting (30 μg/lane) with antibody. Number on the left indicates molecular mass of bands. On the right, the names of proteins of interest are indicated. (Modified from Zilli et al 2008; Criobiology)

6. Conclusions

Cryopreservation, coupled with insemination and short term storage techniques, will lead to an improvement of gamete management of marine fish species. In particular, sperm cryopreservation is considered as a valuable technique for artificial reproduction and genetic improvement since it allows the selection and the storage of gametes of high quality.

However, although seawater fish spermatozoa of marine fish are more resistant than freshwater species to the dynamic changes in osmotic pressure that occur during the process of cryopreservation (Dzuba & Kopeika, 2002), the freezing-thawing procedure, apart from the experimental protocol used and from the fish species considered, determines: a changes of the kinematic characteristics, damages to proteins and DNA, lipid modification and change of the phosphorylation state of proteins involved in sperm motility initiation. The knowledge of the effects of freezing–thawing procedure on spermatozoa is very important to improve cryopreservation techniques for semen of marine fish for the establishment of sperm cryobanks that could play a crucial role in the genetic management and conservation of aquatic resources.

7. References

Aas, G.H., Reftsie, T. & Gjerde, B. (1991). *Evaluation of milt quality of Atlantic salmon.* Aquaculture 95, 125–132

Ashby, J., Tinwell, H., Lefevre, P.A. & Browne, M.A. (1995). *The single cell gel electrophoresis assay for induced DNA damage (comet assay): measurement and tail length and moment.* Mutagenesis 10, 85–90

Babiak, I. & Glogowski, J. (1997). *Effect of individual male variability on cryopreservation of northern pike, Esox lucius L., sperm.* Aquacult Res 28, 191–197

Bailey, J.L. & Buhr, M.M. (1994). *Cryopreservation alters the Ca^{2+} flux of bovine spermatozoa.* Can. J. Anim. Sci. 74, 45–51

Baynes, S.M. & Scott, A.P. (1987). *Cryopreservation of rainbow trout spermatozoa:the influence of sperm quality, egg quality and extender composition on post-thaw fertility.* Aquaculture 66, 53–67

Billard, R. & Cosson, M.P. (1990). *The energetics of fish sperm motility.* In: Gagnon C (ed), Control of Sperm Motility: Biological and Clinical Aspects. CRC Press, Boca Raton, FL, pp. 153–173.

Billard, R. (1986). *Spermatogenesis and spermatology of some teleost fish species.* Reprod. Nutr. Dev. 2, 877–920

Birktoft, J.J., Fu, Z., Carnahan, G.E., Rhodes, G., Roderick, S.L. & Banaszak, L.J. (1989). *Comparison of the molecular structures of cytoplasmic and mitochondrial malate dehydrogenase.* Biochem. Soc.Trans. 17, 301–304

Bleile, D.M., Shultz, J.H. & Harrison, J. (1977). *Investigation of the subunit interactions in malate dehydrogenase.* J. Biol. Chem. 252, 755–758

Braude, P., Bolton, V. & Moore, S. (1988). *Human gene expression first occurs between the four- and eight-cell stages of pre-implantation development.* Nature 332, 459-461

Cao, W.L., Wang, Y.X., Xiang, Z.Q. & Li, Z. (2003) *Cryopreservation-induced decrease in heat-shock protein 90 in human spermatozoa and its mechanism.* Asian J. Androl. 5, 43–46

Chao, N.H. & Liao, I.C. (2001). *Cryopreservation of finfish and shellfish gametes and embryos.* Aquaculture 197, 161–189

Chao, N.H., Chao, W.C., Liu, K.C. & Liao, I.C. (1987). *The properties of tilapia sperm and its cryopreservation.* J. Fish Biol. 30, 107–118

Christen, R., Gatti, J.L. & Billard, R. (1987). *Trout sperm motility. The transient movement of trout sperm motility is related to changes in concentrations of ATP following the activation of flagellar movement.* Eur. J. Biochem. 166, 661–667

Ciereszko, A. & Dabrowski, K. (1994) *Relationship between biochemical constituents of fish semen and fertility: the effect of short-term storage.* Fish Physiol. Biochem. 12, 357–367

Ciereszko, A., Toth, G.P., Christ, S.A. & Dabrowski, K. (1996). *Effect of cryopreservation and theophylline on motility characteristics of lake sturgeon (Acipenser fulvescens) spermatozoa.* Theriogenology 45, 665–672

Collins, A.R., Dobson, V.L., Dusinska, M., Kennedy, G. & Stetina, R. (1997). *The comet assay: what can it really tell us?* Mutat. Res. 29, 183–193

Conget, P., Fernfindez, M., Herrera, G. & Minguell, J.J. (1996). *Cryopreservation of rainbow trout (Oncorhynchus mykiss) spermatozoa using programmable freezing.* Aquaculture 143, 319-329

Cook, S.P., Brokaw, C.J., Muller C.H. & Babcock D.F. (1994). *Sperm chemotaxis: egg peptides control cytosolic calcium to regulate flagellar responses.* Dev. Biol. 165, 10–19

Cosson, J., Linhart, O., Mims, S.D., Shelton, W.L. & Rodina, M. (2000). *Analysis of motility parameters from paddlefish (Polyodon spathula) and shovelnose sturgeon (Scaphirhynchus platorynchus)spermatozoa.* J. Fish Biol. 56, 1348–1367

Drokin, S., Stein, H. & Bartseherer, H. (1998). *Effect of cryopreservation on the fine structure of spermatozoa of rainbow trout (oncorhynchus mykiss) and brown trout (salmo trutta F. fario).* Cryobiology 37, 263–270

Duke, R.C. & Cohen, J.J. (1986). *Endogenous endonuclease-induced DNA fragmentation: an early event in cell-mediated cytolisis.* Proc. Natl. Acad. Sci. USA 80, 61–63

Dzuba, B.B. & Kopeika, E.F. (2002). *Relationship between the changes in cellular volume of fish spermatozoa and their cryoresistance.* Cryo Letters 23, 353-360

Fairbairn, D.W., Olive, P.L. & O-Neill, K.L. (1995). *The comet assay: a comprehensive review.* Mutat. Res. 339, 37–59

Fauvel, C., Suquet, M., Dreanno, C., Zonno, V. & Menu, B. (1998a). *Cryopreservation of sea bass (Dicentrarchus labrax) spermatozoa in experimental and production conditions.* Aquat. Living Resour. 11, 387–394

Fauvel, C., Zonno, V., Suquet, M., Storelli, C. & Dreanno, C. (1998b). *Cryopreservation of sea bass (Dicentrarchus labrax) sperm in both experimental and production conditions.* In: Proceedings of the Eighth International Symposium on Fish Physiology, Upsala, p. 64.

Gwo, J.C. & Arnold, C.R. (1992). *Cryopreservation of Atlantic croaker spermatozoa: evaluation of morphological changes.* J. Exp. Zool. 264, 444–453

Gwo, J.C., Kurokura, H. & Hirano, R. (1993). *Cryopreservation of spermatozoa from rainbow trout, common carp, and marine puffer.* Nippon Suisan Gakkaishi 59, 777–782

Gwo, J.C., Wu, C.Y., Chang, W.S. & Cheng, H.Y. (2003). *Evaluation of damage in Pacific oyster (Crassostrea gigas) spermatozoa before and after cryopreservation using comet assay.* Cryo Lett. 24, 171–180

Harada, K. & Wolf, R.G. (1968). *Malic dehydrogenase. VII. The catalytic mechanism and possible role of identical protein subunits.* J. Biol. Chem. 243, 4131–4137

Helma, C. & Uhl, M. (2000). *A public domain image-analysis program for the single-cell gel-electrophoresis (comet) assay.* Mutat. Res. 466, 9–15

Homma-Takeda, S., Kugenuma, Y., Iwamuro, T., Kumagai, Y. & Shimojo, N. (2001). *Impairment of spermatogenesis in rats by methylmercury: involvement of stage- and cellspecific germ cell apoptotis.* Toxicology 169, 25–35

Huang, S.Y., Kuo, Y.H., Lee, W.C., Tsou, H.L., Lee, Y.P., Chang, H.L., Wu, J.J. & Yang, P.C. (1999). *Substantial decrease of heat-shock protein 90 precedes the decline of sperm motility during cooling of boar spermatozoa.* Theriogenology 51, 1007–1016

Huang, Z. & Vijayaraghavan, S. (2004). *Increased phosphorylation of a distinct subcellular pool of protein phosphatase, PP1gamma2, during epididymal sperm maturation.* Biol. Reprod. 70, 439–447

Hughes, C.M., Lewis, S.E.M., McKelvey-Martin, V.J. & Thomson, W. (1997). *Reproducibility of human sperm DNA measurements using the alkaline single cell gel electrophoresis assay.* Mutat. Res. 374, 261–268

Hwang, P.C. & Idler, D.R. (1969). *A major study on cations, osmotic pressure and pH in seminal components of Atlantic salmon.* J. Fish Res. Board Can. 26, 413–419

Inaba, K., Kagami, O. & Ogawa, K. (1999). *Tctex2-related outer arm dynein light chain is phosphorylated at activation of sperm motility.* Biochem. Biophys. Res. Commun. 256, 177–183

Inoue, K. & Takei, Y. (2003). *Asian medaka fishes offer new models for studying mechanisms of seawater adaptation*. Comp. Biochem. Physiol. B Biochem. Mol. Biol. 136, 635–645

Johnson, L.A. & Ferris, J.A.J. (2002). *Analysis of postmortem DNA degradation by single-cell gel electrophoresis*. Forensic Sci. Int. 126, 43–47

Kopeika, E., Kopeika, J. & Zhang, T. (2007). *Cryopreservation of fish sperm*. Methods Mol. Biol. 368, 203-217

Kopeika, J., Kopeika, E., Zhang, T. & Rawson, D.M. (2003a). *Studies on the toxicity of dimethyl sulfoxide, ethylene glycol, methanol and glycerol to loach (Misgurnus fossilis) sperm and the effect on subsequent embryo development*. Cryo Letters 24, 365-374

Kopeika, J., Kopeika, E., Zhang, T., Rawson, D.M. & Holt, W.V. (2004). *Effect of DNA repair inhibitor (3-aminobenzamide) on genetic stability of loach (Misgurnus fossilis) embryos derived from cryopreserved sperm*. Theriogenology 61,1661-1673

Kopeika, J., Kopeika, E., Zhang, T., Rawson, D.M. & Holt, W.V. (2003b). *Detrimental effects of cryopreservation of loach (Misgurnus fossilis) sperm on subsequent embryo development are reversed by incubating fertilised eggs in caffeine*. Cryobiology 46, 43-52

Krasznai, Z., Morisawa, M., Krasznai, Z.T., Morisawa, S., Inaba, K., Bazsane, Z.K., Rubovszky, B., Bodnar, B., Borsos, A. & Marian, T. (2003a). *Gadolinium, a mechano-sensitive channel blocker, inhibits osmosis-initiated motility of sea- and freshwater fish sperm but does not affect human of ascidian sperm motility*. Cell Motil Cytoskeleton 55, 232–243

Krasznai, Z., Morisawa, M., Morisawa, S., Krasznai, Z.T., Tron, L. & Marian T. (2003b). *Role of ion channel and membrane potential in the initiation of carp sperm motility*. Aquat. Living. Resour. 16, 445–449

Labbe, C., Martoriati, A., Devaux, A., & Maisse, G. (2001). *Effect of sperm cryopreservation on sperm DNA stability and progeny development in rainbow trout*. Mol. Reprod. Dev. 60, 397–404.

Lahnsteiner, F., Berger, B., Wiesmann, T. & Patzner, R.A. (1996b). *Changes in morphology, physiology, metabolism, and fertilization capacity of rainbow trout semen following cryopreservation*. Prog. Fish. Cult. 58, 149-159

Lahnsteiner, F., Weismann, T. & Patzner, R.A. (1992). *Fine structural changes in spermatozoa of the grayling, Thymallus thymallus (Pisces: Teleostei), during routine cryopreservation*. Aquaculture 103, 73-84

Lahnsteiner, F., Berger, B., Horvath, A., Urbanyi, B. & Weismann, T. (2000). *Cryopreservation of spermatozoa in cyprinid fishes*. Theriogenology 54, 1477–1498

Lahnsteiner, F., Berger, B., Weismann, T. & Patzner, R.A. (1996a) *Physiological and biochemical determination of rainbow trout, Onchorhynchus mykiss, semen quality for cryopreservation*. J. Appl. Aquacult. 6, 47–73

Lahnsteiner, F., Patzner, R.A. & Weismann, T. (1994). *The spermatic duct of salmonid fishes. Fine structure, hystochemistry and composition of the secretion*. J. Fish Biol. 42, 79–93

Lahnsteiner, F., Weismann, T. & Patzner, R.A. (1998). *Evaluation of the semen quality of the rainbow trout, Oncorhynchus mykiss, by sperm motility, seminal plasma parameters, and spermatozoal metabolism*. Aquaculture 163, 163–181

Larson, K.L., Brannian, J.D., Singh, N.P., Burbach, J.A., Jost, L.K., Hansen, K.P., Kreger, D.O. & Evenson, D.P. (2001). *Chromatin structure in globozoospermia: a case report*. J. Androl. 22, 424–431

Latham, K.E. & Schultz, R.M. (2001). *Embryonic genome activation*. Front. Biosci. 6, 748–759

Lessard, C., Parent, S., Leclerc, P., Bailey, J.L. & Sullivan, R. (2000). *Cryopreservation alters the levels of the bull sperm surface protein P25b*. J. Androl. 21, 700-707

Lindemann, C.B. & Kanous, K.S. (1989). *Regulation of mammalian sperm motility.* Arch. Androl. 23, 1–22

Lindemann, C.B. (1978). *A cAMP-induced increase in the motility of demembranated bull sperm models.* Cell 13, 9–18

Linhart, O., Rodina, M. & Cosson, J. (2000). *Cryopreservation of sperm in common carp Cyprinus carpio: sperm motility and hatching success of embryos.* Cryobiology 41, 241–250

Linhart, O., Walford, J., Sivaloganathan, B. & Lam, T.J. (1999). *Effects of osmolality and ions on the motility of stripped and testicular sperm of freshwater- and seawater-acclimated tilapia, Oreochromis mossambicus.* J. Fish Biol. 55, 1344–1358

Luconi, M., Carloni, V., Marra, F., Ferruzzi, P., Forti, G. & Baldi, E. (2004). *Increased phosphorylation of AKAP by inhibition of phosphatidylinositol 3-kinase enhances human sperm motility through tail recruitment of protein kinase A.* J. Cell Sci. 117, 1235–1246

Maldjian, A., Pizzi, F., Gliozzi, T., Cerolini, S., Penny, P. & Noble, R. (2005). *Changes in sperm quality and lipid composition during cryopreservation of boar semen.* Theriogenology 63, 411–421

McConnell, E.J., Wagoner, M.J., Keenan, C.E. & Raess, B.U. (1999). *Inhibition of calmodulin-stimulated (Ca21-Mg21)-ATPase activity by dimethyl sulfoxide.* Biochem. Pharmacol. 57, 39–44

McKelvey-Martin, V.J., Green, P., Schmezer, P., Pool-Zobel, B.L., DeMeo, M.P. & Collins, A. (1993). *The single cell gelelectrophoresis assay (comet assay): a European review.* Mutat. Res. 288, 47–63

McLaughlin, E.A. & Ford, W.C.L. (1994). *Effects of cryopreservation on the intracellular calcium-concentration of human spermatozoa and its response to progesterone.* Mol. Reprod. Dev. 37, 241–246

McNiven, M.A., Gallant, R.K. & Richardson, G.F. (1992). *In vitro methods of assessing the viability of rainbow trout spermatozoa.* Theriogenology 38, 679–686

Moore, J.D., Kornbluth, S. & Hunt, T. (2002). *Identification of the nuclear localization signal in Xenopus cyclin E and analysis of its role in replication and mitosis.* Mol. Biol. Cell. 13, 4388–4400

Morisawa, M. & Okuno, M. (1982). *Cyclic AMP induces maturation of trout sperm axoneme to initiate motility.* Nature 295, 703–704

Morisawa, M. & Suzuki, K. (1980). *Osmolality and potassium ions: their roles in initiation of sperm motility in teleosts.* Science 210, 1145–1147

Morisawa, M. (1994). *Cell signaling mechanisms for sperm motility.* Zool. Sci. 11, 647–662

Morisawa, M., Suzuki, K., Shimizu, H., Morisawa, S. & Yasuda, K. (1983). *Effect of osmolality and potassium on motility of spermatozoa from freshwater cyprinid fishes.* J. Exp. Zool. 107, 95–103

Müller, K., Müller, P., Pincemy, G., Kurz, A. & Labbe, C. (2008). *Characterization of sperm plasma membrane properties after cholesterol modification: consequences for cryopreservation of rainbow trout spermatozoa.* Biol. Reprod. 78, 390–399

Nomura, M., Inaba, K. & Morisawa, M. (2000). *Cyclic AMP- and calmodulindependent phosphorylation of 21- and 26-kDa proteins in axoneme is a prerequisite for SAAF-induced motile activation in ascidian spermatozoa.* Dev. Growth Differ. 42, 129–138

Oda, A. & Morisawa, M. (1993). *Rises of intracellular Ca2+ and pH mediate the initiation of sperm motility by hyperosmolality in marine teleosts.* Cell Motil. Cytoskeleton 25, 171–178

Oda, S., Igarashi, Y., Manaka, K., Koibuchi, N., Sawada, M.S., Sasaki, K., Morisawa, M., Ohtake, H. & Shimizu, N. (1998). *Sperm-activating proteins obtained from the herring eggs are homologous to trypsin inhibitors and synthesized in follicle cells.* Dev. Biol. 41, 55–63

Ogier de Baulny, B., Labbé, C. & Maisse G. (1999). *Membrane integrity, mitochondrial activity, ATP content, and motility of the European catfish (Silurus glanis) testicular spermatozoa after freezing with different cryoprotectants.* Cryobiology 39, 177–184

Okamura, N., Tajima, Y., Soejima, A., Masuda, H. & Sugita, Y. (1985). *Sodium bicarbonate in seminal plasma stimulates the motility of mammalian spermatozoa through direct activation of adenylate cyclase.* J. Biol. Chem. 260, 9699–9705

Opresko, L. & Brokaw, C.J. (1983). *cAMP-dependent phosphorylation associated with activation of motility of Ciona sperm flagella.* Gamete Res. 8, 201–218

Ostling, O. & Johanson, K.J. (1984). *Microelectrophoretic study of radiation-induced DNA migration from individual cells.* Biochem. Biophys. Res. Commun. 123, 291–298

Perez-Pe, R., Grasa, P., Fernandez-Juan, M., Peleato, M.L., Cebrian-Perez, J.A. & Muino-Blanco T. (2002). *Seminal plasma proteins reduce protein tyrosine phosphorylation in the plasma membrane of cold-shocked ram spermatozoa.* Mol. Reprod. Dev. 61, 226–233

Piperakis, S.M., Visvardis, E.E. & Tassiou, A.M. (1999). *Comet assay for nuclear DNA damage.* Methods Enzymol. 300, 184-194

Pollack, B.P., Kotenko, S.V., He, W., Izotova, L.S., Barnoski, B.L. & Pestka, S. (1999). *The human homologue of the yeast proteins Skb1 and Hsl7p interacts with Jak kinases and contains protein methyltransferase activity.* J. Biol. Chem. 274, 31531–31542

Pommer, A.C., Rutllant, J. & Meyers, S.A. (2003). *Phosphorylation of protein tyrosine residues in fresh and cryopreserved stallion spermatozoa under capacitating conditions.* Biol. Reprod. 68, 1208–1214

Rana, K.J. (1995). *Preservation of gametes.* In: N.R. Bromage and R.J. Roberts, Editors, Broodstock Management and Eggs and Larval Quality, Cambridge University Press, Cambridge, pp. 53–76

Rawn, J.D. (1983). *Biochemistry.* Harper and Row, New York.

Ritar, A.J. (1999). *Artificial insemination with cryopreserved semen from striped trumpeter (Latris lineata).* Aquaculture 180, 177–187

Rodina, M., Gela, D., Kocour, M., Alavi, S.M., Hulak, M. & Linhart, O. (2007). *Cryopreservation of tench, Tinca tinca, sperm: sperm motility and hatching success of embryos.* Theriogenology 67, 931–940

Ruggia, A., Gelpı´, J.L., Busquets, M., Cascante, M. & Corte´s, A. (2001). *Effect of several anions on the activity of mitochondrial malate dehydrogenase from pig heart.* J. Mol. Catalysis B: Enzymatic 11, 743–755

Rurangwa, E., Volckaert, F.A.M., Huyskens, G., Kime, D.E. & Ollevier, F. (2001). *Quality control of refrigerated and cryopreserved semen using computer-assisted sperm analysis (CASA), viable staining and standardized fertilization in African catfish (Clarias gariepinus).* Theriogenology 55, 751–769

Sakkas, D., Mariethoz, E., Manicardi, G., Bizzarro, D., Bianchi, P.G. & Bianchi, U. (1999). *Origin of DNA damage in ejaculated spermatozoa.* Rev. Reprod. 4, 31–37

Shuyang, H., Jenkins-Keeran, K. & Curry Woods, L. (2004). *Activation of sperm motility in striped bass via a cAMP-independent pathway.* Theriogenology 61, 1487–1498

Steele, E.K., McClure, N. & Lewis, S.E.M. (2000). *Comparison of the effects of two methods of cryopreservation on testicular sperm DNA.* Fertil. Steril. 74, 450–453

Stephens, R.E. & Prior, G. (1992). *Dynein from serotonin-activated cilia and flagella: extraction characteristics and distinct sites for cAMP-dependent protein phosphorylation.* J. Cell. Sci. 103, 999–1012

Stoss, J. (1983). *Fish gamete preservation and spermatozoa physiology.* In: Hoar WS, Randall DJ III, Donaldson EM (eds.), Fish Physiology. Academic Press, New York, pp. 305–350.

Sun, J.G., Jurisicova, A. & Casper, R.F. (2000). *Detection of deoxyribonucleic acid fragmentation in human sperm: correlation with fertilization in vitro.* J. Androl. 21, 33–44

Sun, Y., Clinkebeard, K.D., Clarke, C., Cudd, L., Highlander, S.K. & Dabo, S.M. (1999). *Pasteurella haemolytica leukotoxin induced apoptosis of bovine lymphocytes involves DNA fragmentation.* Vet. Microbiol. 65, 153–166

Suquet, M.D., Dreanno, C., Fauvel, C., Cosson, J. & Billard, R. (2000). *Cryopreservation of sperm in marine fish.* Aquacult. Res. 31, 231–243

Suquet, M.D., Dreanno, C., Petton, B., Normant, Y., Omnes, M.H. & Billard, R. (1998). *Long-term effects of the cryopreservation of turbot (Psetta maxima) spermatozoa.* Aquat. Living Resour. 11, 45–48

Tash, J.S. & Bracho, G.E. (1994). Regulation of sperm motility: emerging evidence for a major role for protein phosphatases. J. Androl. 15, 505–509

Tash, J.S. & Means, A.R. (1983) *Cyclic adenosine 30,50-monophosphate, calcium, and protein phosphorylation in flagellar motility.* Biol. Reprod. 28, 75–104

Tiersch, T.R., Goudie, C.A. & Carmichael, G.J. (1994). *Cryopreservation of channel catfish sperm: storage in cryoprotectants, fertilization trials, and growth of channel catfish produced with cryopreserved sperm.* Trans. Am. Fish Soc. 123, 580–586

Vijayaraghavan, S., Goueli, S.A., Davey, M.P. & Carr, D.W. (1997a). *Protein kinase Aanchoring inhibitor peptides arrest mammalian sperm motility.* J. Biol. Chem. 21, 4747–4752

Vijayaraghavan, S., Trautman, K.D., Goueli, S.A. & Carr, D.W. (1997b). *A tyrosinephosphorylated 55-kilodalton motility-associated bovine sperm protein is regulated by cyclic adenosine 30,50-monophosphates and calcium.* Biol. Reprod. 56, 1450–1457

Wood, D.C., Jurgensen, R.J., Geesin, C. & Harrison, J.H. (1981). *Subunit interactions in mitochondrial malate dehydrogenase. Kinetics and mechanism of reassociation.* J. Biol. Chem. 256, 2377–2382

Yanagimachi, R., Cherr, G.N., Pillard, M.C. & Baldwin, J.D. (1992). *Factors controlling sperm entry into the micropyles of salmonid and herring eggs.* Dev. Growth Differ. 34, 447–461

Zhang, Y.Z., Zhang, S.C., Liu, X.Z., Xu, Y.Y., Wang, C.L.M., Sawant, S., Li, J. & Chen, S. L. (2003). *Cryopreservation of flounder (ParaBchthys olivaceus) sperm with a practical methodology.* Theriogenology 60, 989-996

Zilli, L., Schiavone, R., Storelli, C. & Vilella S. (2008a). *Molecular mechanisms determining sperm motility initiation in two sparids (Sparus aurata and Lithognathus mormyrus).* Biol. Reprod. 79, 356-366

Zilli, L., Schiavone, R., Storelli, C. & Vilella, S. (2008b). *Effect of cryopreservation on phosphorylation state of proteins involved in sperm motility initiation in sea bream.* Cryobiology 57, 150-155

Zilli, L., Schiavone, R., Zonno, V., Rossano, R., Storelli, C. & Vilella S. (2005). *Effect of cryopreservation on sea bass sperm proteins.* Biol. Reprod. 72, 1262-1267

Zilli, L., Schiavone, R., Zonno, V., Storelli, C. & Vilella, S. (2003). *Evaluation of DNA damage in Dicentrarchus labrax sperm following cryopreservation.* Cryobiology 47, 227-235

Zilli, L., Schiavone, R., Zonno, V., Storelli, C. & Vilella, S. (2004). ATP concentration and β-D-glucuronidase activity as indicators of sea bass semen quality. Biol. Reprod. 70, 1679-1684

Part 3

Cryopreservation of Plants

4

Cryopreservation of Plant Genetic Resources

Daisuke Kami

National Agricultural Research Center for Hokkaido Region

Japan

1. Introduction

The plant genetic resources is preserved by pollen, seed, branch, bulb, or tissue culture at the gene bank. Since these need to be updated periodically (several months ~ several years), I point out the problems involving the great labor and space spent for the maintenance of plant genetic resources.

Cryopreservation is a storage method of plant genetic resources at ultra-low temperature, for example, that of liquid nitrogen (LN; -196 °C). It is a preservation method that enables plant genetic resources to be conserved safely, and cost-effectively.

For successful cryopreservation, it is essential to avoid intracellular freezing and induce the vitrification state of plant cells during cooling in LN. In addition, the cryopreservation method should be a simple protocol for everyone to use easily. Since the 1970's, cryopreservation techniques have been researched using different plant organs, tissues and cells. As a result, different cryopreservation procedures have been developed (for example, slow-prefreezing method, vitrification method, dehydration method). With the development of these cryopreservation methods, tissues of tropical plants, which have been conventionally thought to be not cryopreserved, also were successfully preserved in LN (Bajaj, 1995; Towill & Bajaj, 2002). In this Chapter, I describe different types of cryopreservation methods.

In addition, I often ask my colleagues why cryopreservation of plant tissue did not succeed irrespective of the method. As the cause, it is possible that there is a problem in the character of plant species (stress resistance and polyphenol production), or the ways used in the cryopreservation technique. Then, I also would like to present some knowledge about some improvements for making cryopreservation of plant genetic resources more successful in this chapter.

2. Methods of cryopreservation of plant genetic resources

In the section (2.1), I would like to introduce cryopreservation methods of plant genetic resources that have been developed. In the section (2.2), I would like to describe the approach when cryopreserving plant samples from the past literatures or my own experience in order to enhance the regrowth percentage after cryopreservation,

2.1 Cryopreservation method of plant genetic resourses

In this section, I introduce cryopreservation procedures by using figures.

2.1.1 Slow programmed freezing (also known as "prefreezing")

Slow programmed freezing was a major cryopreservation method for plant genetic resources until the 1980's. The procedure in this method is shown in Fig. 1.

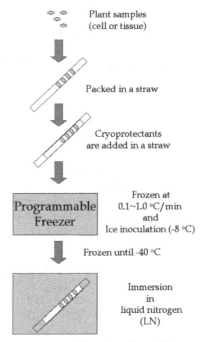

Fig. 1. The protocol of slow programmed freezing (from Kumu et al., 1983).

Plant genetic resources (cells and tissues) were packed in cryotube or straw, and cryoprotectants were added. In this method, dimethyl sulfoxide (DMSO), ethylene glycol (EG) and glucose were utilized as cryoprotectants. In many cases, these were used independently, but Finkle & Ulrich (1979) reported that the regrowth percentage of germplasm after cryopreservation was higher when mixing cryoprotectans in sugar cane cells. Packed specimens were gradually cooled from -20 °C to -100 °C using a programmable freezer or ethanol baths. Processing which freezes cryoprotectant in a tube artificially is performed near -7~-8 °C in the middle of the freezing. In this treatment, ice is made to form out of a cell under gradual cooling. Intracellular moisture penetrates a plasma membrane, and arrives at the surface of the ice besides a cell, and freezes. This is called 'extracellular freezing'. Intracellular moisture is removed and a cytoplasm is contracted by 'extracellular freezing'. Kindly refer to the book of Kartha (1985) to understand the principle of this phenomenon. After making specimens freeze to a predetermined freezing-temperature, they are immersed in LN. The freezing-temperature is arranged by -40 °C in many species. Cryopreserved tubes are warmed using hot water (40 °C) for 1~2 min, and cryoprotectants are removed from a tube. After rewarming, samples are moved from the cryotube, and recultured. The cooling rate in this method is important. It differs from 0.5 ℃/min to 50 ℃/min with plant species and the size of the plant germplasm. However, in the case of the freezing speed of 2 ℃/min or more, the regrowth after preservation tends to fall (Sugawara & Sakai, 1974; Uemura & Sakai, 1980). The disadvantage of this method is that there are many species for which the prefreezing method is not utilized at all. In addition, there are plant tissues which freeze to death partially, and cases in which the decrease in subsequent viability induced also exists (Grout & Henshaw, 1980; Haskins & Kartha, 1980).

2.1.2 Slow unprogrammed freezing (also known as "simple freezing")

This cryopreservation method was reported using samples of several species in the early 1990's. The advantage of this method is that researchers can cryopreserve without a special programmable freezer, compared with slow programmed freezing.

The slow unprogrammed freezing is shown in Fig. 2. Plant tissues are added to the tube containing cryoprotectants. Tubes are treated for about 10 min at room temperature (25 ℃), and are kept at -30 ℃ for 30~120 min. They are then immersed in LN thereafter. Cryopreserved tubes are warmed using hot water (40 ℃) for 1~2 min, and cryoprotectants are removed from a tube. After rewarming, samples are moved from the cryotube, and recultured. In this cryopreservation method, mixtures of glycerol and sucrose or DMSO and sorbitol are used as cryoprotectants (Sakai et al., 1991; Niino et al., 1992; Maruyama et al., 2000). In this cryopreservation method, although 'naked' samples are used, Kobayashi et al. (2005) utilized cells encapsulated with alginate beads in the suspension cells of tobacco.

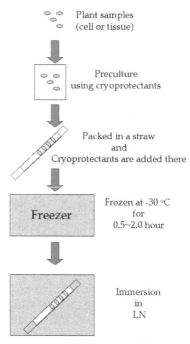

Fig. 2. The protocol of slow unprogrammed freezing (from Sakai et al., 1991).

2.1.3 Vitrification

The vitrification method has been the major cryopreservation method since Uragami et al. (1989) developed it using asparagus culture cells. This cryopreservation method is shown in Fig. 3. Plant tissues are added to the tube containing the loading solution (LS) for the osmoprotection. Beads in tubes are osmoprotected for about 30 min at room temperature (about 25 °C). LS is the liquid culture medium in which sucrose (0.4 mol/L) and the glycerol (2.0 mol/L) were contained. After loading, LS is removed from a tube, and new vitrification solution is added for the dehydration of plant tissues.

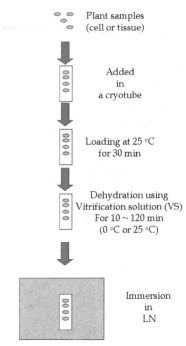

Fig. 3. The protocol of vitirication method (from Sakai et al., 1990).

Many cryoprotectants are dissolved in the vitrification solution, and the optimal dehydration time using the solution changes greatly with treatment temperature.

In many cases, the dehydration using the vitrification solution is performed at 0 °C by the reason of the toxicity to plant cells. Plant Vitrification Solution 2 (PVS2; Sakai et al., 1990) is utilized most as the vitrification solution. Besides PVS2, there are many vitrification solutions. Please refer to Table 1 for the composition. They are immersed in LN after that. Cryopreserved tubes are warmed using hot water (40 °C) for 1~2 min, and the vitrification solution is removed from a tube. After the removal of vitrification solution, unloading solution (the liquid medium supplemented with 1.2 mol/L sucrose) is added to a tube, and cryoprotectants are removed from plant tissues for 30 min at 25 °C. In many cases, the above-mentioned liquid mediums (LS, PVS and unloading solution) were adjusted by pH

5.7~5.8, but without plant growth regulators. After unloading, samples are removed from the cryotube, and recultured.

Conponent (g/L)	PVS1	PVS2	PVS3
Glycerol	220.0	300.0	500.0
Ethylene Glycol	150.0	150.0	
Propylene Glycol	150.0		
Dimethyl Sulfoxide (DMSO)	70.0	150.0	
Sucrose		136.9	500.0
Sorbitol	91.1		

Table 1. Components of major plant vitrification solutions. Components of three plant vitrification solutions are referred from previous reports (Uragami et al., 1989; Sakai et al., 1990; Nishizawa et al., 1993).

2.1.4 Encapsulation-vitrification

The encapsulation-vitrification method was reported first by Matsumoto et al. (1995) using shoot apices of *Wasabia japonica*, and then spread all over the world. The advantage of this method is that regrowth of plant germplasm after cryopreservation is markedly increased by encapsulating plant samples with alginate beads. The encapsulation of plant germplasms makes for less damage to samples during vitrification procedures (loading treatment and dehydration treatment). On the other hand, due to encapsulation-dehydration, treatment time becomes long compared with that of vitrification and the cryopreservation operation becomes complicated (for example, encapsulation).

The procedure for encapsulation-vitrification is shown in Fig. 4. Plant tissues are immersed in the calcium-free liquid medium supplemented with 0.4 mol/L sucrose, 30.0 g/L sodium alginate and glycerol (1.0~2.0 mol/L). The mixture (including a plant cell or tissue) was added drop by drop to the liquid medium containing 0.1 mol/L calcium chloride, forming beads about 5 mm in diameter. The above-mentioned liquid mediums (30.0 g/L sodium alginate and 0.1 mol/L calcium chloride) were adjusted by pH 5.7, but without plant growth regulators. Encapsulated specimens are added to the culture bottle containing LS for osmoprotection. Beads in the bottles are osmoprotected for 16 hours at room temperature (25 °C). LS is the liquid culture medium in which sucrose (0.75~0.8 mol/L) and the glycerol (2.0 mol/L) were contained. After loading, LS is removed from a bottle, and PVS is added newly for the dehydration of plant tissues. The same as with vitrification, the dehydration using PVS is performed at 0 °C in light of the toxicity to plant cells.

After dehydration of PVS, encapsulated samples are moved to a cryotube containing fresh PVS, and immersed in LN. Cryopreserved tubes are warmed using hot water (40 °C) for 1~2 min, and the vitrification solution is removed from the tube. After removal of the solution, unloading solution (supplemented with 1.2 mol/L sucrose; pH 5.7) is added to a tube, and cryoprotectants are removed from plant tissues for 30 min at 25 °C. After unloading, samples are moved from the cryotube, and recultured.

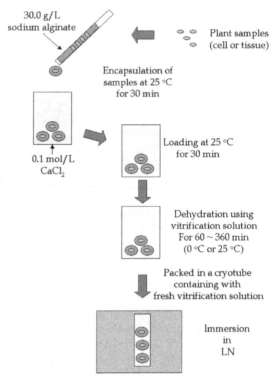

Fig. 4. The protocol of encapsulation-vitirication method (from Matsumoto et al., 1995)

2.1.5 Simplified encapsulation-vitrification

The simplified encapsulation-vitrification method was first reported by Hirai & Sakai (2002) using shoot apices of sweet potato. The operating procedure in this method is the same as encapsulation-vitrification (see Fig. 4), however, the composition of LS differs. LS of simplified encapsulation-vitrification includes high-concentration glycerol (2.0 mol/L) and sucrose (1.6 mol/L), and the viscosity of LS is high. Although this method succeeded with sweet potato, there are some plant species which cannot be cryopreserved using high concentration glycerol (Hirai & Sakai, 1999).

2.1.6 Droplet method

The droplet method was first reported by Schäfer-Menuhr et al. (1994, 1997) using potato apices. The operating procedure is the same for vitrification. However, the LS immersion protocol differs compared with that in the vitrification method. This cryopreservation method is shown in Fig. 5. After treatments by LS and PVS, plant samples are put on aluminum foil which is sterilized and cut small. One drop of PVS is dripped onto plant samples, and the whole aluminum foil is immersed in LN. The aluminum foil after cryopreservation is taken out from LN, and one drop of unloading solution supplemented with 1.0 mol/L sucrose is driiped onto to freezing samples. After rewarming, samples are moved from the cryotube, and recultured. In the droplet method, in order to make a plant sample cool quickly, Wesley-Smith et al. (2001) used not liquid nitrogen but a slush nitrogen (-210 °C) and an isopentane (-160 °C). In addition, the droplet method can reportedly obtain a high regrowth percentage after cryopreservation in tropical plants difficult to cryopreserve (Pennycooke & Towill, 2000, 2001; Leunufna & Keller, 2003; Panis et al., 2005).

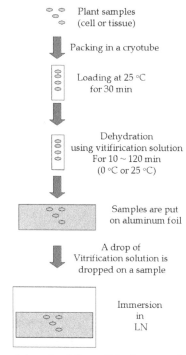

Plant samples
(cell or tissue)

Packing in a cryotube

Loading at 25 °C
for 30 min

Dehydration
using vitifirication solution
For 10 ~ 120 min
(0 °C or 25 °C)

Samples are put
on aluminum foil

A drop of
Vitrification solution is
dropped on a sample

Immersion
in
LN

Fig. 5. The protocol of Droplet method (from Schäfer-menuhr et al., 1997).

2.1.7 Dehydration

Dehydration was first reported by Uragami et al. (1990) using asparagus lateral buds. A dry technique is superior to vitrification in that it does not need to produce PVS. Therefore, there is no influence of medical toxicity at low cost. Problems of dehydration include ready influence of humidity on drying by air flow and dried samples are easily crushed with tweezers.

The cryopreservation procedure is shown in Fig. 6. Plant tissues are put on the filter paper or nylon mesh sterilized and cut small. Samples are dehydrated by silica gel (Uragami et al., 1990) or air flow (Shimonishi et al., 1992; Kuranuki & Yoshida, 1996) before immersion in LN. It is reported that the optimal moisture of the sample is 10%~30% for survival after cryopreservation in the dehydration method (Uragami et al., 1990; Shimonishi et al., 1992; Kuranuki and Yoshida, 1996). After the dehydration, germplasms are moved to a cryotube and immersed in LN. Cryopreserved tubes are warmed at room temperature or using hot water (40 ºC) for 1 ~ 2 min. After rewarming, samples are removed from the cryotube, and recultured.

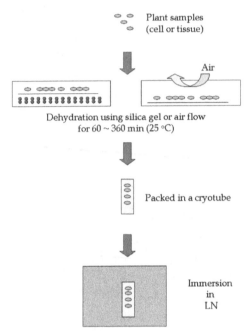

Fig. 6. The protocol of Dehydration method (from Uragami et al., 1990).

2.1.8 Encapsulation-dehydration

The encapsulation-dehydration method was first reported by Fabre & Dereuddre (1990) using shoot apices of potato, and spread worldwide the same way as vitrification and encapsulation-vitrification. This method excels that of dehydration in that regrowth of plant germplasm after cryopreservation is markedly increased by encapsulating plant samples with alginate beads. In addition, encapsulated samples are difficult to be crushed with tweezers compared with the dehydration method.

The encapsulation-dehydration procedure is shown in Fig. 7. Plant tissues are immersed in a calcium-free liquid medium supplemented with 0.4 mol/L sucrose and 30.0 g/L sodium alginate. The mixture (including a plant cell or tissue) was added drop by drop to the liquid medium containing 0.1 mol/L calcium chloride, forming beads about 5 mm in diameter. The above-mentioned liquid mediums (30.0 g/L sodium alginate and 0.1 mol/L calcium chloride) were adjusted by pH 5.7~5.8, but without plant growth regulators. Encapsulated germplasms are added to the culture bottle containing LS for the osmoprotection. Beads in the bottles are osmoprotected for 16 hrs at room temperature (25 ºC). LS is the liquid culture medium in which sucrose (0.75~0.8 mol/L) is contained. After loading, LS is removed from the bottle. Loaded samples are put on sterilized filter papers, and samples are dehydrated by silica gel for 3~7 hours before immersion in LN. After dehydration by silica gel, encapsulated samples are moved to a cryotube, and immersed in LN. Cryopreserved tubes are warmed using hot water (40 ºC) for 1 ~ 2 min. After rewarming, samples are moved from the cryotube, and recultured.

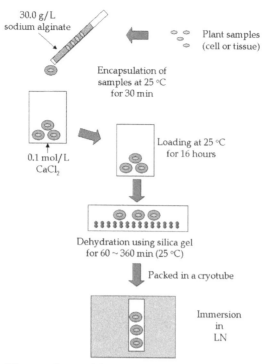

Fig. 7. The protocol of Encapsulation-dehydration method (from Fabre & Dereuddre, 1990).

In encapsulation-dehydration, the addition of glycerol besides sucrose in LS reportedly enhances the regrowth percentage of cryopreserved samples. The optimal concentration of glycerol in LS is 0.5~2.0 mol/L for regrowth of cryopreserved specimens (Matsumoto & Sakai, 1995; Kami et al., 2005, 2007, 2008).

2.1.9 Newly-developed encapsulation-dehydration

A newly developed encapsulation-dehydration method was first reported by Sakai et al. (2000). The operating procedure is the same as for encapsulation-dehydration (see Fig. 7), however, the LS composition differs. LS of the newly developed encapsulation-dehydration includes a high concentration (2.0 mol/L) of glycerol besides sucrose. Therefore, the loading time of this method (1 hour) is shorter than that of encapsulation-dehydration (16 hours).

2.2 Methods of improvement of cryopreservation efficiency

In this section, I introduce some approaches to increase regrowth of samples after rewarming with past reports and actual experimental data I obtained.

2.2.1 Plant material

Before performing cryopreservation of plant samples, it is necessary to grasp the characteristics of the given plant species. For example, it is better to utilize encapsulation-dehydration rather than vitrification for plant species which are subject to toxicity from cryoprotectants. Moreover, if you want to cryopreserve the plant germplasms readily susceptible to toxicity in DMSO with the vitrification method, it is better to use PVS3 rather than PVS2 as the vitrification solution.

Next, I would like to explain this paragraph with actual experimental data I obtained. In cryopreservation, the extracted size of plant material also becomes important. When plant tissues are greatly (3 mm x 3 mm) trimmed, the extraction labor will decline with small tissue size (1 mm x 1 mm). However, the regrowth percentage of large tissues after cryopreservation seems to decrease more than that of small tissues (Fig. 8; Kami et al., 2010). From previous peports, the reason is that the smaller the size of the extracted plant, the more the osmosis cryoprotectant decreases (Kim et al., 2004, 2005).

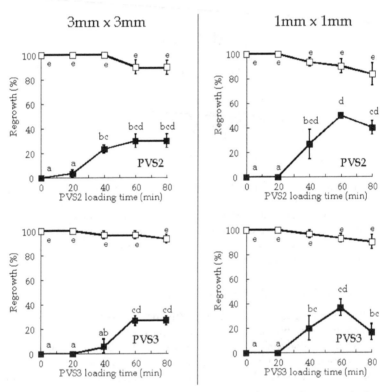

Fig. 8. Effects of excised apex size and exposure time to plant vitrification solution (PVS) on the regrowth of shoot apices immersed in liquid nitrogen (LN) using vitrification. Apices were dehydrated with two types of PVS at 0 °C for various lengths of time prior to cooling (Cryopreserved) or without cooling to -196 °C (Treated Control). The PVS in a cryovial was exchanged just after PVS loading treatment to prevent deterioration of PVS by a loading solution in this study. After cooling for 1hour in LN, rewarming apices were transplanted into regrowth medium. Values represent mean ± SE of three determinations. Differences in mean values of regrowth of treated control and cryopreseved apices with different letters are statistically significant (Tukey's HSD at $p<0.05$) in all data. (from Kami et al., 2010)

2.2.2 Treatment before cryopreservation

Before cryopreservation, cold-acclimation and preculture are done, so survival percentages will increase after cryopreservation.

Cold-acclimation is a treatment by which plantlets are cultured at about 5 ℃ for one week to two months. However, Chang et al. (2000) reported that cold-acclimation was performed at -1 ℃ in grass species (*Zoysia* and *Lolium* sp.). The freezing resistance of plant specimens reportedly increases by cold-acclimation (Chang et al., 2000). However, since cold-acclimation cannot be adapted for a tropical plant, you should not perform this operation. Moreover, optimal acclimation periods differ by plant germplasms. In addition, prolonged cold-acclimation may curve and lower the survival percentage of plant specimens after cryopreservation. Therefore, I recommend that you closely consider the optimal cold acclimation period before trying cryopreservation.

Preculture is the treatment which gives plant cells or tissues dehydration tolerance. In many cases, plant samples are cultivated for 24~48 hours by culture medium supplemented with high-concentration the sucrose (0.3~0.7 mol/L). And some plant species are moved gradually from low to high concentration of sucrose medium (Niino et al., 1992; Niino & Sakai, 1992a,b; Suzuki et al., 1994; Niino et al., 1997). In addition, there are also cases in which glycerol (Matsumoto et al., 1998; Niino et al., 2003), DMSO (Fukai, 1990), or abscisic acid (ABA; Kendal et al., 1993; Tsukazaki et al.,2000) is mixed with a sucrose culture medium, and culture medium containing sorbitol without sucrose are used (Yamada et al., 1991; Maruyama et al., 2000). In many cases, room temperature is used for treatment (20~25 ℃). However, some plant species can be processed by -1 ℃ (Chang et al., 2000) or 5 ℃ (Niino & Sakai, 1992a,b; Kuranuki & Sakai, 1995; Tanaka et al., 2004).

2.2.3 Treatment under cryopreservation

I would like to explain this paragraph with actual experimental data I obtained. In vitrification, I examined the effect of exchange times of fresh PVS2 during a 60-min PVS2 loading treatment on shoot apices (*Cardamine yezoensis* Maxim.) immersed in LN using a vitrification protocol (Fig. 9). The shoot regeneration percentages after cryopreservation was enhanced up to 96.7% when two PVS2 exchanges were used. Moreover, above 80% of shoot regrowth was maintained also by three or more PVS2 exchanges. From this experiment, it became clear that the injury by too much dehydration and medical toxicity are not induced by the exchange of fresh vitrification solution. However, the increase in the exchange time of vitrification solution carries a complex risk of losing the shoot apex and operating. Therefore, I considered that even 2 exchanges during 60-min PVS2 loading treatment on shoot apices of *Cardamine yezoensis* was appropriate (Kami et al., 2010).

Since PVS2 at 0 ℃ has high viscosity and the circulation in the cryobial is poor, it is thought PVS2 around a shoot apex was diluted by the moisture flowing out of the plant tissue. Therefore, by exchanging for fresh PVS2, the dilution of PVS2 around a shoot apex was prevented and the dehydration maintained.

Furthermore, adding an ice blocking agent to PVS reportedly enhances regeneration of cryopreserved sample in recent years (Zhao et al., 2005).

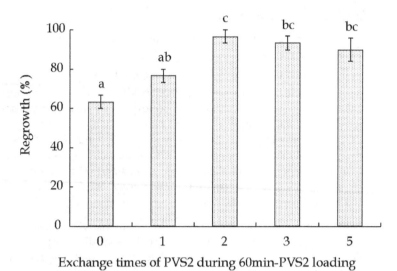

Exchange times of PVS2 during 60min-PVS2 loading

Fig. 9. Effects of exchange times of PVS2 during 60-min PVS2 loading treatment on shoot apices immersed in LN using vitrification. 2 ml of fresh PVS2 were exchanged at 0°C for 60-mins prior to cooling. A PVS2 exchange just after PVS2 loading treatment was not counted as the exchanging time of PVS2 in this study. After cooling for 1hour in LN, rewarming apices were transplanted into 1/4MS. Values represent mean ± SE of three determinations. Differences in mean values of regrowth with different letters are statistically significant (Tukey's HSD at $p<0.05$) in each treatments. (from Kami et al., 2010)

2.2.4 Treatment after cryopreservation

In cryopreservation of plant genetic resources, regeneration after rewarming is the key. Surviving cells or tissues after cryopreservation readily succumb due to different environmental agents because they have been injured by the dehydration or temperature change during the cryopreservation procedure. Moreover, when plant specimens were injured by the cryopreservation process, polyphenol can be produced. Thus, this may threaten the survival of plant specimens after cryopreservation. In that case, regeneration of tissues after preservation reportedly increased when activated charcoal (Bagniol & Engelmann, 1992) and polyvinyl pyrrolidone (Niino et al., 2003), an adsorbent of polyphenol, was mixed with a culture medium.

In recent years, it is also reported that regrowth percentages of rewarming tissues increased by mixing surfactant with regrowth medium (Anthony et al., 1996; Niu et al., 2010). Therefore, special consideration must be given to certain plant species. From previous reports, the regrowth after preservation increases sharply also by decreasing NH_4^+ concentration in a culture medium (Niino et al., 1992a, 1992b; Suzuki et al., 1994; Pennycooke & Towill, 2001).

Next, I would like to explain this paragraph with actual experimental data I obtained. I examined the effects of various nutrient media (Table 2) on regrowth of cryopreserved apices (*Cardamine yezoensis* Maxim.). It was demonstrated that 4-fold dilution of inorganic salts of Murashige and Skoog's medium (1/4MS) or Woody Plant medium (WPM) as basal medium resulted in higher regrowth percentages (both 66.7%) than six other media (Fig. 10; Kami et al., 2010).

Component	Murashige & Skoog (MS)				White	B5	N6	WPM
(mmol/L)	1	1/2	1/4	1/8				
NO_3^-	45.39	22.70	11.35	5.67	0.79	26.77	30.25	6.30
NH_4^+	20.61	10.31	5.15	2.58	0.00	2.33	7.01	5.00
PO_4^{3-}	1.25	0.62	0.31	0.16	0.12	1.09	2.94	1.25
K^+	20.05	10.02	5.01	2.51	1.63	24.73	30.94	12.61
Ca^{2+}	2.99	1.50	0.75	0.37	1.27	1.02	1.13	3.01
Mg^{2+}	3.07	3.07	0.77	0.28	5.98	2.08	1.54	3.07

Table 2. Compositions of eight types of nutrient medium for the regrowth of cryopreserved shoot apices

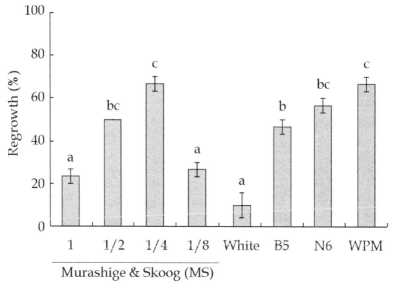

Nutrient media used for regrowth of cryopreserved apices

Fig. 10. Effects of nutrient media on the regrowth of shoot apices immersed in LN using vitrification. Apices were dehydrated with PVS2 at 0°C for 60-mins prior to immersion in LN. The PVS2 in a cryovial was exchanged once just after PVS2 loading treatment. After cooling for 1hour in LN, rewarming apices were transplanted into 8 types of basal medium. Values represent mean ± SE of three determinations. Differences in mean values of regrowth with different letters are statistically significant (Tukey's HSD at $p<0.05$) in each treatments. (from Kami et al., 2010)

3. Conclusion

The cryopreservation technique for plant genetic resources has developed since the 1990s. However, since there are plant species which cannot yet be cryopreserved, improvement of the technology is a pressing need. I have limited my remarks to the introduction of the cryopreservation technique in this section. This seems like a personal comment, not a part of your conclusion. Kartha (1985), in his detailed book on these principles, provides a valuable addition to this chapter, not but provides an explanation of cryopreservation technology.

4. References

Anthony, P.; Jerodar, N.B., Lowe, K.C., Power, J.B. & Davey, M.R. (1996). Pluronic F-68 increases the post-thaw growth of cryopreserved plant cells. *Cryobiology*, Vol.33, No.6, (July 1996), pp.508-514, ISSN 0011-2240

Bagniol, S. & Engelmann, F. (1992). Effect of thawing and recovery conditions on the regrowth of meristems of date palm (*Phoenix dactylifera* L.) after cryopreservation in liquid nitorogen. *Cryo-Letters*, Vol.13, No.3, (January 1992), pp. 253-260, ISSN 0143-2044

Bajaj, Y.P.S. (1995). *Biotecnology Agriculture and Forestry 32 – Cryopreservation of Plant Germplasm I*, Springer-Verlag, ISBN 3-540-57451-4, Berlin Heidelberg, German

Chang, Y.J.; Baker, R.E. & Reed, B.M. (2000). Cold acclimation improves recovery of cryopreserved grass (*Zoysia* and *Lolium* sp.). *Cryo-Letters*, Vol.21, No.2, (November 1999), pp. 107-116, ISSN 0143-2044

Fabre, J. & Dereuddre, J. (1990). Encapsulation Dehydration – A new approach to cryopreservation of *Solanum* shoot-tips. *Cryo-Letters*, Vol.11, No.5, (April 1990), pp. 413-426, ISSN 0143-2044

Finkle, B.J. & Ulrich, J.M. (1979). Efects of cryoprotectants in combination on the survival of frozen sugarcane cells. *Plant Physiology*, Vol.63, No.4, (November 1978), pp. 598-604, ISSN 0032-0889

Fukai, S. (1990). Cryopreservation of chrysanthemum shoot tips. *Scientia Horticulturae*, Vol.45, No.2, (December 1989), pp. 167-174, ISSN0304-4238

Grout, B.W.W. & Henshaw, G.G. (1980). Structural observations on the growth of potato shoot-tip cultures after thawing from liquid nitrogen. *Annals of Botany*, Vol.46, No.2, (January 1980), pp. 243-248, ISSN 0305-7364

Haskins, R.H. & Kartha, K. K. (1980). Freeze preservation of pea meristems: cell survival. *Canadian Journal of Botany*, Vol.58, No.8, (May 1980), pp. 833-840, ISSN 0008-4026

Hirai, D. & Sakai, A. (1999). Cryopreservation of *in vitro*-grown axillary shoot-tip maristems of mint (*Mentha spicata* L.) by encapsulation-vitrification. *Plant Cell Reports*, Vol.19, No.2, (February 1999), pp. 150-155, ISSN 0721-7714

Hirai, D. & Sakai, A. (2002). Simplified cryopreservation of sweet potato (*Ipomoea batatas* (L.) Lam.) by optimizing conditions for osmoprotection. *Plant Cell Reports*, Vol.21, No.10, (February 2002), pp. 961-966, ISSN 0721-7714

Kami, D.; Suzuki, T. & Oosawa, K. (2005). Cryopreservation of blue honeysuckle *in vitro*-cultured tissues using encapsulation-dehydration and vitrification. *Cryobiology and Cryotechnology*, Vol.51, No.2, (December 2005), pp.63-68, ISSN 1340-7902

Kami, D.; Uenohata, M., Suzuki, T. & Oosawa, K. (2008). Cryopreservation of black chokeberry *in vitro* shoot apices. *Cryo-Letters*, Vol.29, No.3, pp.209-216, (January 2008), ISSN 0143-2044

Kami, D.; Shi, L., Sato, T., Suzuki, T. & Oosawa, K. (2009). Cryopreservation of shoot apices of hawthorn *in vitro* cultures originating from East Asia. *Scientia Horticulturae*, Vol.120, No.2, (August 2008), pp. 84-88, ISSN0304-4238

Kami, D.; Kido, S., Otokita, K., Suzuki, T., Sugiyama, K. & Suzuki, M. (2010). Cryopreservation of shoot apices of *Cardamine yezoensis in vitro*-cultures by vitrification Method. *Cryobiology and Cryotechnology*, Vol.56, No.2, (September 2010), pp.119-126, ISSN 1340-7902 (in Japanese with English summary)

Kartha,K.K. (1985). *Cryopreservation of Plant Cells and Organs*, CRC Press, ISBN 0-8493-6102-8, Florida, U.S.A.

Kendall, E.J.; Kartha, K.K., Qureshi, J.A. & Chermak, P. (1993). Cryopreservation of immature spring wheat zygotic embryos using an abscisic acid pretreatment. . *Plant Cell Reports*, Vol.12, No.2, (Octobar 1992), pp. 89-94, ISSN 0721-7714

Kim, H.H.; Kim, J.B., Baek, H.J., Cho, E.G., Chae, Y.A. & Engelmann, F. (2004). Evolution of DMSO concentration in garlic shoot tips during a vitrification procedure, *Cryo-Letters*, Vol.25, No.1, (October 2003), pp. 91-100, ISSN 0143-2044

Kim, J.B.; Kim, H.H., Baek, H.J., Cho, E.G., Kim, Y.H. & Engelmann, F. (2005). Changes in sucrose and glycerol content in garlic shoot tips during freezing using PVS3 solution, *Cryo-Letters*, Vol.26, No.2, (December 2004), pp. 103-112, ISSN 0143-2044

Kobayashi, T.; Niino, T. & Kobayashi, M. (2005). Simple cryopreservation protocol with an encapsulation technique for tabacco BY-2 supension cell cultures. *Plant Biotechnology*, Vol.22, No.2, (March 2005), 105-112, ISSN 1342-4580

Kumu, Y.; Harada, T. & Yakuwa, T. (1983). Development of a whole plant from a shoot tip of *Asparagus officinalis* L. frozen to –196ºC. *Journal of the Faculty of Agriculture, Hokkaido University*, Vol.61, No.3, (May 1983), pp. 285-294, ISSN 0018-344X

Kuranuki, Y. & Sakai, A. (1995). Cryopreservation of *in vitro*-grown shoot tips of tea (*Camellia sinensis*) by vitrification. *Cryo-Letters*, Vol.16, No.4, (April 1995), pp. 345-352, ISSN 0143-2044

Kuranuki, Y. & Yoshida, S. (1996). Differential responses of embryogenic axes and cotyledons from tea seeds to desiccation and cryoexposure. *Breeding Science*, Vol. 46, No.2, pp. 149-154, (March 1996), ISSN 1347-3735

Leunufna, S. & Keller, E.R.J. (2003). Investigating a new cryopreservation protocol for yams (*Discorea* spp.). *Plant Cell Reports*, Vol.21, No.12, (April 2003), pp. 1159-1166, ISSN 0721-7714

Maruyama, E., Tanaka, T., Hosoi, Y., Ishii, K. & Morohoshi, N. (2000). Embryogenic cell culture, protoplast regeneration, cryopreservation, biolistic gene transfer and plant renegeration in Japanese cedar (*Cryptomeria japonica* D. Don). *Plant Biotechnology*, Vol.17, No.4, (April 2000), 281-296, ISSN 1342-4580

Matsumoto, T.; Sakai, A., Takahashi, C. & Yamada, K. (1995). Cryopreservation of *in vitro*-grown apical meristems of wasabi (*Wasabia japonica*) by encapsulation-vitrification method. *Cryo-Letters*, Vol.16, No.2, (November 1994), pp. 189-196, ISSN 0143-2044

Matsumoto, T. and A. Sakai. 1995. An approach to enhance dehydration tolerance of alginate-coated dried meristems cooled to -196°C. *Cryo-Letters*, Vol.16, No.3, (January 1995), pp. 299-306, ISSN 0143-2044

Matsumoto, T.; (1998). A novel preculturing for enhancing the survival of *in vitro*-grown meristems of wasabi (*Wasabia japonica*) cooled to -196°C by vitrification. *Cryo-Letters*, Vol.19, No.1, (September 1997), pp. 27-36, ISSN 0143-2044

Niino, T. & Sakai, A. (1992a). Cryopreservation of alginate-coated *in vitro*-grown shoot tips of apple, pear and mulberry. *Plant Science*, Vol.87, No.2, (August 1992), pp. 199-206, ISSN 0168-9452

Niino, T. & Sakai, A. (1992b). Cryopreservation of *in vitro*-grown shoot tips of apple and pear by vitrification. *Plant Cell, Tissue and Organ Culture*. Vol.28, No.3, (October 1991), pp. 261-266, ISSN 0167-6857

Niino, T.; Sakai, A., Yakuwa, H. & Nojiri, K. (1992). Cryopreservation of *in vitro*-grown shoot tips of mulberry by vitrification. *Cryo-Letters*, Vol.13, No.4, (April 1992), pp. 303-312, ISSN 0143-2044

Niino, T.; Tashiro, K., Suzuki M., Ochiai, S., Magoshi, J. & Akihama, T. (1997). Cryopreservation of *in vitro*-grown shoot tips of cherry and sweet cherry by one-step vitrification. *Scientia Horticulturae*, Vol.70, No.2, (March 1997), pp. 155-163, ISSN0304-4238

Niino, T.; Tanaka, D., Ichikawa, S., Takano, J., Ivette, S., Shirata, K. & Uemura, M. (2003). Cryopreservation of *in vitro*-grown apical shoot tips of strawberry by vitrification. *Plant Biotechnology*, Vol.20, No.1, (Nobember 2002), 75-80, ISSN 1342-4580

Nishizawa, S.; Sakai, A., Amano, Y. & Matsuzawa, T. (1993). Cryopreservation of asparagus (*Asparagus officinalis* L.) embryogenic suspension cells and subsequent plant regeneration by vitrification. *Plant Science*, Vol.91, No.1, (March 1993), pp. 67-73, ISSN 0168-9452

Niu, Y.L.; Zhang, Y.F., Zhang, Q.L. & Luo, Z.R. (2010). A preliminary study on cryopreservation protocol applicable to all types of *Diospyros Kaki* Thunb. *Biotecnology & Biotechnological Equipment*, Vol.24, No.3, (April 2010), pp. 1960-1964, ISSN 1310-2818

Panis, B.; Piette, B. & Swennen, R. (2005). Droplet vitrification of apical meristems: a cryopreservation protocol applicable to all Musaceae. *Plant Science*, Vol.168, No.1, (July 2004), pp. 45-55, ISSN 0168-9452

Pennycooke, J.C. & Towill, L.E. (2000). Cryopreservation of shoot tips from *in vitro* plants of sweet potato [*Ipomoea batatas* (L.) Lam.] by vitrification. *Plant Cell Reports*, Vol.19 No. 8, (September 1999), pp. 733–737, ISSN 0721-7714

Pennycooke, J.C. & Towill, L.E. (2001). Medium alternations improve regrowth of sweet potato (*Ipomoea batatas* {L.} Lam.) shoot tips cryopreserved by vitrification and encapsulation–dehydration. *Cryo-Letters* Vol.22, No.4, (February 2001), pp. 381–389, ISSN 0143-2044

Sakai, A.; Kobayashi, S. & Oiyama, I. (1990). Cryopreservatioin of nucellar cells of navel orange (*Citrus sinensis Osb.* var *Brasiliensis* Tanaka) by vitrification. *Plant Cell Reports*, Vol.9, No.1, (January 1990), pp. 30-33, ISSN 0721-7714

Sakai, A.; Kobayashi, S. & Oiyama, I. (1991). Cryopreservatioin of nucellar cells of navel orange (*Citrus sinensis Osb.* var *Brasiliensis* Tanaka) by a simple freezing method. *Plant Science,* Vol.74, No.3, (October 1990), pp. 243-248, ISSN 0168-9452

Sakai, A.; Matsumoto, T., Hirai, D. & Niino, T. (2000). Newly developed encapsulation-dehydration protocol for plant cryopreservation. *Cryo-Letters,* Vol.21, No.1, (October 1999), pp. 53-62, ISSN 0143-2044

Schäfer-Menuhr, A.; Schumacher, H.M. & Mix-Wagner, G. (1994). Langzeitlagerung alter Kartoffelsorten durch Kryokonservierung der Meristeme in flüssigem Stickstoff. *Landbauforschung Völkenrode,* Vol.44, No.4, (March 1994), pp. 301-313, ISSN 0376-0723 (in German with English summary)

Schäfer-Menuhr, A.; Schumacher, H.M. & Mix-Wagner, G. (1997). Long-term storage of old potato varieties by cryopreservation of shoot-tips in liquid nitrogen. *Plant Genetic Resources Newsletter,* (October 1998), Vol.111, No.1, pp. 19-24, ISSN 1020-3362

Shimonishi, M.; Ishikawa, M., Suzuki, S. & Oosawa, K. (1991). Cryopreservation of melon somatic embryos by desiccation method. *Japanese Journal of Breeding,* Vol.41, No.2, (January 1991), 347-351, ISSN 0536-3683

Sugawara, Y. & Sakai, A. (1974). Survival of suspension-cultured sycamore cells cooled to the temperature of liquid nitrogen. *Plant Physiology,* Vol.74, No.5, (August 1974), pp. 722-724, ISSN 0032-0889

Suzuki, M.; Niino, T. & Akihama, T. (1994). Cryopreservation of shoot tips of kiwifruit seedlings by the alginate encapsulation-dehydration technique. *Plant Tissue Culture Letters,* Vol.11, No2., pp. 122-128, (November 1993), ISSN 0289-5773

Tanaka, D.; Niino, T., Isuzugawa, K., Hikage, T. & Uemura, M. (2004). Cryopreservation of shoot apices of *in-vitro* grown gentian plants: comparison of vitrification and encapsulation-vitrification protocols. *Cryo-Letters,* Vol.25, No.3, (January 2004), pp. 167-176, ISSN 0143-2044

Tsukazaki, H.; Mii, M., Tokuhara, K. & Ishikawa, K. (2000). Cryopreservation of *Doritaenopsis* suspension culture by vitrification. *Plant Cell Reports,* Vol.19, No.12, (June 2000), pp. 1160-1164, ISSN 0721-7714

Towill, L.E. & Bajaj, Y.P.S. (2002). *Biotecnology Agriculture and Forestry 50 – Cryopreservation of Plant Germplasm II,* Springer-Verlag, ISBN 3-540-41676-5, Berlin Heidelberg, German

Uemura, M. & Sakai, A. (1980). Survival of carnation (*Dianthus caryophyllus* L.) shoot apices frozen to the temperature of liquid nitrogen. *Plant & Cell Physiology,* Vol.21, No.1, (September 1979), pp. 85-94, ISSN 1471-9053

Uragami, A.; Sakai, A., Nagai, M. & Takahashi, T. (1989). Survival of cultured cells and somatic embryos of *Asparagus officinalis* cryopreserved by vitrification. *Plant Cell Reports,* Vol.8, No.5, (August 1989), pp. 418-421, ISSN 0721-7714

Uragami, A.; Sakai, A. & Nagai, M. (1990). Cryopreservation of dried axillary buds from plantlets of *Asparagus officinalis* L. grown *in vitro. Plant Cell Reports.* Vol.9, No.4, pp. 328-331, (August 1990), ISSN 0721-7714

Yamada, T.; Sakai, A., Matsumura, T. & Higuchi, S. (1991). Cryopreservation of apical meristems of white clover (*Trifolium repens* L.) by vitrification. *Plant Science* Vol.78, No.1, (September 1990), pp. 81–87, ISSN 0168-9452

Zhao, M.A.; Xhu, Y.Z., Dhital, S.P., Song, Y.S., Wang, M.Y. & Lim, H.T. (2005). Anefficient cryopreservation procedure for potato (*Solanum tuberosum* L.) utilizing the new ice blocking agent, Supercool X1000. *Plant Cell Reports*. Vol.24, No.5, (April 2005), pp. 477-481, ISSN 0721-7714

5

Current Issues in Plant Cryopreservation

Anja Kaczmarczyk[1,2], Bryn Funnekotter[1,2], Akshay Menon[1,2],
Pui Ye Phang[1,2], Arwa Al-Hanbali[1,2], Eric Bunn[2,3] and Ricardo L. Mancera[1]
[1]Curtin Health Innovation Research Institute, Western Australian Biomedical Research Institute, Curtin University, Perth
[2]Botanic Gardens and Parks Authority, Fraser Avenue, West Perth
[3]School of Plant Biology, Faculty of Natural and Agricultural Sciences, University of Western Australia
Australia

1. Introduction

Plant cryopreservation involves the storage of plant tissues (usually seed or shoot tips) in liquid nitrogen (LN) at -196°C or in the vapour phase of LN at -135°C in such a way that the viability of stored tissues is retained following re-warming (Day et al., 2008; Hamilton et al., 2009). Cryopreservation is usually applied to species with recalcitrant (i.e. dehydration sensitive) seeds that are not storable by any other means, or preservation of specific cultivars of vegetatively propagated crop plants like banana or potato, or for unique ornamental genotypes (Halmagyi et al., 2004; Kaczmarczyk et al., 2011a; Panis et al., 2005). Another reason to utilise cryostorage is to conserve endangered plant species, particularly where seeds may be extremely scarce or of doubtful quality and/or the species is threatened with imminent extinction (Decruse et al., 1999; Mallon et al., 2008; Mandal & Dixit-Sharma, 2007; Paunescu, 2009; Sen-Rong & Ming-Hua, 2009; Touchell et al., 2002).

The main advantage of cryopreservation is that once material has been successfully cooled to LN temperatures, it can be conserved in principle indefinitely, because at these ultra-low temperatures no metabolic processes occur. Replenishing a small volume of LN weekly in cryo-dewars is the only on-going maintenance operation usually required in cryostorage. There are further advantages to this approach: the low costs of storage, minimal space requirements and reduced labour maintenance compared to living collections and even when compared to maintenance of tissue cultures at room temperature. Once in storage, there is no risk of new contamination by fungus or bacteria, and cryogenically stored material has been reported to retain genetic stability (Harding, 2004). Depending on the species, small cryopreserved samples may take several weeks to re-establish shoot cultures, and several months to a year may be required to produce micropropagated plants capable of transfer to soil under greenhouse conditions and (following weaning) into the field.

Shoot tips (containing the apical meristem) are the most commonly used plant material for cryostorage. The apical meristem is composed of small unvacuolated cells served with a relatively small vascular system. The organised structure of apical meristems generally results in direct shoot formation after re-warming, thereby maintaining the genetic integrity of the resulting propagated material. While callus tissue (unorganised wound tissue) can also be cryostored, the risk of occurrence of genetic deviations may be higher when utilising the indirect organogenesis pathway. Besides shoot tips, callus cultures, cell cultures, somatic embryos, pollen or plant buds as well as recalcitrant and orthodox seeds can be used as explants in plant cryopreservation (Reed, 2008).

Plant cryopreservation began with research on the freezing of mulberry twigs in LN (Sakai, 1965). Since then, a huge variety of plants and genotypes have been successfully cryostored for conservation of agriculture and horticultural genotypes, as well as for endangered and threatened plant species (Gonzalez-Arnao et al., 2008; Hamilton et al., 2009; Kaczmarczyk et al., 2011b; Mycock et al., 1995; Reed, 2008; Sakai & Engelmann, 2007). This chapter reviews and gives examples of different plant cryopreservation protocols that have been successfully applied. It will focus on free radical damage and membrane structure, both important topics in the cryopreservation of biological tissues. The topic of genetic and epigenetic stability in plant cryopreservation is also discussed. Recent reviews of plant cryopreservation have been written by Benson (2008), Day & Stacey (2007), Hamilton et al. (2009) and Reed (2008).

2. Plant cryopreservation methods

In the last three decades a number of different cryopreservation protocols, such as classical slow-cooling, vitrification, droplet vitrification, encapsulation/dehydration and encapsulation/vitrification protocols have been developed and utilised for germplasm storage (Reed, 2008). The choice of cryopreservation method to attain the highest survival rates is largely dependent on the plant species and tissue type that is being cryostored.

2.1 Slow cooling or controlled rate cooling

This technique involves the simple dehydration of plant material before cryogenic storage in LN. This is can be done by slow cooling of the plant tissue to a temperature of approximately -40°C (Reinhoud et al., 2000). This forces the formation of extracellular ice ahead of intracellular ice, thus causing an outflow of water from the cells due to the resulting osmotic imbalance and, consequently, dehydration. Dehydration can also be brought about by incubation of tissue material on media containing a relatively high concentration of an osmo-regulant, commonly sucrose, although other compounds can also be used (Panis et al., 2002). Usually water concentrations must be decreased to between 10% and 20% of the fresh weight for optimal cryogenic survival (Engelmann, 2004). This has the aforementioned effects of reducing the extent of ice crystal formation due to the reduced water concentration and assisting in the achievement of the vitrified state of water as a result of the increased solute concentration. These techniques do not necessarily make use of cryoprotective agents (CPAs), however they can be used in conjunction with them to further improve dehydration (Reinhoud et al., 2000), though these agents can be toxic to plant cells at high concentrations (Arakawa et al., 1990). Rapid re-warming rates are used after cryogenic storage to prevent ice crystal formation during thawing (Reinhoud et al., 2000). This approach can result in extreme rates of dehydration, which can cause cell volume reductions that are potentially lethal (Day et al., 2008). It has been suggested that slow-cooling is only suitable for non-organised tissues, as sufficient dehydration is more difficult to achieve in tissues with complex structures due to the different rates of water movement between and within plant cells with different characteristics (Gonzalez-Arnao et al., 2008).

2.2 Encapsulation-dehydration

This method, developed by Fabre and Dereuddre (1990), involves encapsulating shoot tips, somatic embryos or callus cells within alginate beads. This is followed by incubation in media with high sugar concentrations in order to raise intracellular solute concentrations and promote desiccation. Finally, silica gel or airflow is used to dehydrate the beads until the moisture content drops to 20-30%, before they are immersed in LN (Fabre & Dereuddre, 1990; Hamilton et al., 2009; Reinhoud et al., 2000). The encapsulating material is thought to promote a vitrified state in the tissue regardless of the cooling and re-warming rates, thus reducing damage from ice crystal formation (Scottez et al., 1992). Mechanical stress is also reduced because the bead protects the explants from damage during handling. The benefits of this method include avoiding the use of high concentrations of (potentially toxic) CPAs (Reinhoud et al., 2000) and the presence of a nutritive bead, which may enhance post-regeneration survival or re-growth of the material following immersion in LN and re-warming (Panis & Lambardi, 2005).

2.3 Vitrification

Vitrification involves the treatment of tissues in a mixture of highly concentrated penetrating and non-penetrating CPAs applied at non-freezing temperatures, followed by rapid cooling in LN (Gonzalez-Arnao et al., 2008; Panis & Lambardi, 2005). The combination of high intracellular solute concentrations (due to dehydration and some CPA penetration) and rapid cooling prevents the nucleation of water and the formation of ice crystals, both intracellularly and extracellularly, thus promoting the vitrification of water (Kreck et al., 2011; Mandumpal et al., 2011; Reinhoud et al., 2000).

For plants sensitive to direct exposure to vitrification solutions, due to dehydration intolerance and osmotic stresses, a loading step of 10-20 minutes can be incorporated prior to incubation within the CPA solution. This is done by incubation of the samples within a less toxic/concentrated CPA solution (a media containing 0.4 M sucrose and 2 M glycerol proved highly effective [Nishizawa et al., 1993; Sakai et al., 1990]), thereby improving dehydration tolerance. The CPAs used in vitrification usually contain high concentrations of glycerol, dimethyl sulfoxide (DMSO), ethylene glycol and various sugars (Day et al., 2008). The most commonly used mixture of CPAs for vitrification is plant vitrification solution 2 (PVS2), which consists of 30% (w/v) glycerol, 15% (w/v) ethylene glycol and 15% (w/v) DMSO in basal culture medium containing 0.4 M sucrose (Sakai et al., 1990, 2008).

Exposure time to cryoprotective solutions is a vital step in the cryostorage process. Volk and Walters (2006) demonstrated that the extent of penetration of PVS2 into mint and garlic shoot tips was directly proportional to exposure time. The water content of the shoot tips also significantly decreased with an increase in exposure time to PVS2 (Volk & Walters, 2006). Greater penetration of CPAs can be useful as it helps to increase the internal solute concentration and may contribute to maintaining cell volume, thus preventing damage to the cells (Meryman, 1974). However, overexposure to CPAs may cause damage to the cells owing to the toxic nature of the CPAs or excessive dehydration.

The vitrification protocol is a more widely applied cryopreservation method than slow cooling due to its ease of use, high reproducibility and the wide range of species with complex tissue structure (such as shoot tips and embryos) that have been successfully cryopreserved with this procedure (Takagi et al., 1997; Touchell et al., 1992; Vidal et al., 2005).

2.4 Droplet-vitrification

The droplet-vitrification technique is a modification of the basic vitrification protocol that involves placing the sample within a droplet of 1-10 µl of cryoprotective solution on a piece of aluminium foil before immersion in LN (as opposed to 1-2 mL of cryoprotective solution in the original protocol), as shown in Fig. 1.

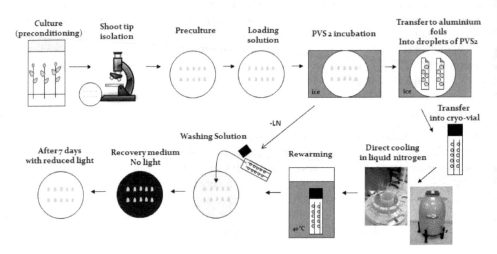

Fig. 1. Cryopreservation procedure for the droplet vitrification method.

This approach achieves higher cooling and re-warming rates, as the small volume of liquid allows a higher rate of heat transfer to and from the sample (Sakai & Engelmann, 2007). Cooling rates are increased to >130°C/sec (Panis & Lambardi, 2005), therefore facilitating the direct transition of intracellular water from a liquid state to a glassy state far more rapidly, thus minimising water crystallisation. The droplet-vitrification protocol has been successfully applied in the cryopreservation of garlic and chrysanthemum (Kim et al., 2009), yams (Leunufna & Keller, 2005), lily (Chen et al., 2011), potato (Yoon et al., 2006) and other plants (Sakai & Engelmann, 2007).

2.5 Encapsulation-vitrification

Another modification of the vitrification approach termed encapsulation/vitrification combines elements of the encapsulation/dehydration method with the vitrification method. As with the standard encapsulation method, shoot tips or calluses are first encapsulated in alginate beads and then the encapsulated material is incubated in a vitrification solution to promote sufficient dehydration and vitrification rather than dehydration under a constant airflow, which is very time consuming and relatively imprecise (Hirai & Sakai, 1999). Successful protocols have been established for potato (Hirai & Sakai, 1999), gentian (Tanaka et al., 2004), strawberry (Hirai et al., 1998) and pineapple (Gamez-Pastrana et al., 2004).

3. Free radical and oxidative damage in plant cryopreservation

While achieving an optimum cryopreservation protocol that successfully avoids ice damage is important, there are various other factors that can affect post-cryogenic survival (Fig. 2). During the cryopreservation procedure plant tissues are susceptible to a variety of stresses, including oxidative stresses (Benson & Bremner, 2004).

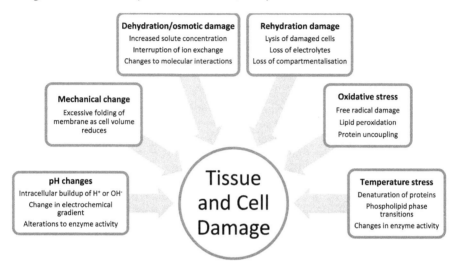

Fig. 2. Main causes of damage to plant tissues during cooling and cryopreservation (modified from [Turner, 2001]).

3.1 Reactive oxygen species (ROS)

The formation of reactive oxygen species (ROS) during cryopreservation can occur during the many steps involved in this process. For example, ROS formation has been detected in photo-oxidative stress during tissue culture, during excision of shoot apices, osmotic injury and desiccation following application of CPAs, as well as during the rapid changes in temperature when the samples are first cryostored in LN and then re-warmed (Benson & Bremner, 2004; Roach et al., 2008). ROS are highly reactive molecules that can cause a wide range of damage in cells. There is a large variety of molecules that are classified as ROS, some of which include oxygen-free radical species and reactive oxygen non-radical derivatives (Table 1).

Radicals	Non-Radicals
Superoxide ($O_2^{\bullet-}$)	Hydrogen peroxide (H_2O_2)
Hydroxyl ($OH\bullet$)	Peroxynitrite ($ONOO^-$)
Hydroperoxyl ($OOH\bullet$)	Peroxynitrous acid ($ONOOH$)
Peroxyl ($ROO\bullet$)	Hypochlorous acid ($HOCl$)
Alkoxyl ($RO\bullet$)	Hypobromous acid ($HOBr$)
Carbonate ($CO_3^{\bullet-}$)	Ozone (O_3)
Carbon dioxide ($CO_2^{\bullet-}$)	Singlet oxygen ($^1\Delta g$)
Singlet oxygen ($^1\Sigma g^+$)	

Table 1. Common reactive oxygen species (ROS) (Halliwell & Gutteridge, 2007).

Many of the more active ROS are free radicals, which are molecules that contain an unpaired electron, thus being able to react non-specifically with neighbouring molecules by removing electrons and causing a self-propagating chain reaction of radical formation. The removal of electrons can lead to a loss of function and structural alterations in macromolecules like proteins, lipids and DNA (Benson, 1990; Halliwell & Gutteridge, 2007). ROS are frequently produced as by-products during cellular metabolism. The electron transport chain used in respiration and photosynthesis are the major producers of ROS, caused by the leakage of free electrons onto molecular oxygen (O_2), resulting in the formation of superoxide (Benson & Bremner, 2004; Benson, 1990; Halliwell & Gutteridge, 2007). The formation of ROS is controlled by a high concentration of antioxidants and proteins that can quench the ROS and fix the damage in these regions. Nevertheless, if there is a sudden increase in ROS formation then cellular repair processes can be overwhelmed and excessive damage can occur.

Temporarily reducing cryo-sample exposure to light immediately after cryopreservation has been shown to increase post-cryogenic survival due to the removal of photo-oxidative stress in the plants (Senula et al., 2007; Touchell et al., 2002). Photo-oxidative stress in plants can result in high levels of singlet oxygen ($^1\Sigma g^+$) and superoxide ($O_2^{\cdot-}$) being produced, either from direct UV radiation on oxygen or the leakage of light energy onto oxygen from chlorophyll when the carotenoid pigments become saturated (Wise, 1995). Plants are highly susceptible to photo-oxidative stresses at low temperatures when exposed to strong light conditions. This can be demonstrated in alpine plants, many of which display adaptations (especially to leaves, i.e. the production of carotenoid pigments) that reduce photo-oxidation damage from enhanced UV-B radiation at high altitudes (Streb et al., 1998). When the ability of antioxidants to quench the formation of ROS and the recycling of antioxidants is reduced, greater damage occurs to the chloroplast through lipid peroxidation, inactivation of photosynthetic proteins and loss of pigments (i.e. bleaching) (Wise, 1995; Wise & Naylor, 1987). Damage to the chloroplast during chilling stress has been shown to severely impede growth rates (Partelli et al., 2009); therefore, reducing the damage that occurs to plant cells due to low temperature oxidative stress is vitally important for improving survival and recovery in cryopreservation.

The most reactive ROS commonly found in plants include superoxide ($O_2^{\cdot-}$), the hydroxyl radical ($OH\bullet$), hydroperoxyl ($OOH\bullet$) and singlet oxygen ($^1\Sigma g^+$). Superoxide and singlet oxygen are often formed as by-products of the electron transport chain from both metabolism and photosynthesis, while hydroperoxyl and hydroxyl radicals are commonly formed in a process called Fenton's reaction, where hydrogen peroxide is converted into the hydroxyl or hydroperoxyl radical (1).

$$Fe^{2+} + H_2O_2 \rightarrow Fe^{3+} + OH\bullet + OH^-$$

$$Fe^{3+} + H_2O_2 \rightarrow Fe^{2+} + OOH\bullet + H^+ \tag{1}$$

The formation of the hydroxyl radical is the major cause of lipid peroxidation in membranes, but can also cause a wide range of damage to all cellular components, including proteins and DNA (Halliwell & Gutteridge, 2007). The addition of specific chelating agents (such as desferrioxamine) has been shown to reduce the levels of iron in cryopreserved tissues, with subsequent decreased levels of the hydroxyl radical observed (Benson et al., 1995; Fleck et al., 2000; Obert et al., 2005).

Damage caused by ROS is difficult to quantify as these molecules are non-specific in their interactions, reacting freely with lipids, proteins and DNA. ROS are highly reactive (and therefore short-lived) and thus, direct measurement of the ROS present in cells is difficult due to time constraints, and does not reflect the damage that may be done prior to the ROS being quenched by antioxidants. Consequently, it is easier to measure the formation of by-products of oxidative damage or the antioxidant status of the cells. The ratio of oxidised to reduced antioxidants is a good indication of ROS formation and the ability of cells to regulate oxidative stress. Identifying end products of ROS oxidation is an indication of the damage caused and is a sign that cells have been unable to satisfactorily quench ROS activity.

3.2 Lipid peroxidation

The cell membrane represents one of the major areas where cryo-injury can occur. Any damage to the cell membrane can alter the delicate balance between intra and extracellular solutes, leading to cell death (Anchordoguy et al., 1987; Dowgert & Steponkus, 1984; Gordon-Kamm & Steponkus, 1984; Lynch & Steponkus, 1987). Lipid peroxidation of fatty acids (FA) in phospholipids can cause extensive damage to the cell membrane if the chain reaction is not controlled, leading to large areas where the semi-permeability of the membrane is altered and thus can no longer function normally (Benson et al., 1992; Halliwell & Gutteridge, 2007). Lipid peroxidation is caused when specific ROS (hydroxyl radical, peroxyl radical and singlet oxygen) interact with a FA. Polyunsaturated fatty acids (PUFA) are the most susceptible to peroxidation (Møller et al., 2007; Young & McEneny, 2001). Glutathione peroxidase can detoxify lipid peroxides by reducing them back to their lipid alcohol form and, in the process, oxidising glutathione (Benson, 1990). The formation of the toxic end-products of lipid peroxidation, such as malondialdehyde (MDA) and 4-hydroxylnonenal (HNE), can also cause damage to cells (Halliwell & Gutteridge, 2007). The toxicity of MDA is debatable (Halliwell & Gutteridge, 2007); however, there is evidence that MDA can interact with proteins and DNA, causing loss of function in proteins and mutations in DNA (Halliwell & Gutteridge, 2007; Hipkiss et al., 1997; Marnett, 1999). The formation of HNE has shown greater toxicity to cells as it can damage mitochondria, inhibit synthesis of DNA and proteins, and interfere with the action of repair proteins such as chaperones (Halliwell & Gutteridge, 2007). Identifying the formation of MDA and HNE is commonly used in cryopreservation as an indicator of oxidative stress. High levels of MDA or HNE detected have correlated with decreased survival rates in rice cell suspensions, olive somatic embryos, flax and blackberry shoot tips (Benson et al., 1992; Obert et al., 2005; Uchendu et al., 2010; Lynch et al., 2011).

Volatile headspace sampling (VHS) measures the formation of volatile compounds released from a sample in a non-destructive and non-invasive assay method. This provides an important tool for measuring oxidative stress. Free radical damage can cause the formation of volatile compounds such as methane, ethane, ethylene and pentane. Quantification of these compounds can be indicative of oxidative damage. The detection of ethylene in plants is of particular importance as ethylene is a vital hormone. Fang et al. (2008) observed that decreased levels of ethylene production correlated with decreased survival and growth. The production of the other volatile compounds is indicative of lipid peroxidation, where increased levels correlate to excessive oxidative damage. Benson et al. (1987) observed a large increase in volatile compounds produced after thawing that has since been observed in other species.

The use of DMSO as a free radical scavenger and probe for the hydroxyl radical can be utilised with VHS. The formation of methane when DMSO interacts with the hydroxyl radical can be measured if the sample is placed in an airtight container, with the levels of methane detected correlating with the formation of the hydroxyl radical. This technique has been used as a measurement of oxidative stress in multiple different plant species, such as rice, cocoa, *Daucus carota* and flax (Benson & Withers 1987; Benson et al., 1995; Obert et al., 2005; Fang et al., 2008). The production of methane is particularly strong during the initial phase of recovery, where it is predicted that antioxidant activity and production is reduced, resulting in increased ROS (Fang et al., 2008). The use of chelating agents to reduce the formation of hydroxyl radicals from the Fenton reaction has shown significant benefits in plant cryopreservation. Desferrioxamine is the most common chelating agent used. It binds to and reduces the amount of free iron. The addition of desferrioxamine to rice cells during cryopreservation showed a decreased recovery period after cryopreservation (Benson et al., 1995). Detection of methane formation was delayed in flax tissue when exposed to desferrioxamine (Obert et al., 2005), indicating a delayed production of hydroxyl radicals.

3.3 Antioxidants

An antioxidant can be defined as any molecule that "delays, prevents or removes oxidative damage to a target molecule" (Halliwell & Gutteridge, 2007). Antioxidants can be classified into two major groups: enzymes that catalytically remove ROS, or sacrificial antioxidant molecules that are preferentially oxidised to protect more important molecules by quenching ROS (Halliwell & Gutteridge, 2007). The addition of exogenous antioxidants during cryopreservation has been shown to result in increased survival in some cases (Uchendu et al., 2010; Wang & Deng, 2004). It is possible that the addition of exogenous antioxidants may also aid in reducing oxidative stresses immediately following re-warming, when cellular metabolism has not been restored to its original state.

Glutathione (GSH) is one of the main water soluble sacrificial antioxidants in plants (Kranner et al., 2006). This low molecular weight thiol is converted to its oxidised form (GSSG) upon interaction with ROS. The ratio of reduced to oxidised GSH is a good indicator of oxidative stress, as an increased amount of GSSG indicates the inability of the plant to control oxidative damage, thereby triggering premature cell death (Kranner et al., 2006). GSH is recycled by the enzyme GSH reductase using NADPH as the electron donor (Halliwell & Gutteridge, 2007). Ascorbic acid is the most abundant water-soluble antioxidant in plant cells (Foyer & Noctor, 2009). This antioxidant is able to quench free radicals, forming the stable radical semidehydroascorbate, which can be further oxidised to dehydroascorbate by another free radical. *Arabidopsis* mutants, which were unable to express ascorbic acid activity, demonstrated the important role this antioxidant plays in reducing oxidative stress, as these plants were not viable following exposure to photo-oxidative stress (Dowdle et al., 2007). Tocopherol (Vitamin E) is a lipophilic antioxidant. This is the main antioxidant involved in membrane protection against lipid peroxidation, as it is preferentially oxidised instead of PUFAs (Halliwell & Gutteridge, 2007; Uchendu et al., 2010). This antioxidant is thought to be reduced to its original, functional state by ascorbic acid (Packer et al., 1979). Carotenoid pigments are the main defence in chloroplasts, where large amounts of singlet oxygen can be produced if the activated chlorophylls transfer their energy onto oxygen. The carotenoid pigments can absorb the energy directly from the chlorophylls, thus suppressing the formation of singlet oxygen, and they can also quench the singlet oxygen directly (Halliwell & Gutteridge, 2007).

The enzyme superoxide dismutase (SOD) catalytically removes the ROS superoxide, producing oxygen and hydrogen peroxide. The removal of superoxide is more important than the formation of hydrogen peroxide, as superoxide is a more reactive species, causing wider damage in the cells. There is potential for SOD to cause formation of ROS if levels of hydrogen peroxide are not controlled. SOD contains a metal cofactor that can cause Fenton reactions and the formation of the hydroxyl radical. Catalase is the main enzyme involved in removing hydrogen peroxide, resulting in the decomposition of hydrogen peroxide to water and oxygen. This enzyme is vital for the removal of hydrogen peroxide before it can damage the cell or be converted to the highly reactive hydroxyl radical through Fenton's reaction.

4. Plant cryopreservation and membrane structure

Membrane systems within cells are usually the site of freezing injury in plants (Steponkus, 1984). Membrane stability is therefore important for reducing such injury. There are four types of injury: (i) expansion-induced lysis, where the cells overexpand as a result of increased extracellular osmotic pressure during warming/thawing; (ii) loss of osmotic responsiveness, where there is no osmotic change during warming due to a slow cooling rate (cells remain dehydrated); (iii) altered osmotic behaviour, where cells membranes turn "leaky", resulting in the release of water and solutes into the surroundings; and (iv) intracellular ice formation, where rapid cooling causes membrane disruption due to the formation of ice crystals (Steponkus, 1984).

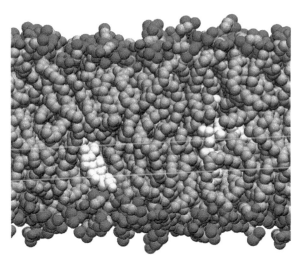

Fig. 3. Typical cell membrane structure consisting of a phospholipid bilayer with embedded sterols. Phospholipid chains are shown in grey, choline groups in blue and phosphate groups in red, while sterol molecules are shown in yellow.

The cell membrane is a bilayer consisting of different lipids and associated proteins (Fig. 3), where the lipids define the cell membrane structure and fluidity, and have a function in signal transduction (Foubert et al., 2007; Furt et al., 2011). The three main classes of lipids found in cell membranes are glycerolipids (mostly phospholipids), sterols and sphingolipids (Furt et al., 2011).

4.1 Phospholipids

Phospholipids are amphiphilic molecules that form the core structure of the cell membrane lipid bilayer. They consist of a polar head (containing a phosphate group and simple organic molecule) and a (mostly) non-polar fatty acid tail. The most common phospholipid components of membranes include phosphatidylcholine (PC), phosphatidylethanolamine (PE), phosphatidylglycerol (PG), phosphatidylinositol (PI), phosphatidylserine (PS) and phosphatidic acid (PA). Phospholipids essentially maintain the structure, fluidity and permeability of the membrane, determined to some extent by the type of phospholipid present and the level of unsaturation in their fatty acid tails (Emmelot & Vanhoeven, 1975; van Meer et al., 2008).

4.2 Sterols

Sterols are steroid alcohols that are integral components of living cells. They are formed as intermediate molecules during the metabolic production of hormones and are an integral component of cell membranes (Hartmann, 1998; Hodzic et al., 2008). Sterols are non-polar molecules with a polar hydroxyl (-OH) side chain that allows them to interact with the polar and non-polar groups of the phospholipid bilayer. They restrict the motion of the fatty acid chains in phospholipids, thus controlling the fluidity of the cell membrane (Hartmann, 1998). The major plant sterols are cholesterol, campesterol, sitosterol and stigmasterol (Grunwald, 1975). Sitosterol is usually found in higher concentration in apex tissue than stigmasterol (Grunwald & Saunders, 1978). The composition and concentration of these sterols within the membrane modify its permeability and fluidity (Grunwald, 1975; Grunwald & Saunders, 1978; Nes, 1974). Cholesterol has the greatest stabilising effect on membranes due to its small side chain (Grunwald & Saunders, 1978).

The ability of sterols to alter stability and permeability of membranes can have a large effect on post-cryopreservation survival of plant tissue. The ratio of stigmasterol to sitosterol has been found to increase after sucrose preculture and was unfavourable to obtaining high percentages of shoot regeneration after cryopreservation of banana meristems (Zhu et al., 2006). Marsan et al. (1998) investigated the interactions of sitosterol and stigmasterol with phosphatidylcholine molecules in soybean and discovered, using neutron scattering experiments, that sitosterol has a greater effect than stigmasterol on the ordering of the fatty acyl chains of PC and increasing the hydrophobic thickness of PC bilayers. Cold acclimation of winter rye (*Secale cereale*) seedlings showed an increase in free sterol content, with β-sitosterol having the largest increase (Lynch & Steponkus, 1987). Uemura and Steponkus (1994) also found an increase of β-sitosterol with cold acclimation in winter rye; however, these results were not mimicked in spring oats (*Avena sativa* L. cv Ogle), where there was no significant change in β-sitosterol, but the stigmasterol proportion increased, whilst cholesterol decreased.

4.3 Soluble sugars

Intracellular soluble sugars and sugar alcohols, such as the ones used in preculture media, reduce damage sustained by cell membranes when the cells undergo desiccation and can improve membrane stability (Crowe et al., 1988; Steponkus, 1984). Smaller sugar molecules help membranes to osmotically retain water and may enter the interlamellar space to maintain a degree of hydration and increase the separation between membranes, thus reducing compressive stresses and, consequently, reducing the chance of a fluid-gel phase transition (Wolfe & Bryant, 1999). Furthermore, the polar hydroxyl (-OH) groups on sugars and sugar alcohols have been thought to interact with the polar membrane phospholipid headgroups and stabilise the membranes by maintaining the separation of the phospholipid molecules (Steponkus, 1984; Turner et al., 2001; Wolfe & Bryant, 1999). Turner et al. (2001) tested several sugars and sugar polyalcohols and determined that the small size of molecules such as glycerol and erythritol allowed them to 'pack' around membranes and have better bonding abilities. Additionally, the stereochemical arrangement of the -OH groups, particularly their orientation along one side of the molecules, imparted more stable hydrogen bonds with the membrane phospholipids and, consequently, resulted in more stable membranes (Turner et al., 2001). Nonetheless, recent biophysical studies have established that specific interactions of sugar molecules to phospholipid headgroups are not necessary to explain the stabilising effect of sugars on membrane gel to fluid transition temperatures (Lenne et al. 2006, 2007, 2009) and that sugars do not in fact insert between phospholipid headgroups but instead a solvation layer of water molecules separates them from the phospholipid headgroups (Kent et al., 2010).

4.4 Preconditioning and cold acclimation

Cold acclimation is the process in which plants being exposed to low non-freezing temperatures increase their freezing tolerance (Thomashow, 1999; Sakai & Engelmann, 2007). Preconditioning of raspberry and blueberry plants at 22/-1°C alternating temperatures for four weeks was essential for optimal post-cryopreservation shoot-tip re-growth using encapsulation-dehydration and vitrification protocols (Gupta & Reed, 2006). Similarly, better recovery rates were seen in mint (Senula et al., 2007) and yams (Leunufna & Keller, 2005) when they were cold acclimated for several weeks before cryostorage.

Cold acclimation is thought to activate genes that improve plant survival at low temperatures by improving stability in membranes (Guy et al., 1985; Thomashow, 1999). Cell membrane phospholipids and sterols have been observed to increase in proportion upon cold acclimation in winter rye, with a particular increase in di-unsaturated PC and β-sitosterol (Uemura & Steponkus, 1994). Changes in phospholipid and sterol composition were found in *Arabidopsis thaliana* after cold acclimation at 2°C for one week, which increased its freezing tolerance from -2°C down to -10°C (Uemura et al., 1995). Cell membrane phospholipid changes were also observed in oat leaves (Uemura & Steponkus, 1994). These changes may be related to improved membrane stability in these plants.

Preconditioning of plants has also been shown to increase antioxidant levels in plants prior to cryopreservation (Baek & Skinner, 2003; Dai et al., 2009; Harding et al., 2009; Zhao & Blumwald, 1998). Baek and Skinner (2003) analysed the expression of antioxidant genes in wheat species after cold acclimation and found increased expression of antioxidant enzymes, such as SOD and catalase. Increased levels of antioxidants may allow plants to better tolerate oxidative stress. Lynch and Steponkus (1987) observed an increase in the di-unsaturated levels of PC and PE in winter rye seedlings. Sucrose pre-treatment of banana meristems prior to cryopreservation increased survival after warming and was related in most cases to a decrease of the double bond index in phospholipids, free fatty acids, glycolipids and sphingolipids (Zhu et al., 2006).

5. Genetic and epigenetic stability

The aim of successful cryopreservation is to maintain genetically stable plant material. While cryopreservation is now recognised as the method of choice for the long term preservation of plant material, the usefulness of cryostorage only applies if it does not lead to genetic changes in the plant species of interest (Zarghami et al., 2008). It is thus recommended to avoid the use of tissue in a non-organised dedifferentiated state, such as callus, and to use organised tissue like shoot tips instead to reduce the likelihood of non-desirable genetic mutations (Benson et al., 1996; Harding, 2004) as well as due to their higher regrowth rates (Bunn et al., 2007). Cryopreservation can cause injury at cellular level, but it is not clear if this injury can change the genetic composition of plants. The genetic stability of cryopreserved plants has nonetheless been confirmed for most of the analysed samples at morphological, histological and molecular level (Harding, 2004). Where differences between control and cryopreserved genotypes were found, it was suggested that the genetic changes might not be associated with the cryopreservation process itself, but rather that they are caused by the overall tissue culture, cryoprotection and regeneration process (Harding, 2004).

Comparisons of morphological development and analyses of the characteristics of control and cryopreserved plants have shown no differences in many species, such as *Prunus*, sugarcane, onion, kiwi, *Eucalyptus*, coffee, *Dendrobium* and *Cosmos* (Harding, 2004), as well as in the hybrid aspen (*Populus tremula*), an economically important woody plant and widely used model forest plant (Jokipii et al., 2004). Alterations in phenotype related to flower colouring have been observed in *Chrysanthemum*, which might be related to the chimeric structure of the plant (Harding, 2004). Morphological and phenotypic studies in potato, where shoot tips were used for cryopreservation, showed stable genetic integrity (Kaczmarczyk et al., 2010). Biometric studies examining morphological characters, agronomic traits or vegetal development descriptors in *Dioscorea floribunda*, sugarcane and banana revealed no significant differences between cryopreserved-derived and control plants (Harding, 2004). No abnormalities in chromosome number or cell structure were observed in cryopreserved *Vanda pumila* (*Orchidaceae*), with the regrown shoot primordia being able to induce new meristematic tissues like those of the non-cryopreserved controls (Na & Kondo, 1996). Long term storage of cryopreserved plants, such as strawberry, pea, *Rubus* and potato, did not result in any overall changes in regeneration capability and phenotype when regenerated explants were compared at the time point of storage as well as 10, 12 and 28 years later (Castillo et al., 2010; Caswell & Kartha, 2009; Keller et al., 2006; Mix-Wagner et al., 2003).

Histological studies and cytological analysis using flow cytometry have confirmed the genetic stability of plant species such as pea, oil palm, silver birch, *Rubus*, *Solanum tuberosum* and rice (Harding, 2004). Biochemical analyses have compared products of gene expression such as the formation and concentration of secondary metabolites. Examples of compounds compared in cryopreserved and control plants have been diosgenin in *Dioscorea floribunda*, chlorophyll and pyrethrin in *Chrysanthemum*, hypericin production in *Hypericum perforatum*

L. (Skyba et al., 2010; Urbanova et al., 2006), betalanin pigments in *Beta vulgaris* and nicotin alkaloids in *Nicotiana rustica*, which were all unchanged in cryopreserved plants and thus confirmed the integrity of metabolic functions after cryostorage (Harding, 2004). Similar stability was observed upon comparison of proteins and enzymes (Marin et al., 1993; Paulet et al., 1993; Wu et al., 2001).

A variety of different techniques and markers have been applied to compare genomic DNA patterns, such as restriction fragment length polymorphism (RFLP), randomly amplified polymorphic DNA (RAPDs) fragments, simple sequence repeat (SSR) analysis and amplified fragment length polymorphism (AFLP). Most studies have confirmed the presence of genetic stability (Castillo et al., 2010; Helliot et al., 2002) and where changes in the genome have been found, such as in sugarcane and potato with RFLP markers, the changes could not be related to the process of cryopreservation itself (Castillo et al., 2010; Harding, 2004).

In contrast to genetic variations manifested by DNA nucleotide sequence alterations, epigenetic changes do not change the original DNA sequence (Boyko & Kovalchuk, 2008) but can result in heritable changes of gene expression. Typical features of epigenetic characteristics are DNA methylation, histone modification and changes in chromatin structure (Boyko & Kovalchuk, 2008). Epigenetic gene regulatory mechanisms have a function in plant development and might be influenced or changed by environmental conditions and osmotic stress during tissue culture and cryopreservation. Some recent studies have analysed epigenetic characteristics like DNA methylation in tissue culture and cryopreserved plants. Modifications in DNA methylation have been found in almond (Channuntapipat et al., 2003), papaya (Kaity et al., 2008), chrysanthemum (Martín & González-Benito, 2006), *Ribes* (Johnston et al., 2009), strawberry (Hao et al., 2002a), citrus (Hao et al., 2002b) and potato (Kaczmarczyk et al., 2010). Changes in methylation might be caused by stressful *in vitro* conditions, osmotic dehydration and the application of cryoprotective agents (Harding, 2004).

6. Conclusion

Many plant species have been successfully cryopreserved through the development of various cryopreservation methods. As a standard protocol, vitrification and droplet vitrification are widely applied. Shoot tips are the preferred material for cryostorage as they contain the meristem and an organised structure, with direct shoot development avoiding unstructured phases, which could lead to mutations. Preconditioning of plants (cold acclimation or sugar preculture) can have positive effects on survival and regeneration after cryopreservation, which could be due to increased membrane stability. Cryopreserved plants have been found to be genetically stable in most cases, but epigenetic changes have been detected, although most of the molecular analyses have only compared fractions of the genome.

Success in cryopreservation cannot be guaranteed for all plants, as some species are recalcitrant to tissue culture or the cryopreservation process. Fundamental studies looking at membrane composition, membrane damage and repair are likely to help to elucidate why some species are cryosensitive and how cryopreservation protocols can be improved for those species.

7. Acknowledgment

The authors thank ALCOA of Australia Ltd and BHP Billiton Worsley Alumina Pty Ltd for their generous support. AK is supported through the Australian Research Council Linkage Project Grant LP0884027. BF is supported through an Australian Postgraduate Award and a Curtin Research Scholarship. Dr. Zak Hughes is gratefully acknowledged for producing Fig. 3.

8. References

Anchordoguy, T.J., Rudolph, A.S., Carpenter, J.F. & Crowe, J.H. (1987). Modes of interaction of cryoprotectants with membrane phospholipids during freezing. *Cryobiology* 24:pp. 324-331.

Arakawa, T., Carpenter, J.F., Kita, Y.A. & Crowe, J.H. (1990). The basis for toxicity of certain cryoprotectants - a hypothesis. *Cryobiology* 27:pp. 401-415.

Baek, K-H. & Skinner, D.Z. (2003). Alteration of antioxidant enzyme gene expression during cold acclimation of near isogenic wheat lines. *Plant Science* 165:pp. 1221-1227.

Benson, E.E. (1990). *Free Radical Damage in Stored Plant Germplasm*, International Board for Plant Genetic Resources, ISBN 92-9043-196-2, Rome, Italy.

Benson, E.E. (2008). Cryopreservation of phytodiversity: A critical appraisal of theory & practice. *Critical Reviews in Plant Sciences* 27:pp. 141-219.

Benson, E.E. & Bremner, D. (2004). Oxidative stress in the frozen plant: a free radical point of view. In: *Life in the Frozen State*, Fuller, B.J., Lane, N. & Benson, E.E. (Eds.), pp. 206–241, CRC Press, ISBN 0-415-24700-4, Boca Raton, Florida.

Benson, E.E., Lynch, P.T. & Jones, J. (1992). The detection of lipid peroxidation products in cryoprotected and frozen rice cells: consequences for post-thaw survival. *Plant Science* 85:pp. 107-114.

Benson, E.E., Lynch, P.T. & Jones, J. (1995). The use of the iron chelating agent desferrioxamine in rice cell cryopreservation: a novel approach for improving recovery. *Plant Science* 110:pp. 249-258.

Benson, E.E., Reed, B.M., Brennan, R.M., Clacher, K.A. & Ross, D.A. (1996). Use of thermal analysis in the evaluation of cyropreservation protocols for *Ribes nigrum* L. germplasm. *CryoLetters*. 17:pp. 347–362.

Benson, E.E. & Whithers, L.A. (1987). Gas-chromatographic analysis of volatile hydrocarbon production by cryopreserved plant-tissue cultures - a nondestructive method for assessing stability. *CryoLetters* 8:pp. 35-46.

Boyko, A. & Kovalchuk, I. (2008). Epigenetic control of plant stress response. *Environmental and Molecular Mutagenesis* 49:pp. 61-72.

Bunn, E., Turner, S., Panaia, M. & Dixon, K.W. (2007). The contribution of in vitro technology and cryogenic storage to conservation of indigenous plants, *Australian Journal of Botany* 55:pp. 345-355.

Castillo, N.R.F., Bassil, N.V., Wada, S. & Reed, B.M. (2010). Genetic stability of cryopreserved shoot tips of *Rubus* germplasm. *In Vitro Cellular & Developmental Biology-Plant* 46:pp. 246-256.

Caswell, K.L. & Kartha, K.K. (2009). Recovery of plants from pea and strawberry meristems cryopreserved for 28 years. *CryoLetters* 30:pp. 41-46.

Channuntapipat, C., Sedgley, M. & Collins, G. (2003). Changes in methylation and structure of DNA from almond tissues during in vitro culture and cryopreservation. *J. Amer. Soc. Hort. Sci.* 128:pp. 890-897.

Chen, X.-L., Li, J.-H., Xin, X., Zhang, Z.-E., Xin, P.-P. & Lu, X.-X. (2011). Cryopreservation of in vitro-grown apical meristems of *Lilium* by droplet-vitrification. *South African Journal of Botany* 77:pp. 397-403

Crowe, J.H., Crowe, L.M., Carpenter, J.F., Rudolph, A.S., Wistrom, C.A., Spargo, B.J. & Anchordoguy, T.J. (1988). Interactions of sugars with membranes. *Biochimica et Biophysica Acta* 947:pp. 367-84.

Dai, F., Huang, Y., Zhou, M. & Zhang G. (2009). The influence of cold acclimation on antioxidant enzymes and antioxidants in sensitive and tolerant barley cultivars. *Biologia Plantarum* 153:pp. 257-262.

Day, J.G., Harding, K.C., Nadarajan, J. & Benson, E.E. (2008). Cryopreservation, conservation of bioresources at ultra low temperatures. In: *Molecular Biomethods Handbook*. Walker J.M & Rapley R. (Eds.), pp. 917-947, Humana Press, ISBN 978-1-60327-374-9, Totowa, NJ.

Day, J.G. & Stacey, G.N. (2007). *Cryopreservation and Freeze-Drying Protocols*. Humana Press Inc., ISBN 978-1-58829-377-0, Totowa, NJ.

Decruse, S.W., Seeni, S. & Pushpangadan, P. (1999). Cryopreservation of alginate coated shoot tips of in vitro grown *Holostemma annulare* (Roxb.) K. Schum, an endangered medicinal plant: Influence of preculture and DMSO treatment on survival and regeneration. *CryoLetters* 20:pp. 243-250.

Dowdle, J., Ishikawa, T., Gatzek, S., Rolinski, S. & Smirnoff, N. (2007). Two genes in *Arabidopsis thaliana* encoding GDP 1 galactose phosphorylase are required for ascorbate biosynthesis and seedling viability. *The Plant Journal* 52:pp. 673-689.

Dowgert, M.F. & Steponkus, P.L. (1984). Behavior of the plasma membrane of isolated protoplasts during a freeze-thaw cycle. *Plant Physiol* 75:pp. 1139-1151.

Emmelot, P. & Vanhoeven, R.P. (1975). Phospholipid unsaturation and plasma membrane organization. *Chemistry and Physics of Lipids* 14:pp. 236-246.

Engelmann, F. (2004). Plant cryopreservation: Progress and prospects. *In Vitro Cellular & Developmental Biology-Plant* 40:pp. 427-433.

Fabre, J. & Dereuddre, J. (1990). Encapsulation-dehydration: A new approach to cryopreservation of *Solanum* shoot-tips. *CryoLetters* 11:pp. 423-426.

Fang, J.Y., Wetten, A. & Johnston, J. (2008). Headspace volatile markers for sensitivity of cocoa (*Theobroma cacao* L.) somatic embryos to cryopreservation. *Plant Cell Reports* 27: pp. 453-461.

Fleck, R.A., Benson, E.E., Bremner, D.H. & Day, J.G. (2000). Studies of free radical-mediated cryoinjury in the unicellular green alga *Euglena gracilis* using a non-destructive hydroxyl radical assay: a novel approach for developing protistan cryopreservation strategies. *Free Radic Res* 32:pp. 157-70.

Foubert, I., Dewettinck, K., Van de Walle, D., Dijkstra, A.J. & Quinn, P.J. (2007). Physical properties: structural and physical characteristics, In: *The Lipid Handbook with CD-ROM*, Gunstone, F.D., Harwood J.L. & Dijkstra A.J. (Eds.), pp. 471-534, CRC Press, Taylor & Francis Group, ISBN 0-8493-9688-3, LLC, Boca Raton, FL.

Foyer, C.H. & Noctor, G. (2009). Redox regulation in photosynthetic organisms: signaling, acclimation, and practical implications. *Antioxidants & Redox Signaling* 11:pp. 861-905.

Furt, F., Simon-Plas, F. & Sebastien, M. (2011). Lipids of the plat plasma membrane, In: *The Plant Plasma Membrane*. Murphy, A.S., Peer, W. & Schulz, B. (Eds.), pp. 3-30, Springer-Verlag, ISBN 978-3-642-13430-2, Berlin Heidelberg.

Gamez-Pastrana, R., Martinez-Ocampo, Y., Beristain, C.I. & Gonzalez-Arnao, M.T. (2004). An improved cryopreservation protocol for pineapple apices using encapsulation-vitrification. *CryoLetters* 25:pp. 405-414.

Gonzalez-Arnao, M.T., Panta, A., Roca, W.M., Escobar, R.H. & Engelmann, F. (2008). Development and large scale application of cryopreservation techniques for shoot and somatic embryo cultures of tropical crops. *Plant Cell Tissue and Organ Culture* 92:pp. 1-13.

Gordon-Kamm, W.J. & Steponkus, P.L. (1984). Lamellar-to-hexagonal II phase transitions in the plasma membrane of isolated protoplasts after freeze-induced dehydration. *Proc Natl Acad Sci USA* 81:pp. 6373-6377.

Grunwald, C. (1975). Plant sterols. *Annual Review of Plant Physiology and Plant Molecular Biology* 26:pp. 209-236.

Grunwald, C. & Saunders, P.F. (1978). Function of sterols [and discussion]. *Philosophical Transactions of the Royal Society of London B Biological Sciences* 284:pp. 541-558.

Gupta, S. & Reed, B.M. (2006). Cryopreservation of shoot tips of blackberry and raspberry by encapsulation-dehydration and vitrification. *CryoLetters* 27:pp. 29.42.

Guy, C.L., Niemi, K.J. & Brambl, R. (1985). Altered gene expression during cold acclimation of spinach. *Proc Natl Acad Sci USA* 82: pp. 3673-3677.

Halliwell, B. & Gutteridge, J.M.C. (2007). *Free Radicals in Biology and Medicine* (Fourth edition), Oxford University Press Inc., ISBN 978-0-19-856868-1, Oxford.

Halmagyi, A., Fischer-Klüver, G., Mix-Wagner, G. & Schumacher, H.M. (2004). Cryopreservation of *Chrysanthemum morifolium* (*Dendranthema grandiflora* Ramat.) using different approaches. *Plant Cell Rep* 22:pp. 371-375.

Hamilton, K.N., Turner, S.R. & Ashmore, S.E. (2009). Cryopreservation, In: *Plant Germplasm Conservation in Australia: Strategies and Guidelines for Developing, Managing and Utilising Ex Situ Collections*, Offord, C.A & Meagher, P.F. (Eds.), pp. 129-128, Australian Network for Plant Conservation Inc., ISBN 978-0-9752191-1-9, Canberra.

Hao, Y.J., You, C.X. & Deng, X.X. (2002a). Analysis of ploidy and the patterns of amplified fragment length polymorphism and methylation sensitive amplified polymorphism in strawberry plants recovered from cryopreservation. *CryoLetters* 23:pp. 37-46.

Hao, Y.J., You, C.X. & Deng, X.X. (2002b). Effects of cryopreservation on developmental competency, cytological and molecular stability of citrus callus. *CryoLetters* 23:pp. 27-35.

Harding, K. (2004). Genetic integrity of cryopreserved plant cells: a review. *CryoLetters* 25:pp. 3-22.

Harding, K.N., Johnston, J.W. & Benson, E.E. (2009). Exploring the physiological basis of cryopreservation success and failure in clonally propagated in vitro crop plant germplasm. *Agricultural and Food Science* 18:pp. 103-116.

Hartmann, M.A. (1998). Plant sterols and the membrane environment. *Trends in Plant Science* 3:pp. 170-175.

Helliot, B., Madur, D., Dirlewanger, E. & De Boucaud, M.T. (2002). Evaluation of genetic stability in cryopreserved *Prunus*. *In Vitro Cellular & Developmental Biology-Plant* 38:pp. 493-500.

Hipkiss, A.R., Preston, J.E., Himswoth, D., Worthington, V.C. & Abbot, N.J. (1997). Protective effects of carnosine against malondialdehyde-induced toxicity towards cultured rat brain endothelial cells. *Neuroscience Letters* 238:pp. 135-138.

Hirai, D. & Sakai, A. (1999). Cryopreservation of in vitro-grown meristems of potato (*Solanum tuberosum* L.) by encapsulation-vitrification. *Potato Research* 42:pp. 153-160.

Hirai, D., Shirai, K., Shirai, S. & Sakai, A. (1998). Cryopreservation of in vitro grown meristems of strawberry (*Fragaria* x *ananassa* Duch.) by encapsulation-vitrification. *Euphytica* 101:pp. 109-115.

Hodzic, A., Rappolt, M., Amenitsch, H., Laggner, P. & Pabst, G. (2008). Differential modulation of membrane structure and fluctuations by plant sterols and cholesterol. *Biophysical Journal* 94:pp. 3935-3944.

Johnston, J.W., Benson, E.E. & Harding, K. (2009). Cryopreservation induces temporal DNA methylation epigenetic changes and differential transcriptional activity in *Ribes* germplasm. *Plant Physiology and Biochemistry* 47:pp. 123-131.

Jokipii, S., Ryynänen, L., Kallioc, P.T., Aronen, T. & Häggmana, H. (2004). A cryopreservation method maintaining the genetic fidelity of a model forest tree, *Plant Science*. 166:pp. 799-806.

Kaczmarczyk, A., Houben, A., Keller, E.R.J. & Mette, M.F. (2010). Influence of cryopreservation on the cytosine methylation state of potato genomic DNA. *CryoLetters* 31:pp. 380-391.

Kaczmarczyk, A., Rokka, V.-M. & Keller, E.R.J. (2011a). Potato shoot tip cryopreservation, a review. *Potato Research* 54:pp. 45-79.

Kaczmarczyk, A., Turner, S.R., Bunn, E., Mancera, R.L. & Dixon, K.W. (2011b). Cryopreservation of threatened native Australian species - what have we learned and where to from here? *In Vitro Cellular & Developmental Biology-Plant* 47:pp. 17-25.

Kaity, A., Ashmore, S.E., Drew, R.A. & Dulloo, M.E. (2008). Assessment of genetic and epigenetic changes following cryopreservation in papaya. *Plant Cell Reports* 27:pp. 1529-1539.

Keller, E.R.J., Senula, A., Leunufna, S. & Grube, M. (2006). Slow growth storage and cryopreservation - tools to facilitate germplasm maintenance of vegetatively propagated crops in living plant collections. *International Journal of Refrigeration-Revue Internationale Du Froid* 29:pp. 411-417.

Kent, B., Garvey, C.J., Lenne, T., Porcar, L, Garamus, V.M. & Bryant, G. (2010). Measurement of glucose exclusion from the fully hydrated DOPE inverse hexagonal phase. *Soft Matter* 6:pp. 1197-1202.

Kim, H.H., Lee, Y.G., Shin, D.J., Ko, H.C., Gwag, J.G., Cho, E.G. & Engelmann, F. (2009). Development of alternative plant vitrification solutions in droplet-vitrification procedures. *CryoLetters* 30:pp. 320-34.

Kranner, I., Birtic, S., Anderson, K.M. & Pritchard, H.W. (2006). Glutathione half-cell reduction potential: A universal stress marker and modulator of programmed cell death? *Free Radical Biology and Medicine* 40:pp. 2155-2165.

Kreck, C.A., Mandumpal, J.B. & Mancera, R.L. (2011). Prediction of the glass transition in aqueous solutions of simple amides by molecular dynamics simulations. *Chemical Physics Letters* 501:pp. 273-277.

Lenne, T., Bryant, G., Garvey, C.J., Kelderling, U. & Koster, K.L. (2006). Location of sugars in multilamellar membranes at low hydration. *Physica B-Condensed Matter* 385-86:pp. 862-864.

Lenne, T., Bryant, G., Holcomb, R. & Koster, K.L. (2007). How much solute is needed to inhibit the fluid to gel membrane phase transition at low hydration? *Biochimica et Biophysica Acta-Biomembranes* 1768:pp. 1019-1022.

Lenne, T., Garvey, C.J., Koster, K.L. & Bryant, G. (2009). Effects of sugars on lipid bilayers during dehydration - SAXS/WAXS measurements and quantitative model. *Journal of Physical Chemistry B*, 113:pp. 2486-2491.

Leunufna, S. & Keller, E.R.J. (2005). Cryopreservation of yams using vitrification modified by including droplet method: effects of cold acclimation and sucrose. *CryoLetters* 26:pp. 93-102.

Lynch, D.V. & Steponkus, P.L. (1987). Plasma membrane lipid alterations associated with cold acclimation of winter rye seedlings (*Secale cereale* L. cv Puma). *Plant Physiol* 83:pp. 761-767.

Lynch, P.T., Siddika, A., Johnston, J.W., Trigwell, S.M., Mehra, A., Benelli, C., Lambardi, M. & Benson, E.E. (2011). Effects of osmotic pretreatments on oxidative stress, antioxidant profiles and cryopreservation of olive somatic embryos. *Plant Science* 181:pp. 47-56.

Mallon, R., Bunn, E., Turner, S.R. & Gonzalez, M.L. (2008). Cryopreservation of *Centaurea ultreiae* (Compositae) a critically endangered species from Galica (Spain). *CryoLetters* 29:pp. 363-370.

Mandal, B.B. & Dixit-Sharma, S. (2007). Cryopreservation of in vitro shoot tips of *Dioscorea deltoidea* Wall., an endangered medicinal plant: effect of cryogenic procedure and storage duration. *CryoLetters* 28:pp. 461-470.

Mandumpal, J.B., Kreck, C.A. & Mancera, R.L. (2011). A molecular mechanism of solvent cryoprotection in aqueous DMSO solutions. *Physical Chemistry Chemical Physics* 13:pp. 3839-3842.

Marin, M.L., Gogorcena, Y., Ortiz, J. & Duranvila, N. (1993). Recovery of whole plants of sweet orange from somatic embryos subjected to freezing-thawing treatments. *Plant Cell Tissue and Organ Culture* 34:pp. 27-33.

Marnett, L.J. (1999). Chemistry and biology of DNA damage by malondialdehyde. *IARC scientific publications* 150:pp. 1-17.

Marsan, M.P., Bellet-Amalric, E., Muller, I., Zaccai, G. & Milon, A. (1998). Plant sterols: a neutron diffraction study of sitosterol and stigmasterol in soybean phosphatidylcholine membranes. *Biophysical Chemistry* 75:pp. 45-55.

Martin, M.C. & González-Benito, M. E. (2006) Sequence comparison in somaclonal variant of cryopreserved *Dendranthema grandiflora* shoot apices. *Cryobiology* 53:pp. 424.

Meryman, H.T. (1974). Freezing injury and its prevention in living cells *Annual Review of Biophysics and Bioengineering* 3:pp. 341-363.

Mix-Wagner, G., Schumacher, H.M. & Cross, R.J. (2003). Recovery of potato apices after several years of storage in liquid nitrogen. *CryoLetters* 24:pp. 33-41.

Møller, I.M., Jensen, P.E. & Hansson, A. (2007). Oxidative modifications to cellular components in plants. *Annu. Rev. Plant Biol.* 58:pp. 459-481.

Mycock, D.J., Wesley-Smith, J. & Berjak, P. (1995). Cryopreservation of somatic embryos of four species with and without cryoprotectant pre-treatment. *Annals of Botany* 75: pp. 331-336.

Na, H.-Y. & Kondo, K. (1996). Cryopreservation of tissue-cultured shoot primordia from shoot apices of cultured protocorms in *Vanda pumila* following ABA preculture and desiccation. *Plant Science* 118:pp. 195-201.

Nes, W.R. (1974). Role of sterols in membranes. *Lipids* 9:pp. 596-612.

Nishizawa, S., Sakai, A., Amano, Y. & Matsuzawa, T. (1993). Cryopreservation of *Asparagus officinalis* L. embryogenic suspension cells and subsequent plant regeneration by vitrification. *Plant Science* 91:pp. 67-73.

Obert, B., Benson, E.E., Millam, S., Pretová, A. & Bremner, D.H. (2005). Moderation of morphogenetic and oxidative stress responses in flax in vitro cultures by hydroxynonenal and desferrioxamine. *Journal of Plant Physiology* 162:pp. 537-547.

Packer, J.E., Slater, T. & Willson, R. (1979). Direct observation of a free radical interaction between vitamin E and vitamin C. *Nature* 278:pp. 737-738.

Panis, B., Strosse, H., Van den Hende, S. & Swennen, R. (2002). Sucrose preculture to simplify cryopreservation of banana meristem cultures. *CryoLetters* 23:pp. 375-384.

Panis, B. & Lambardi, M. (2005). Status of cryopreservation technologies in plants (crops and forest trees). Proceedings of the International Workshop on 'The Role of Biotechnology', pp. 43-54, Villa Gualino, Turin, Italy, 5.-7. March 2005.

Panis, B., Piette, B. & Swennen, R. (2005). Droplet vitrification of apical meristems: a cryopreservation protocol applicable to all Musaceae. *Plant Science* 168:pp. 45-55.

Partelli, F.L., Vieira, H.D., Pais, I.P., Quartin, V.L., Campos, P.S., Fortunato, A., Eichler, P., Viana, A.P., Ribeiro, A. & Ramalho, J.C. (2009). Chloroplast membrane lipids from *Coffea* sp. under low positive temperatures. Proceedings of 22nd International Conference on Coffee Science, pp. 891-898, ASIC 2008, ISBN 2900212219, Campinas, SP, Brazil, September, 2008.

Paulet, F., Engelmann, F. & Glaszmann, J.C. (1993). Cryopreservation of apices of in vitro plantlets of sugarcane (*Saccharum* sp hybrids) using encapsulation dehydration. *Plant Cell Reports* 12:pp. 525-529.

Paunescu, A. (2009). Biotechnology for endangered plant conservation: a critical overview *Romanian Biotechnological Letters* 14:pp. 4095-4103.

Reed, B.M. (2008). *Plant Cryopreservation: A Practical Guide*, Springer, ISBN 978-0-387-72275-7, New York.

Reinhoud, P.J., Versteege, I., Kars, I., van Iren, F. & Kijne, J.W. (2000). Physiological and molecular changes in tobacco suspension cells during development of tolerance to cryopreservation by vitrification, In: *Cryopreservation of Tropical Plant Germplasm. Current Research Progress and Application,* Engelmann, F. & Takagi, H. (Eds.), pp. 57-66, International Plant Genetic Resources Institute, ISBN 92-9043-428-7, Rome.

Roach, T., Ivanova, M., Beckett, R.P., Minibayeva, F.V., Green, I., Pritchard, H.W. & Kranner, I. (2008). An oxidative burst of superoxide in embryonic axes of recalcitrant sweet chestnut seeds as induced by excision and desiccation. *Physiologia Plantarum* 133:pp. 131-139.

Sakai, A. (1965). Survival of plant tissue at super-low temperatures III. Relation between effective prefreezing temperatures and the degree of frost hardiness. *Plant Physiol.* 40:pp. 882-887.

Sakai, A. & Engelmann, F. (2007). Vitrification, encapsulation-vitrification and droplet-vitrification: a review. *CryoLetters* 28:pp. 151-172.

Sakai, A., Hirai, D. & Niino, T. (2008). Development of PVS-based vitrification and encapsulation-vitrification protocols, In: *Plant Cryopreservation: A Practical Guide*, Reed, B. (Ed.), pp. 33-57, Springer, ISBN 978-0-378-72275-7, New York.

Sakai, A., Kobayashi, S. & Oiyama, I. (1990). Cryopreservation of nucellar cells of navel orange (*Citrus sinensis* Osb. var. *brasiliensis* Tanaka) by vitrification. *Plant Cell Reports* 9:pp. 30-33.

Scottez, C., Chevreau, E., Godard, N., Arnaud, Y., Duron, M. & Dereuddre, J. (1992). Cryopreservation of cold acclimated shoot tips of pear in vitro cultures after encapsulation-dehydration. *Cryobiology* 29:pp. 691-700.

Sen-Rong, H. & Ming-Hua, Y. (2009). High-efficiency vitrification protocols for cryopreservation of in vitro grown shoot tips of rare and endangered plant *Emmenopterys henryi* Oliv. *Plant Cell Tissue and Organ Culture* 99:pp. 217-226.

Senula, A., Keller, E.R.J., Sanduijav, T. & Yohannes, T. (2007). Cryopreservation of cold-acclimated mint (*Mentha* spp.) shoot tips using a simple vitrification protocol. *CryoLetters* 28:pp. 1-12.

Skyba, M., Urbanova, M., Kapchina-Toteva, V., Kosuth, J., Harding, K. & Cellarova, E. (2010). Physiological, biochemical and molecular characteristics of cryopreserved *Hypericum perforatum* L. shoot tips. *CryoLetters* 31:pp. 249-260.

Steponkus, P.L. (1984). Role of the plasma membrane in freezing injury and cold acclimation. *Annual Review of Plant Physiology* 35:pp. 543-584.

Streb, P., Shang, W., Feierabend, J. & Bligny, R. (1998). Divergent strategies of photoprotection in high-mountain plants. *Planta* 207:pp. 313-324.

Takagi, H., Thinh, N.T., Islam, O.M., Senboku, T. & Sakai, A. (1997). Cryopreservation of in vitro-grown shoot tips of taro (*Colocasia esculenta* (L) Schott) by vitrification 1. Investigation of basic conditions of the vitrification procedure. *Plant Cell Reports* 16:pp. 594-599.

Tanaka, D., Niino, T., Isuzugawa, K., Hikage, T. & Uemura, M. (2004). Cryopreservation of shoot apices of in-vitro grown gentian plants: Comparison of vitrification and encapsulation-vitrification protocols. *CryoLetters* 25:pp. 167-176.

Thomashow, M.F. (1999). Plant cold acclimation: Freezing tolerance genes and regulatory mechanisms. *Annual Review of Plant Physiology and Plant Molecular Biology* 50:pp. 571-599.

Touchell, D.H., Dixon, K.W. & Tan, B. (1992). Cryopreservation of shoot-tips of *Grevillea scapigera* (Proteaceae) - a rare and endangered plant from Western-Australia. *Australian Journal of Botany* 40:pp. 305-310.

Touchell, D.H., Turner, S.R., Bunn, E. & Dixon, K.W. 2002. Cryostorage of somatic tissues of endangered Australian species, In: *Cryopreservation of Plant Germplasm*. Towill, L.E. & Bajaj, Y.P.S. (Eds.), pp. 357-372, Springer Verlag, ISBN 7055-8630, Berlin Heidelberg, Germany.

Turner, S. 2001. *Cryopreservation of somatic germplasm of selected Australian monocodyledonous taxa (Haemodoraceae)*, PhD Thesis, Curtin University of Technology, Perth, Western Australia.

Turner, S., Senaratna, T., Touchell, D., Bunn, E., Dixon, K. & Tan, B. (2001). Stereochemical arrangement of hydroxyl groups in sugar and polyalcohol molecules as an important factor in effective cryopreservation. *Plant Science* 160:pp. 489-497.

Uchendu, E.E., Leonard, S.W., Traber, M.G. & Reed, B.M. (2010). Vitamins C and E improve regrowth and reduce lipid peroxidation of blackberry shoot tips following cryopreservation. *Plant Cell Reports* 29:pp. 25-35.

Uemura, M. & Steponkus, P.L. (1994). A contrast of the plasma membrane lipid composition of oat and rye leaves in relation to freezing tolerance. *Plant Physiology* 104:pp. 479-496.

Uemura, M., Joseph, R.A. & Steponkus, P.L. (1995). Cold-acclimation of *Arabidopsis thaliana* - effect on plasma membrane lipid composition and freeze-induced lesions. *Plant Physiology* 109:pp. 15-30.

Urbanova, M., Kosuth, J. & Cellarova, E. (2006). Genetic and biochemical analysis of *Hypericum perforatum* L. plants regenerated after cryopreservation. *Plant Cell Reports* 25:pp. 140-147.

van Meer, G., Voelker, D.R. & Feigenson, G.W. (2008). Membrane lipids: where they are and how they behave. *Nature Reviews Molecular Cell Biology* 9:pp. 112-124.

Vidal, N., Sanchez, C., Jorquera, L., Ballester, A. & Vieitez, A.M. (2005). Cryopreservation of chestnut by vitrification of in vitro-grown shoot tips. *In Vitro Cellular & Developmental Biology-Plant* 41:pp. 63-68.

Volk, G.M. & Walters, C. (2006). Plant vitrification solution 2 lowers water content and alters freezing behavior in shoot tips during cryoprotection. *Cryobiology* 52:pp. 48-61.

Wang, Z.C. & Deng, X.X. (2004). Cryopreservation of shoot-tips of citrus using vitrification: effect of reduced form of glutathione. *CryoLetters* 25:pp. 43-50.

Wise, R.R. (1995). Chilling-enhanced photooxidation: the production, action and study of reactive oxygen species produced during chilling in the light. *Photosynthesis Research* 45:pp. 79-97.

Wise, R.R. & Naylor, A.W. (1987). Chilling-enhanced photooxidation: evidence for the role of singlet oxygen and superoxide in the breakdown of pigments and endogenous antioxidants. *Plant Physiology* 83:pp. 278.

Wolfe, J. & Bryant, G. (1999). Freezing, drying, and/or vitrification of membrane-solute-water systems. *Cryobiology* 39:pp. 103-129.

Wu, Y.J., Zhao, Y.H., Engelmann, R. & Zhou, M.D. (2001). Cryopreservation of kiwi shoot tips. *CryoLetters* 22:pp. 277-284.

Yoon, J.-W., Kim, H.-H., Ko, H.-C., Hwang, H.-S., Hong, E.-S., Cho, E.-G. & Engelmann, F. (2006). Cryopreservation of cultivated and wild potato varieties by droplet vitrification: effect of subculture of mother-plants and of preculture of shoot tips. *CryoLetters* 27:pp. 211-222.

Young, I. & McEneny, J. (2001). Lipoprotein oxidation and atherosclerosis. *Biochemical Society Transactions* 29:pp. 358-362.

Zarghami, R., Pirseyedi, S.M., Hasrak, Sh. & Pakdaman, B.S. (2008). Evaluation of genetic stability in cryopreserved *Solanum tuberosum*. *African Journal of Biotechnology* 7:pp. 2798-2802.

Zhao, S. & Blumwald, E. (1998). Changes in oxidation reduction state and antioxidant enzymes in the roots of jack pine seedlings during cold acclimation. *Physiologia Plantarum* 104: pp.134-142.

Zhu, G.Y., Geuns, J.M.C., Dussert, S., Swennen, R. & Panis, B. (2006). Change in sugar, sterol and fatty acid composition in banana meristems caused by sucrose-induced acclimation and its effects on cryopreservation. *Physiologia Plantarum* 128:pp. 80-94.

6

Cryopreservation of Spices Genetic Resources

K. Nirmal Babu[1], G. Yamuna[1], K. Praveen[1], D. Minoo[2],
P.N. Ravindran[1] and K.V. Peter[1]
[1]Indian Institute of Spices Research, Kerala
[2]Providence Women's College, Kerala
India

1. Introduction

Plant genetic resources - constituting genotypes or populations of cultivars (landraces, advance/improved cultivars), genetic stocks, wild and weedy species, which are maintained in the form of plants, seeds, tissues, etc. - hold key to food security and sustainable agricultural development (Iwananga, 1994). They are non-renewable and are among the most essential of the world's natural resources. Due to deforestation, spread of superior varieties and selection pressure, genetic variability is gradually getting eroded. This demands priority action to conserve germplasm be it at species, genepool or ecosystem level, for posterity (Frankel, 1975).

Whilst ecologists focused on *in situ* conservation might argue that *ex situ* conserved germplasm cannot offer the advantages afforded by selection and adaptation as a result of environmental pressures, there is no denying that if species are under threat—or worse, near extinction—then *ex situ* conservation of even limited germplasm is preferable to extinction. The opportunities offered by conservation biotechnology should not be missed or restricted by lack of interconnectivity between traditional and contemporary conservation practitioners.

2. Spices and germplasm conservation

Spices and herbs are aromatic plants–fresh or dried plant parts like foliage, young shoots, roots, bark, buds, seeds, berries and other fruits of which are mainly used to flavour our culinary preparations, confectionary. They are also major ingredients in indigenous medicine and perfumery. Spices and herbs are grown throughout the world–different plant species in different regions. Peninsular India is a rich repository of spices and over 100 species of spices and herbs are grown. The other major spice growing countries are Brazil, China, Guatemala, Indonesia, Madagascar, Nigeria, West Indies, Malaysia, Sri Lanka, Spain, Turkey, Mediterranean region and the Central America. Black pepper, cardamom, ginger, turmeric, vanilla, capsicum, cinnamon, clove, nutmeg, tamarind, coriander, cumin, fennel, fenugreek, dill, caraway, anise and herbs like saffron, lavender, thyme, oregano, celery, anise, sage and basil are important as spices. India being the native home of many spices, their conservation and characterization are one of the priority programmes. Deforestation, habitat degradation and overexploitation caused considerable loss of diversity in spices.

In many spices, conventional seed storage can satisfy most of the conservation requirements. But in crops with recalcitrant seeds and those having conservation needs cannot be satisfied by seed storage, have to be stored *in vitro*. Most field gene banks are prone to high labour cost, vulnerable to hazards like natural disasters, pests and pathogens attack (especially viruses and systemic pathogens), to which they are continuously exposed and require large areas of space. This supports *in vitro* and cryo conservation. In addition, other resources like continuous supply of standard stock cultures for experiments to examine physiological and biochemical processes, cell and callus lines developed for *in vitro* synthesis of valuable secondary products, flavours and other important compounds will benefit strongly from *in vitro* cultures. Most of the spice crops are either vegetatively propagated or have recalcitrant seeds. The spices germplasm is mostly conserved in field gene banks. Most of the spices are plagued by destructive and epidemic diseases caused by viruses, bacteria and fungi. This makes germplasm conservation in field gene bank risky. Thus *in vitro* and cryo storage system becomes important in the overall strategy of conserving genepool. Each technology should be chosen on the basis of utility, security and complementarily to other components of the strategy. A balance needs to be struck between seed, field gene bank, *in vitro* and cryo conservation of propagules, tissues, pollen, cell lines and DNA storage for overall objective of conserving gene pool.

3. Methodologies

3.1 Micropropagation

Plant regeneration and successful cloning of genetically stable plantlets in tissue culture is an important pre-requisite in any conservation effort of recalcitrant species. These techniques form the base for establishing tissue cultures and developing *in vitro* and cryo conservation technology for conservation. Simultaneously these tissue-cultured plants should be evaluated for their morphological and genetic stability in culture. The *in vitro* storage experiments, as much as possible, use growth regulators free media to reduce the rate of multiplication which in turn will reduce the extent of variation.

Micropropagation (culture initiation, multiplication, plant regeneration and *in vitro* rooting) form the cycle of events that form the backbone of cryopreservation studies. For initial culture establishment earlier protocols developed by Nirmal Babu *et al.*, 1997 can be used.

Murashige and Skoog (1962), Woody Plant (McCown and Amos, 1979) and Schenk and Hildebrandt (1972) media can be used depending upon the crop for micropropagation Table 1. The miniaturized *in vitro* grown shoots can be used for cryopreservation.

Micropropagation protocols for stable cloning of elite genotypes of spice crops were standerdised. Protocols were available for black pepper and its related species cardamom, ginger, turmeric and related genera, large cardamom, kasturi turmeric, mango ginger, *Kaempferia galanga*, *K. rotunda*, *Alpinia* spp, large. Cardamom, vanilla and related species, cinnamon, camphor, cassia seed and herbal spices like lavender, celery, thyme, mint, anise, savory, spearmint and oregano (Nirmal Babu *et al.*, 1997, 2005, Minoo 2002). These techniques form the base for establishing tissue cultures and developing *in vitro* technology for conservation. The basal media used are MS (Murshige and Skoog, 1962) for crops like cardamom, ginger, turmeric, kasturi turmeric, mango ginger, large cardamom, *Kaempferia*, *Vanilla* spp. seed and herbal spices and WPM-Woody Plant Medium (Mc Cown and Amos,

1979) for black pepper and its related species, cinnamon, camphor and cassia. Simultaneously these tissue-cultured plants are being evaluated for their morphological and genetic stability in culture (Luckose *et al*, 1993, Chandrappa *et al*, 1997, Nirmal babu *et al* 2003, Madhusoodanan *et al* 2005). Though micropropagation protocols were standardized using growth regulators, all the *in vitro* storage experiments were carried out using growth regulators free media to reduce the rate of multiplication which in turn will reduce the extent of variation.

Composition	Molecular formula	Concentration (mgl⁻¹) MS	Concentration (mgl⁻¹) WPM	Concentration (mgl⁻¹) SH
Macronutrients				
Ammonium nitrate	NH_4NO_3	1650.00	400.00	-
Ammonium phosphate	$NH_4H_2PO_4$	-	-	300.00
Potassium nitrate	KNO_3	1900.00	-	2500.00
Calcium chloride	$CaCl_2.2H_2O$	440.00	-	-
Calcium chloride	$CaCl_2$	-	72.50	151.00
Calcium nitrate	$Ca(NO_3)_2.4H_2O$	-	386.00	-
Potassium di hydrogen orthophosphate	KH_2PO_4	170.00	170.00	-
Potassium sulfate	K_2SO_4	-	990.00	-
Magnesium sulphate	$MgSO_4.7H_2O$	370.00	180.70	195.40
Micronutrients				
Sodium EDTA	Na_2EDTA	37.30	37.30	20.00
Ferrous sulphate	$FeSO_4.7H_2O$	27.80	27.800	15.00
Boric acid	H_3BO_3	6.20	6.20	5.00
Manganese sulphate	$MnSO_4.4H_2O$	22.30	22.30	10.00
Potassium iodide	KI	0.83	-	1.00
Zinc sulphate	$ZnSO_4.7H_2O$	8.60	8.60	1.00
Sodium molybdate	$Na_2MoO_4.2H_2O$	0.25	0.25	0.10
Copper sulphate	$CuSO_4.5H_2O$	0.025	0.25	0.20
Cobalt chloride	$CoCl_2.6H_2O$	0.025	-	0.10
Vitamins				
Myo-inositol	$C_6H_{12}O_6$	100.00	100.00	1000
Thiamine HCl	$C_{12}H_{17}CIN_4OS.HCl$	0.10	0.50	100
Nicotinic acid	$C_6H_5NO_2$	0.50	0.025	1.00
Pyridoxine HCl	$C_6H_{11}NO_3.HCl$	0.50	0.025	1.00
Amino acid				
Glycine	$C_2H_5NO_2$	2.00	1	-

*Murashige and Skoog, 1962, McCown and Amos, 1979, Schenk and Hildebrandt 1972

Table 1. Composition of MS*, WPM* and SH* basal media

Protocols are available for micropropagation and multiplication of many endangerd species like *Piper hapnium, P. silent vallyensis, P.schmidtii, P. wightii, P. barberi , Vaniilla aphylla, V. pilifera, V. walkyrie, V. wightiana, K. rotunda and Alpinia galanga* are available (Peter et al 2002, Minoo 2002, Nirmal Babu et al 1999, 2005).

Bertaccini *et al* (2004), Du *et al* (2004) reported micropropagation and establishment of mite-brone virus-free garlic.

3.2 Callus and cell culture systems

Quatrano (1968) and Nag and Street (1973) reported the first successful experiments on cryopreservation of plant cells. Since then a large number of cell suspension and calli cultures have been successfully cryopreserved (Engelmann et al 1994). In general, callus cultures are more difficult to cryopreserve than cell suspensions, because of the relative volume of the callus, its slow growth rate and the cellular heterogeneity (Withers 1987). One successful cryopreservation procedure that is applicable to all different cell suspensions or calli cultures has not been developed yet. Research focuses on optimizing the factors on which successful cryopreservation of plant organs cells suspensions and calli depends, such as: (i) starting material, (ii) pretreatment, (iii) cryopreservation procedure, and (iv) post-thaw treatment.

Plant cells cultured *in vitro* produce wide range of primary and secondary metabolites of economic value. Production of phytochemicals from plant cell cultures has been presently used for pharmaceutical products. Production of flavour components and secondary metabolites *in vitro* using immobilised cells is an ideal system for spices crops. Production of saffron and capsaicin was reported using such system (Ravishankar *et al.*, 1988; 1993, Johnson *et al.*, 1996; Venkataraman and Ravishankar 1997). Johnson *et al* (1996) reported biotransformation of ferulic acid vanillamine to capsacin and vanillin in immobilised cell cultures of *Capsicum frutescens*. Reports on the *in vitro* synthesis of crocin, picrocrocin and safranel from saffron stigma (Himeno and Sano, 1995) and colour components from cells derived from pistils (Hori *et al*, 1988) are available for further scaling up. Callus and cell cultures were established in nutmeg, clove, camphor, ginger, lavender, mint, thyme, celery etc. Cell immobilization techniques have been standardized in ginger, sage, anise and lavender (Ilahi and Jabeen, 1992; Ravindran *et al*, 1996; Sajina *et al*, 1997).

Studies on conservation of cell lines is yet o become popular in spices. Suspensions of embryogenic cell lines of fennel, conserved at 4 ^{0}C for up to 12 weeks produced normal plants upon transfer to normal laboratory conditions (Umetsu *et al*, 1995).

3.3 Somatic embryogenesis and plant regeneration

In black pepper primary embryogenic cultures can be established as per the method described by Nair and Dutta Gupta (2003). Culture the surface sterilized seeds on agar gelled full-strength, PGR-free SH (Schenk and Hildebrandt, 1972) medium containing 3.0% (W/V) sucrose under darkness. Primary somatic embryos (PEs) derived from micropylar tissues of germinating seeds after 90 days could be utilized for inducing secondary somatic embryogenic cultures.

Primary somatic embryo clumps having pre-globular to torpedo shaped embryos (5–6 visible embryos per seed) were carefully detached and inoculated on half strength PGR-free SH medium containing 1.5 % sucrose and gelled with 0.8% agar (Bacteriological grade, Hi-media). The pH of the medium was adjusted to 5.9 prior to autoclaving. Cultures were maintained at darkness at a temperature of 25±2°C. The culture conditions remained the same for all further experiments unless otherwise specified. While inoculating, the PEs were uniformly spread on the surface of the medium. Secondary embryogenic cultures were further maintained by subculturing on SH medium containing 1.5% sucrose at intervals of 20 d. The proliferating SEs were spread periodically on the surface of the medium, to facilitate proliferation.

3.4 Pollen storage

Pollen storage can be considerable value supplementing the germplasm conservation strategy by facilitating hybridisation between plants with different time of flowering and to transport pollen across the globe for various crop improvement programmes in addition to developing haploid or homozygous lines. No significant work was done in India, except a few initial reports.

The technique of pollen storage is comparable with that of seed storage, since pollen can be dried (less than 5% moisture content on a dry weight basis) and stored below 0°C. There are limited reports on the survival and fertilizing capacity of cryopreserved pollen more than five years old. Pollen might represent an interesting alternative for the long-term conservation of problematic species (IPGRI, 1996). However, pollen has a relatively short life compared with seeds (although this varies significantly among species) and viability testing can be time-consuming and uneconomical. Other disadvantages of pollen storage are the small amount produced by many species, the lack of transmission of organelle genomes via pollen, the loss of sex-linked genes in dioecious species and the general inability to regenerate into plants. Pollen, therefore, has been used to a limited extent in germplasm conservation (Hoekstra ,1995). An advantage is that pests and diseases are rarely transferred by pollen (excepting some virus diseases). This allows safe movement and exchange of germplasm as pollen.

3.5 Cryo preservation

For long-term conservation of the problem species, cryopreservation is the only method currently available. Dramatic progress has been made in recent years in the development of new cryopreservation techniques and cryopreservation protocols have been established for over 100 different plant species.

Cryopreservation is an attractive option for long-term storage. Liquid nitrogen (–196°C) is routinely used for cryogenic storage, since it is relatively cheap and safe, requires little maintenance and is widely available. Below –120°C the rate of chemical or biophysical reactions is too slow to cause biological deterioration (Kartha 1985). Only in the long term might there be a small risk of ionising radiation causing genetic changes in materials stored at cryogenic temperatures (Grout 1995).

An array of plant material could be considered for cryopreservation as dictated by the actual needs vis-a-vis preservation. These include meristems, cell, callus and protoplast cultures, somatic and zygotic embryos, anthers, pollen or microspores and whole seeds (Withers, 1985; Kartha, 1985).

Plant germplasm stored in liquid nitrogen (-196°C) does not undergo cellular divisions. In addition, metabolic and most physical processes are stopped at this temperature. As such, plants can be stored for very long time periods and both the problem of genetic instability and the risk of loosing accessions due to contamination or human error during subculturing are overcome. Most cryopreservation endeavours deal with recalcitrant seeds, *in vitro* tissues from vegetatively propagated crops, species with a particular gene combination (elite genotypes) and dedifferentiated plant cell cultures. Care must be taken to avoid ice crystallisation during the freezing process, which otherwise would cause physical damage to the tissues. The existing cryogenic strategies rely on air-drying, freeze dehydration, osmotic dehydration, addition of penetrating cryoprotective substances and adaptive metabolism (hardening), encapsulation, vitification or combinations of these processes.

Cryopreservation methods have been developed for more than 80 different plant species in various forms like cell suspensions, calluses, apices, somatic and zygotic embryos (Kartha and Engelmann, 1994; Engelmann, 1997, 2000, Engelmann *et al* 1994, 1995). However, their routine utilisation is still restricted almost exclusively to the conservation of cell lines in research laboratories

For small volumes, long-term storage is practicable through storage of cultures in cryopreservation at ultra-low temperature, usually by using liquid nitrogen (-196°C). At this temperature all cellular divisions and metabolic processes are virtually halted and consequently, plant material can be indefinitely stored without alteration or modification.

The normal approach of tissue culture is to find a medium and set of conditions that favour the most rapid rate of growth with a subculture interval of 20 – 30 days. For cryopreservation storage biological materials are stored in liquid nitrogen for long term with out subculturing. Cryopreservation, i.e., the storage of biological material at ultra low temperature usually that of liquid nitrogen (-196°C) can be achieved by different techniques like direct freezing, encapsulation- dehydration, encapsulation- vitrification and vitrification.

3.5.1 Encapsulation - Dehydration

A simplified methodology for vitrification is given below (Yamuna 2007).

The *in vitro* plants already established were used as mother plants for source of explants. This in turn facilitates the reduction in size of the plantlets and smaller somatic embryos which made them suitable for cryopreservation.

1. Suspend *in vitro* grown shoots/ somatic embryos in MS basal medium supplemented with 4% (w/v) Na alginate, 2M Glycerol and 0.4 M sucrose.
2. Drop the mixturecontining microshoots, with a sterile pipette into 0.1M $CaCl_2$ solution containing 2M Glycerol and 0.4M sucrose and left for 20 min to form beads about 4 mm in diameter, each bead containing at least one shoot.
3. Preculture the encapsulated shoots – stepwise - on MS medium enriched with different concentration of 0.3, 0.5, 0.75 and 1.0M for four days with one day on each.
4. Place the precultured beads on sterile fitter paper in Petridishes (diameter 90mm) and dehydrated by air drying on a flow bench (at room temperature and humidity) for periods of 0-10 h to determine the optimal dehydration time.
5. Measure the water content of the beads was by weighing them prior and after drying in an oven at 80°C for 48h.
6. Transfer the dehydrated beads into a 2 ml cryovial (ten beads per tube) and directly immerse in liquid nitrogen for 24h.

3.5.2 Vitrification

A simplified methodology for vitrification is given below (Yamuna 2007).

1. Shoots (1-2mm)/ somatic embryos were excised and cultivated on MS medium supplemented with 0.3 M sucrose for 24h at 25°C.

2. The treated explants were then cultured on MS medium supplemented with sucrose at 0.75 M for 1 day in the same conditions.
3. After pretreatments explants were transferred to a cryovial with 1.8 ml of loading solution (2 M Glycerol + 0.4 M sucrose) and kept for 15 min.
4. Different incubation periods in PVS2 (40-100 minutes) were tested for osmoprotected explants
5. Cryovials containing 8-10 explants were directly immersed in liquid nitrogen and kept for 24 h.

3.5.3 Encapsulation – Vitrification

A simplified methodology for encapsulation - vitrification is given below (Yamuna 2007).

1. Suspend pre-cultured shoots (1-2mm)/ somatic embryos with 2-3 apical domes on 0.3M sucrose for 16h in MS basal medium supplemented with 4% sodium alginate and 0.3 M sucrose.
2. Dispense the mixture including shoots, were with a sterile pipette into MS medium supplemented with 0.1M $CaCl_2$ and 0.4 to 1.0M sucrose, with or without 2M Glycerol gently shaken (20 rpm) on a rotary shaker for 1h at 25°C.
3. The encapsulated and osmo-protected shoots were dehydrated with 20 ml PVS2 in a 100 ml Erlenmeyer flask at 25°C and plunged into LN and held for at least 24 h at -196°C.

3.6 Thawing and recovery of conserved materials

After LN storage, cryovials warm rapidly in a 40 °C water bath for 2-3 minutes. The solution was drained from the cryovials and replace twice at 10 min intervals with 1 ml 1.2 M sucrose solution in the case of encapsulation- vitrification and vitrification methods. The composition of recovery medium was MS/WPM/SH basal medium supplemented with 2.22 – 4.44 µM and BA, 2.69- 5.37 µM NAA.

In the Encapsulation - dehydration, Encapsulation - vitification and vitrification procedures, surviving shoots can be identified by greening of explants following 2 weeks of post culture. Regrowth can be defined as the shoots that regenerated to shoots in 6 weeks of postculture. Elongated shoots can be used for micropropagation and rooting and subculture was done every 4 weeks. For rooting well grown shoots can be transferred to solid MS medium used for multiplication.

3.7 Genetic stability of conserved materials

An important prerequisite for any conservation technique is that the regenerants produced from the conserved material should be true-to-type. There are ample evidences to indicate that under certain culture conditions the materials undergo genetic changes (somaclonal variations) and as a consequence lose their integrity and uniformity. This would be highly undesirable in spices varieties where the purpose is not only to conserve a genotype but also retain its specific quality traits. Thus testing for the genetic stability of *in vitro* conserved materials is of utmost importance. Besides morphology, cytology and isozyme profiling sophisticated biochemical and DNA-based techniques have enabled more critical analysis of the genetic stability of *in vitro* materials.

RAPD, ISSR and SSR analysis can be done to evaluate genetic fidelity of the cryopreserved lines of Spices. DNA isolation can be done as per CTAB method (Ausubel et al., 1995 or Sambrook et al. 1989). RAPD and ISSR, SSR profiles were developed as per the method suggested by Williams et al., (1990), Nirmal babu et al., (2003, 2007) and Ravindran et al., (2004).

Morphological characters coupled with RAPD profiles using 24 operon primers have indicated genetic fidelity among randomly selected micropropagated plants of Subhakara and Aimpiriyan, indicating that micropropagation protocol can be used for commercial cloning of black pepper (Nirmal Babu et al., 2003). Genetic uniformity of micropropagated Piper longum using RAPD profiling was reported by Ajith (1997) and Parani et al. (1997) for conservation.

Peter et al (2001) and Ravindran et al (2004) reported that the conserved materials of all the species conserved by them showed normal rate of multiplication when transferred to multiplication medium after storage. The normal sized plantlets when transferred to soil established with over 80% success. They developed into normal plants without any deformities and were morphologically similar to mother plants. RAPD profiling of these conserved plants also showed their genetic uniformity.

Ravindran et al (2004), Yamuna et al (2007) and Yamuna (2007) reported genetic uniformity was observed in cryo preserved and recovered plants of cardamom, ginger, black pepper and endangered species of Piper, P. barberi based on RAPD and ISSR profiling.

4. Status of cryo conservation in spices

Reports on cryopreservation of spices are meager and limited. The present status of cryo preservation in major spices is given Table 2. The number of accessions conserved in cryo genebank at the National Bureau of Plant Genetic Resources (NBPGR), New Delhi are given in Table 3.

Application	Technique	Reference
Black pepper (*Piper nigrum*) and related species	Meristem culture	Philip et al., 1992
Disease eradication propagation	Shoot Culture Leaf/root	Broome and Zimmerman, 1978
Cryopreservation	Seeds	Chaudhury and Chandel ,1994
Cryopreservation	synseeds	Ravindran et al., 2004; Nirmal Babu et al., 2007; Yamuna 2007
Cryopreservation	Seed	Decruse and Seeni, 2003
Slow growth storage and cryopreservation	Plantlets and shoot tips	Ravindran et al., 2004; Nirmal Babu et al., 2007; Yamuna 2007
Allium Spp		
Disease eradication	Meristem culture and themotherapy	Conci and Nome, 1991
Cryopreservation	Shoot culture,	Keller 1991

Application	Technique	Reference
	microbullbets	
Cryopreservation	Shoot tips	Niwata 1995
Cardamom (*Elettaria cardamomum* Maton)		
Disease eradication	Meristem culture	Nadagauda *et al.* 1983
Cryopreservation	Seeds	Chaudhury and Chandel, 1995
Slow growth storage and cryopreservation	Plantlets and shoot tips	Ravindran *et al.*, 2004; Nirmal Babu *et al.*, 2007; Yamuna 2007
Zingiber spp.		
Disease eradication	Shoot cultures, shoot buds	Balachandran *et al* .,1990
Propagation	Somatic embryo regeneration	Hosoki and Sagawa ,1977, Nirmal Babu, 1997
Cryopreservation	Synseeds	Sharma *et al.*, 1994
Slow growth storage and cryopreservation	Plantlets and shoot tips	Ravindran *et al.*, 2004; Nirmal Babu *et al.*, 2007; Yamuna *et al* 2007 ; Yamuna 2007
Curcuma spp		
Slow growth storage and cryopreservation	Plantlets and shoot tips	Ravindran *et al.*, 2004 ; Nirmal Babu *et al.*, 2007
Vanilla spp.		
Disease eradication	Apical meristem	Cereveta and Madrigal, 1981
Cryo preservation	Synthetic seeds	Ravindran *et al.*, 2004
Pollen Cryo preservation	Pollen	Minoo, 2002; Minoo *et al* 2011
Slow growth storage and cryopreservation	Plantlets and shoot tips	Ravindran *et al.*, 2004; Nirmal Babu *et al.*, 2007, Minoo and Babu 2009
Herbal spices		
Slow growth storage	*In vitro* plantlets	Nirmal Babu *et al.*1996
Capsicum		
Cryopreservation	Seed	Peter *et al* 2002 ; Ravindran *et al* 2004
Cryopreservation	Pollen	Alexander *et al.*, 1991
Cryopreservation	Pollen	Rajasekharan and Ganeshan, 2003
Fennel (*Foeniculum vulgare*)		
Cold storage	Embryogenic	Umetsu *et al.*, 1995

Application	Technique	Reference
	suspension cells	
Coriander (*Coriandrum sativum*)		
Cryopreservation	somatic embryos	Elena *et al.*, (2010)
Mint (*Mentha* spp.)		
Cryopreservation	Somatic embryos	Leigh and Remi 2003
Ocimum spp		
Slow growth	Encapsulated beads	Mandal *et al* (2000)
Syzygium francissi		
	Shoot tips	Shatnawi *et al* (2004).
Armoracia rusticana		
Cryopreservation	Hairy root cultures	Phunchindawan *et al*
Crocus spp.		
	Encapsulated calluses	Chand *et al* (2000); Baghdadi et al., (2010)

*Ashmore, 1997, 2002 and Nirmal Babu *et al*, 1999, 2007, Yamuna 2007; Yamuna et al 2007

Table 2. Present status of information on cryo conservation of spices

Species	No.of accessions
Maintained as in vitro cultures	
Spices and industrial crops	380 accessions (7 genera, 27 species)
Medicinal and Aromatic plants	169 accessions (21 genera, 28 species)
Maintained in cryo bank	
Spices and Condiments	148 accessions
Medicinal and Aromatic plants	5 accessions
Total	702

Source: Annual Report NBPGR 2010-11

Table 3. Present status of Spices in *in vitro* and Cryo genebank at NBPGR

4.1 Black pepper and related species

Cryopreservation of black pepper (*Piper nigrum* L.)seeds in liquid nitrogen (LN_2) was reported by Choudhary and Chandel, (1994), and Choudhury and Malik (2004). Pepper seeds are recalcitrant and the seed viability decreases with reduction in moisture content. Seeds desiccated to 12% & 6%moisture contents were successfully cryopreserved in liquid nitrogen at -196^0C, with a survival rate of 45% & 10.5% respectively (Chaudhury and Chandel 1994).

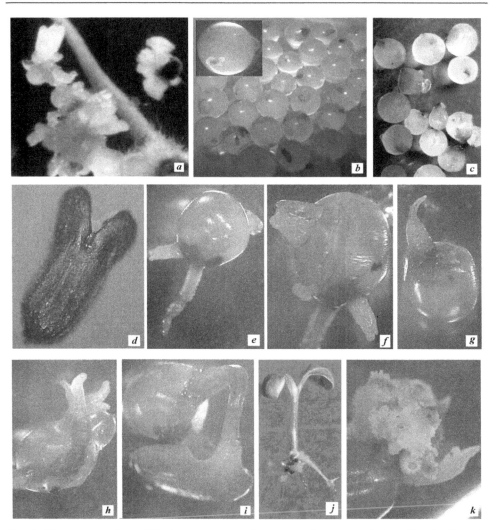

Fig. 1. Cryopreservation of black pepper somatic embryos by encapsulation dehydration. a) Somatic embryos used for cryopreservation, b) Somatic embryos encapsulated in Na-alginate, c) Encapsulated and dehydrated somatic embryos, d) Viable somatic embryo stained in red colour after cryopreservation, e), f), g), h) & i) Various stages of development of somatic embryos to plantlet after cryopreservation, j) Fully developed plantlet from a somatic embryo cryopreserved by encapsulation dehydration, k) A cluster of somatic embryos at different stages of development, originated from an embryogenic line after cryopreservation

Yamuna (2007) reported the effect of encapsulation-dehydration and vitrification methods on survival of cryo preserved somatic embryos in black pepper. In encapsulation dehydration treatment, the best survival rates (62 %) of somatic embryos was obtained after freezing, by preculturing in 0.7 M sucrose (direct) for 1 day, followed by dehydration in the laminar air flow for 6 h which resulted in 21 % moisture content. In the vitrification procedure, the somatic embryos were precultured for 3 days on SH basal medium containing 0.3 M sucrose and subjected to vitrification treatment for 60 minutes at 25°C resulted in 71 % survival after cryopreservation. The study concluded that the embryogenic lines of *Piper nigrum* cultivar karimunda can be successfully cyopreserved following an encapsulation dehydration/desiccation procedure (62 % success). This success rate can be enhanced to 71 % using a vitrification/one step freezing in liquid nitrogen (Fig. 1).This was mainly because of the nature of somatic embryos which is more suitable to cryopreservation compared to shoot buds. The genetic stability of the conserved somatic embryos was proved by RAPD and ISSR profiling. Cryopreservation of encapsulated shoot buds of endangered *Piper barberi* was reported by Peter *et al* (2001) and Ravindran *et al* (2004).

Encapsulated shoot tips of *Piper barberi* were cryopreserved with 60% success using vitrification technique. In encapsulation vitrification the encapsulated shoot tips were precultured on MS medium, supplemented with 0.3 M, 0.5 M and 0.7 M sucrose (pH 5.8) for three days followed by dehydration with PVS2 solution (100%) at 0° C for 3 hours. After dehydration the beads (10 encapsulated shoot tips in 0.8 ml PVS2 solution per 1.5 ml cryotube) were frozen rapidly by direct immersion in to liquid nitrogen (- 196 °C) and kept for one hour (Peter *et al* 2001 and Ravindran *et al* 2004). Yamuna 2007 also reported that studies on cryopreservation of endangered *P.barberi* shoot tips revealed that, the encapsulation- vitrification procedure produced higher survival (70 %) of cryopreserved shoot tips (Fig. 2) compared to encapsulation - dehydration which gave 40 % survival. Genetic fidelity studies showed that the regenerated plants were similar to the controls. Thus encapsulation - vitrification as a simple and efficient method for long term preservation of *P.barberi* propagules.

4.2 Cardamom and related species

Choudhary and Chandel (1995) attempted cryo-conservation of cardamom (*Elettaria cardamomum* Maton.) seed. They tried to conserve seeds at ultra-low temperature by suspending seeds in cryovials in vapor phase of liquid nitrogen (-150°C) by slow freezing and also by direct immersion in liquid nitrogen (-196°C) by fast freezing. The result showed that seeds possessing 7.7-14.3% moisture content could be successfully cryo-preserved with 80% germination when tested after one-year storage in vapor phase of liquid nitrogen (at-150°C).

Shoot tips(1.0-2.0mm) from *in vitro* grown plantlets of cardamom were subjected to progressive increase of sucrose concentrations (0.1, 0.3, 0.5, 0.7, 0.9, and 1.0) for two days each under the same cultural conditions as the parent plantlets. These shoot tips were transferred to 1.8ml cryotube containing ice cold PVS2 solution (30%(v/v) glycerol + 15% (v/v) ethylene glycol + 15% (v/v) DMSO in culture medium with 0.4 M sucrose, pH (5.8)) at 0°C for 3 hours. After 3 hours equilibration at 0°C, the shoot tips were directly immersed into liquid nitrogen for 1 hour. Vials were thawed in 40°C water for 1 minute. The cryoprotectant was removed and the shoot tips were washed 2-3 times in 1.2M sucrose solution. About 70%Shoot tips were recovered on MS medium supplemented with BAP and NAA. But the encapsulation vitrification method gave only 60% success (Ravindran *et al* 2004).

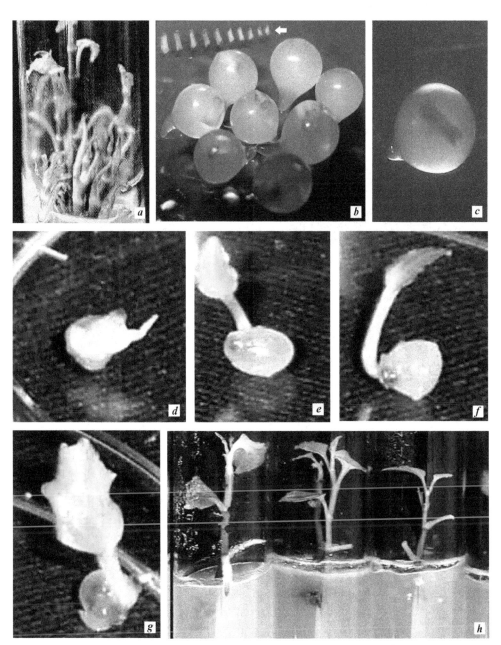

Fig. 2. Cryopreservation of *Piper barberi* by encapsulation vitrification. a) *In vitro* culture of *P. barberi*, b) & c) Shoot tips encapsulated in Na-alginate, arrow indicates shoot tip used as explants, d), e), f) & g) Various stages of development of cryopreserved shoot tips after post culturing, h) Regenerated plantlets after 3 months of post culturing

Yamuna (2007) tested the effect of encapsulation – dehydration, encapsulation vitrification and vitrification methods on cryopreservation of cardamom. In the vitrification treatment, to enhance tolerance to vitrification solution (PVS2), a two step sucrose preculture with 0.3 M and 0.75 M sucrose for one day each and an osmo protection step with a loading solution (LS) of 2 M glycerol and 0.4 M sucrose were performed prior to PVS2 treatment. The shoots

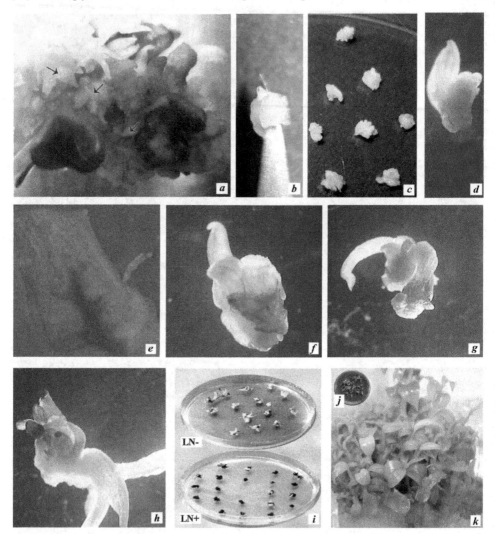

Fig. 3. Plant regeneration from cryopreserved miniature shoots of cardamom by vitrification. a) Cardamom culture with miniature shoots, b) & c) Excised meristematic clumps used for cryopreservation, d) Explant turned brown after cryopreservation, e) Viable tissues stained in TTC after cryopreservation, f), g), h), & i) Shoot development after 10, 14 and 25 days of post culturing , j) regenerating shoot buds in a petridish, k) Development of multiple shoots after 4 months of post culturing

dehydrated with PVS2 for 60 min retained a high level of shoot formation (70 %). The vitrification procedure resulted in higher regrowth (70 %) (Fig.3) when compared to encapsulation vitrification (62 %) and encapsulation dehydration (60 %). In all the three cryopreservation procedures tested, shoots grew after cryopreservation without intermediary callus formation. The genetic stability of cryopreserved cardamom shoots were confirmed using ISSR and RAPD profiling.

Fig. 4. Plant regeneration from cryopreserved shoot buds of ginger by encapsulation vitrification. a) *In vitro* culture, b) A typically excised shoot bud used for cryopreservation, c) & d) Shoot buds encapsulated in Na-alginate, e) & f) Shoot buds turned brown after thawing, g) Viable apical dome stained in red colour after liquid nitrogen storage (TTC staining), h) Regenerating shoot bud 20 days after post culturing, i) & j) Elongated shoot with no intermediary callus formation, k) & l) Regenerating shoot buds in petriplates, m) Plantlets regenerating from cryopreserved shoot bud

4.3 Ginger, turmeric and related species

Cryopreservation of Ginger (*Zingiber officinale* Rosc) and turmeric (*Curcuma longa* L.) shoot tips was successfully done with 80% of recovery using vitrification method. But the rate of recovery was only 40% when encapsulated shoot tips were dehydrated in progressive increase of sucrose concentration together with 4- 8 hrs. of desiccation (Peter *et al* 2001 and Ravindran *et al* 2004).

Efficient cryopreservation techniques were developed for *in vitro* grown shoots of ginger based on encapsulation dehydration, encapsulation vitrification and vitrification procedures (Yamuna et al 2007 and Yamuna 2007. The vitrification procedure resulted in higher re-growth (80 %) when compared to encapsulation vitrification (66 %) and encapsulation dehydration (41 %). The genetically stability of shoot apices was confirmed by molecular profiling. The RAPD and ISSR assays performed suggested that no genetic aberrations originated in ginger plants during culture and cryopreservation (Fig. 4).

4.4 Vanilla and related species

Technology for cryopreservation of vanilla germplasm - using encapsulation and vitrification methods – were available. Encapsulated *in vitro* grown shoot tips of vanilla could be cryo preserved with 70% success when pretreated with progressive increase of sucrose concentration (0.1M-1.0M) for one day each and dehydrated for 8 hrs (Peter *et al* 2001; Minoo 2002 and Ravindran *et al* 2004) (Fig. 5).

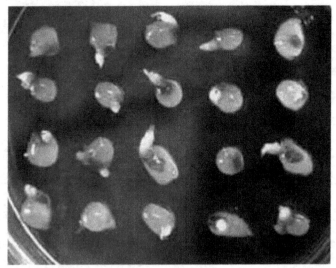

Fig. 5. Germination of cryopreserved encapsulated shoot tips protocorms of vanilla

Ginzalez-Arnao, *et al.*, (2009) attempted to cryo-preserve *V. planifolia* Andr. using *in vitro* fragmented explants (IFEs) and the apices derived from them. Cryopreservation of apices from *in vitro* grown plants was achieved using the droplet vitrification protocol. Maximum survival (30%) and further regeneration (10%) of new shoots were obtained for apices derived from clusters of *in vitro* plantlets produced from microcuttings through a three-step droplet vitrification protocol: 1-d preculture of apices on solid MS medium with 0.3 M sucrose; loading with a 0.4 M sucrose + 2 M glycerol solution for 20–30 min; and exposure to plant vitrification solution PVS3 for 30 min at room temperature.

Minoo (2002) reported cryopreservation of vanilla pollen for conservation (Fig. 6) of haploid genome as well as assisted pollination between species that flower at different seasons and successful fertilisation using cryopreserved pollen (Minoo, 2002, Minoo *et al* 2011). Pollen from two asynchronously flowering species of *Vanilla viz.*, cultivated *V. planifolia* and its wild relative *V. aphylla,* were cryopreserved after desiccation to 12 % moisture content, pretreated with cryoprotectant Dimethyl sulphoxide (5%) and cryopreserved -196°C in Liquid Nitrogen. This cryopreserved pollen was latter thawed and tested for their viability both *in vitro* and *in vivo*. A germination percentage of 82.1% and 75.4% in *V. planifolia* and *V.aphylla* pollen respectively were observed indicating their viability(Fig.6). This cryopreserved pollen of *V. planifolia* was used successfully to pollinate *V.aphylla* flowers resulting in fruit set (Fig.7). The seeds thus obtaines were sussfully cultured to develop hybrid plantlets. This system is of great importance and can be used for conserving the haploid gene pool of *Vanilla* in cryobanks and their subsequent utility in crop improvement (Fig. 6 and 7)

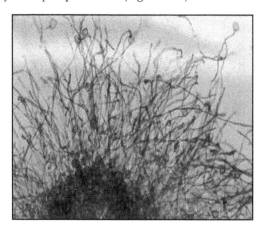

Fig. 6. Germination of cryopreserved Vanilla pollen

Fig. 7. Fruit set after pollination with cryopreserved pollen

4.5 Capsicum

Plants could be successfully regenerated (Fig 8) from cryopreserved seeds of capsicum (Peter *et al* 2001 and Ravindran *et al* 2004). Alexander *et al* (1991) and Rajasekharan and Ganeshan. (2003) reported freeze preservation of capsicum pollen (*Capsicum annuum*) in liquid nitrogen (–196°C) for 42 months.

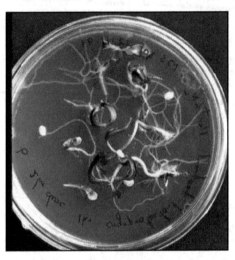

Fig. 8. Successful germination of cryopreserved seeds of capsicum

4.6 Seed herbal and other spices

Elena *et al.,* (2010) successfully cryopreserved coriander (*Coriandrum sativum* L.) somatic embryos using sucrose pre-culture and air desiccation procedure utilized embryo clumps (ECs). The regrowth after cryopreservation and average number of new embryos developed from cryopreserved ECs were retained at the level of the untreated control (98% and 13 embryos per clump, respectively). Both normal and abnormal plants were produced from control and cryopreserved cultures, indicating that appearance of abnormalities was not related to cryopreservation. The regenerants with normal phenotype showed the same peaks of relative DNA content regardless of cryopreservation. The results suggest that simple desiccation method is effective for cryopreservation of coriander somatic embryos with subsequent regeneration. Plants could be regenerated from cryopreserved seeds of Anise.(Peter *et al* 2001).

Successful Cryopreservation of seeds, meristems, somatic or zygotic embryos were reported in *Allium* Spp (Niwata, 1995, Hyung *et al* 2003, Haeng *et al* 2003, 2004.2005, Jung *et al* 2005, Gayle *et al* 2004). Preliminary success was reported in cryo preservation of Mint (Leigh and Remi 2003).

Most of the reports are confined to a few genotypes and hence the techniques standardized needs to be extended to more genotypes before adopting them for routine conservation. Reports of cryoconservation of spices like *Ocimum, Lavendula, Salvia* are available from National Bureau of Plant Genetic Resources (NBPGR), New Delhi.

Mandal *et al* (2000) reported propagation and conservation of four pharmaceutically important herbs, *Ocimum americanum* L. syn. *O. canum* Sims. (hoary basil); *O basilicum* L. (swett basil); *O. gratissimum* L. (shrubby basil); and *O. sanctum* L. (sacred basil) using synthetic seed technology. Synthetic seeds were produced by encapsulating axillary vegetative buds harvested from garden-grown plants of these four *Ocimum* species in calcium alginate gel. The gel contained Murashige and Skoog (MS) nutrients and 1.1-4.4 μM benzyladenine (BA). Shoots emerged from the encapsulated buds on all six planting media tested. However, the highest frequency shoot emergence and maximum number of shoots per bud were recorded on media containing BA. Of the six planting media tested, both shoot and root emergence from the encapsulated buds in a single step was recorded on growth regulator-free MS medium as well as on vermi-compost moistened with halfstrength MS medium. Rooted shoots were retrieved from the encapsulated buds of *O. americanum, O. basilicum,* and *O. sanctum* on these two media, whereas shoots of *O. gratissimum* failed to root. The encapsulated buds could be stored for 60 d at 4°C. Plants retrieved from the encapsulated buds were hardened off and established in soil.

An efficient procedure for the *in vitro* propagation and cryogenic conservation of *Syzygium francissi* was developed by Shatnawi *et al* (2004). Shoot tips excised from *in vitro*-grown plants were successfully cryostoraged at −196°C by the encapsulation-dehydration method. A preculture of formed beads on MS medium containing 0.75 *M* sucrose for 1 d, followed by 6 h dehydration (20% moisture content) led to the highest survival rate after cryostorage for 1h. This method is a promising technique for *in vitro* propagation and cryopreservation of shoot tips from *in vitro*-grown plantlets of *S. francissi* germplasm.

Hairy root cultures of *Armoracia rusticana* Gaertn. Mey. et Scherb. (horseradish) were successfully cryopreserved by two cryogenic procedures (Phunchindawan *et al.,* 1997). Encapsulated shoot primordia were precultured on solidified Murashige-Skoog medium supplemented with 0.5M sucrose for 1 day and then dehydrated with a highly concentrated vitrification solution (PVS2) for 4 h at 0°C prior to a plunge into liquid nitrogen. The survival rate of encapsulated vitrified primordia amounted to 69%. In a revised encapsulation-dehydration technique, the encapsulated shoot primordia were precultured with a mixture of 0.5M sucrose and 1M or 1.5M glycerol for 1 day to induce dehydration tolerance and then subjected to air-drying prior to a plunge into liquid nitrogen. The survival rate of encapsulated dried primordia was more than 90%, and the revived primordia produced shoots within 2 weeks after plating. A long-term preservation of shoot primordia was also achieved by the technique. Thus, this revised encapsulation-dehydration technique appears promising as a routine method for the cryopreservation of shoot primordia of hairy roots

The effect of sucrose concentration and dehydration period on survival and regrowth of encapsulated calluses were also studied in 2 species of Crocus (Chand et al 2000). Highest survival (83.3; 88.9%) and regrowth (77.6; 83.3%) rates were obtained when encapsulated unfrozen calluses of *Crocus hyemalis* and *C. moabiticus* precultured with 0.1 M sucrose for two days without further air dehydration. After cryopreservation, the highest survival (55.6; 61.1%) and regrowth (16.7; 27.8%) rates were achieved when calluses of *C. hyemalis* and *C. moabiticus* were pretreated with 0.5 M sucrose for two days after two hours of dehydration. Viability of crocus decreased with increased sucrose concentration and dehydration period. Dehydration of encapsulated calluses of *C. hyemalis* and *C. moabiticus* with silica gel for one hour prior to freezing resulted in maximum rates of survival (77.8; 83.3%) and re-growth

(33.3; 72.1%). However, further studies should be initiated to improve regrowth of surviving embryogenic calluses and to study genetic stability after cryopreservation.

5. DNA bank

Concurrent with the advancements in gene cloning and transfer has been the development of technology for the removal and analysis of DNA. DNAs from the nucleus, mitochondrion and chloroplast are now routinely extracted and immobilized onto nitrocellulose sheets where the DNA can be probed with numerous cloned genes. In addition, the rapid development of polymerase chain reaction (PCR) now means that one can routinely amplify specific oligonucleotides or genes from the entire mixture of genomic DNA.These advances, coupled with the prospect of the loss of significant plant genetic resources throughout the world,have led to the establishment of DNA bank for the storage of genomic DNA.

The conserved DNA will have numerous uses viz, molecular phylogenetics and systematics of extinct taxa, production of previously characterized secondary compounds in transgenic cell cultures, production of transgenic plants using genes from gene families, *in vitro* expression and study of enzyme structure and function and genomic probes for research laboratories.

The vast resources of dried specimens in the world's herbaria may hold considerable DNA that would be suitable for PCR. It seems likely that the integrity of DNA would decrease with the age of specimens. Because there are many types of herbarium storage environments, preservation and collections, there is a need for systematic investigations of the effect of modes of preparation, collection and storage on the integrity of DNA in the world's major holdings.

The advantage of storing DNA is that it is efficient and simple and overcomes many physical limitations and constraints that characterize other forms of storage (Adams 1988, 1990, 1997, Adams and Adams 1991, Adams *et al* 1994). The disadvantage lies in problems with subsequent gene isolation, cloning and transfer but, most importantly, it does not allow the regeneration of live organisms (Maxted *et al.*, 1997). DNA banking is yet to catch up in spices. DNA samples of over 600 genotypes of spices is stored in the DNA bank of Indian Institute of Spices Research (IISR), Calicut.

6. Future focus

In contrast to the prevailing attitude among conservation biologists, globally there is considerable interest among cryobiologists in the use of in vitro, cold and ultra-cold technology for germplasm conservation. The procedures for plant material are given in-depth coverage by Reed et al. (2004) who stress equally the ecological and plant/germplasm health aspects preceeding and following storage. Panis and Lambardi (2006) discussed the evolution of technologies for plant material, covering cell suspensions and callus cultures of herbaceous species, pollen, shoot meristems, woody species, as well as seed and embryonic axes. The *ex situ* gene bank at Gatersleben in Germany houses 986 potato accessions are cryopreserved and trials on other species are performed (Börner 2006). The National Bureau of Plant Genetic Resources (NBPGR), New Delhi has over 702 accesion of various spices, medicinal and aromatic crops in its cryo gene bank (Table. 3). Keller *et al.* (2008) make the point that cryopreservation affords the best of conditions for the long-term maintenance of plant material, particularly for vegetatively propagated species. Cryopreservation is the only viable method available for long-term preservation of the both plant and animal origin species. As an ultimate aim of cryoconservation is the reintroduction of preserved material into the field, it is appropriate at this point to consider the concept of restoration a little more closely. In terms of ultimate ecosystem restoration, the possibilities raised by in vitro conservation, including cryoconservation, do not mean that species selection should merely take random advantage of what germplasm has or can be conserved as there are many genetic, physiological and phenotypic considerations to be taken into account (Kramer and Havens 2009).

The establishment and maintenance of biological resource centers (BRCs) or germplasm conservatories requires careful attention to implementation of reliable preservation technologies and appropriate quality control to ensure that recovered cultures and other biological materials perform in the same way as the originally isolated culture or material. There are many types of BRC that vary both in the kinds of material they hold and in the purposes for which the materials are provided. All BRCs are expected to provide materials and information of an appropriate quality for their application and work to standards relevant to those applications. There are important industrial, biomedical, and conservation issues that can only be addressed through effective and efficient operation of BRCs in the long term. This requires a high degree of expertise in the maintenance and management of collections of biological materials at ultra-low temperatures, or as freeze dried material, to secure their long-term integrity and relevance for future research, development, and conservation. The application of cryogenic preservation in biotechnology and medicine has recently been a topic of interest. The use of cryogenic preservation in this area has given new horizon to this field of applications.

7. References

Adams, R.P. and Adams, J.E., 1991. Conservation of Plant Genes: DNA Banking and In Vitro Biotechnology. Academic Press, New York.

Adams, R.P., 1988. The preservation of genomic DNA: DNA Bank Net. *Amer. J. Bot.*, 75: 156.

Adams, R.P., 1990. The preservation of Chihuahuan plant genomes through *in vitro* biotechnology:DNA Bank-Net, a genetic insurance policy. In: *Third Symposium on Resources of the Chihuahuan Desert Region*, (Eds.) Powell, A.M., Hollander, R.R., Barlow, J.C., McGillivray, W.B. and Schmidly, D.J. Printech Press, Lubbock, TX, pp. 1–9.

Adams, R.P., 1997.Conservation of DNA: DNA banking. In: *Biotechnology and Plant Genetic Resources: Conservation and Use*, (Eds.) Callow, J.A. Ford-Loyd, B.V. and Newbury, H.J. Biotechnology in Agriculture Series, No. 19. CAB International, pp. 163–174.

Adams, R.P., Miller, J.S., Golenberg, E.M. and Adams, J.E., 1994. *Conservation of Plant Genes 11: Utilization of Ancient and Modern DNA*. Missouri Botanical Garden Press. St. Louis, MI, 276 pp. 227.

Ajith, A., 1997. Micropropagation and genetic fidelity studies in *Piper longum* L. In: *Biotechnology of Spices, Medicinal and Aromatic Plants*, (Eds.) Edison, S., Ramana,

K.V., Sasikumar, B., Nirmal Babu K. and Santhosh J. Eapen. Indian Society for Spices, Calicut, India, p. 94–97.

Alexander, M.P., Ganeshan, S. and Rajasekharan, P.E., 1991. Freeze preservation of capsicum pollen (*Capsicum annuum*) in liquid nitrogen (–196°C) for 42 months: Effect on viability and fertility. *PlantCell Incompatibility Newsletter*, 23: 1–4. NBPGR 2011 Annual Report, National Bureau of Plant Genetic Resources, 2010-11

Ashmore, S.E., 1997. Status report on the development and application of *in vitro* techniques for conservation and use of plant genetic resources. International Plant Genetic Resources Institute, Rome, Italy. 67 p.

Ausubel F M, Brent R, Kingston R E, Moore D D, Seidman, J G, Smith J A & Struhl K 1995 Short Protocols in Molecular Biology, John Wiley & Sons, New York, Ch. 2.4.

Balachandran, S.M., Bhat, S.R. and Chandel, K.P.S., 1990. *In vitro* clonal multiplication of turmeric (*Curcuma longa*) and ginger (*Zingiber officinale* Rosc.). *Plant Cell Reports*, 3: 521–524.

Bertaccini, A., Botti, S., Tabanelli, D., Dradi, G., Fogher, C., Previati, A., da Re, F., Nicola, S., Nowak, J. and Vavrina, C.S., 2004. Micropropagation and establishment of mite-borne virus-free garlic *Allium sativum*), *Acta Horticulturae*, 631: 201–206.

Börner A. Preservation of plant genetic resources in the biotechnology era. Biotech J 1: 1393–1404; 2006.

Broome, O.C. and Zimmerman, R.N., 1978. *In-vitro* propagation of black pepper. *Hort. Sci.*, 43: 151–153.

Chand, P.K., Mandal, J. and Pattnaik, S., 2000. Alginate encapsulation of axillary buds of *Ocimum americanum* L. (hoary basil), *O.basilicum* L. (sweet basil), *O. gratissimum* L. (shrubby basil), and *O. sanctum* L. (sacred basil). *In vitro Cellular and Developmental Biology* (Plant), 36(4): 287–292.

Chandrappa, H.M., Shadakshari, Y.G., Sudharshan, M.R. and Raju, B. ,1997. Preliminary yield trial of tissue cultured cardamom selections. In: *Biotechnology of Spices, Medicinal and Aromatic Plants,*(Eds.) Edison, S., Ramana, K.V., Sasikumar, B., Nirmal Babu K. and Santhosh J. Eapen. Indian Society for Spices, Calicut, India, p. 102–105.

Chaudhury, R. and Chandel, K.P.S., 1994. Germination studies and cryopreservation of seeds of black pepper (*Piper nigrum* L.): A recalcitrant species. *Cryoletters*, 15: 145–150.

Choudhary, R. and Chandel, K.P.S., 1995. Studies on germination and cryopreservation of cardamom (*Elletaria cardamomum* Maton.) seeds. *Seed Science and Biotechnology*, 23(1): 235–240.

Choudhury, R. and Malik, S.K., 2004. Genetic conservation of plantation crops and spices using cryopreservation. *Ind. J. Biotechnology*, 3: 348–358.

Conci, V.C. and Nome, S.F., 1991. Virus free garlic (*Allium sativum* L.) plants obtained by thermotherapy and meristem tip culture. *J. Phytopathol.*, 132: 186–192.

Decruse, S W and Seeni, S 2003. Seed cryopreservation is a suitable storage procedure for a range of Piper species. Seed Sci. Technol. 31, 213-217.

Du, Y.Q., Zhu, J.Z. and Shen, W.P., 2004. Study on virus elimination technology of garlic in Jiading by shoot tip culture. *Acta Agriculturae Shanghai*, 20(1): 9–12.

Elena Popova, Haeng-Hoon Kim and Kee-Yoeup Paek 2010. Cryopreservation of coriander (*Coriandrum sativum* L.) somatic embryos using sucrose pre-culture and air desiccation *Scientia Horticulturae* 124 (4), 522-528.

Engelmann F. Use of biotechnologies for conserving plant biodiversity. *Acta Hort* 812: 63–82; 2009.

Engelmann, F., 1997. *In vitro* conservation methods. In: *Biotechnology and Plant Genetic Resources:Conservation and Use,* (Eds.) Callow, J.A., Ford-Loyd, B.V. and Newbury, H.J. Biotechnology in Agriculture Series, No. 19. CAB International, UK, pp. 119–161.

Engelmann, F., 2000. Importance of cryopreservation for the conservation of plant genetic resources. In: *Cryopreservation of Tropical Plant Germplasm,* (Eds.) Engelmann, F. and Takagi, H. Current Research Progress and Application. Japan International Research Centre for Agricultural Sciences, Japan/International Plant Genetic Resources Institute, Rome, Italy, pp. 8–20.

Engelmann, F., Benson, E.E., Chabrillange, N., Gonzalez-Arnao, M.T., Mari, S., Michaux-Ferriere, N., Paulet, Glazmann, J.C. and Charrier, A., 1994. Cryopreservation of several tropical plant species using encapsulation/dehydration of apices. In. *Proc. VIIIth IAPTC Meeting,* Firenze, Italy.

Engelmann, F., Dumet, D., Chabrillange, N., Abdelnour-Esquivel, A., Assy-Bah, B., Dereuddre, J. and Duval, Y., 1995. Cryopreservation of zygotic and somatic embryos from recalcitrant and intermediate-seed species. *Plant Genet. Resources Newsletter,* 103: 27–31.

Frankel, O.H., 1975. Genetic resources survey as a basis for exploration. In: *Crop Genetic Resources for Today and Tomorrow,* (Eds.) O.H. Frankel and J.H. Hawkes. Cambridge University Press, Cambridge. pp. 99.

Gayle, M., Volk, Nicholas Maness and Kate Rotindo, 2004. Cryopreservation of Garlic (*Allium sativum* L) using Plant Vitrification Solution 2. *CryoLetters,* 25: 219–226.

Gonzalez-Arnao M.T, Claudia Esther Lazaro-Vallejo, Florent Engelmann, Roberto Gamez-Pastrana, Yolanda Maria Martinez-Ocampo, Miriam Cristina Pastelin-Solano and Carlos Diaz-Ramos 2009. Multiplication and cryopreservation of vanilla (*Vanilla planifolia* 'Andrews') *In Vitro Cellular & Developmental Biology - Plant* 45 (5): 574-582.

Grout B W W 1995 Introduction to the in vitro preservation of plant cells, tissues and organs. In: Grout, B.W.W. (Ed.), Genetic Preservation of Plant Cells in Vitro. Springer–Verlag, Berlin. pp : 1–20.

Haeng-Hoon Kim, Eun-Gi Cho, Hyung-Jin Baek, Chang-Yung Kim, ER Joachim Keller and Florent Engelmann, 2004. Cryopreservation of garlic shoot tips by vitrification: Effects of dehydration, Rewarming, unloading and regrowth conditions *CryoLetters,* 25: 59–70.

Haeng-Hoon Kim, Jung-Bong Kim, Hyung-Jin Baek, Eun-Gi Cho, Young-Am Chaand Florent Engelmann, 2004. Evalution of DMSO Concentration in Garlic Shoot tips during a Vitrification Procedure. *CryoLetters,* 25: 91–100.

Haeng-Hoon Kim, Ju-Won Yoon, Jung-Bong Kim, Florent Engelmann and Eun-Gi Cho, 2005. Thermal analysis of Garlic Shoot tips during a Vitrification Procedure. *CryoLetters*, 26(1): 33–44.

Himeno, H. and Sano, K., 1995. Synthesis of crocin, picrocrocin and safranal by saffron stigma like structures proliferated *in vitro*. *Agricultural Biology and Chemistry*, 51 (9): 2395–2400.

Hoekstra, F.A., 1995. Collecting pollen for genetic resources conservation. In: *Collecting Plant Genetic Diversity: Technical Guidelines*, (Eds.) Guarino, L., Rao, V.R. and Reid, R. CAB International, Wallingford, UK, pp. 527–550.

Hori, H., Enomoto, K. and Nakaya, H., 1988. Induction of callus from pistils of *Crocus sativus* L. and production of colour components in the callus. *Plant Tissue Culture Letters*, 5: 72 –77.

Hosoki, T. and Sagawa, Y., 1977. Clonal propagation of ginger (*Zingiber officinale* Rosc.) through tissue culture. *Horticulture Science*, 12: 451-452.

Hyung-Jin Baek, Haeng-Hoon Kim, Eun-Gi Cho, Young-Am Chae and Florent Engelmann, 2003. Importance of explant size and origin and of preconditioning treatments for cryopreservation of garlic shoot apices by vitrification. *CryoLetters*, 24: 381–388.

Ilahi, I. and Jabeen, M., 1992. Tissue culture studies for micropropagation and extraction of essential oils from *Zingiber officinale* Rosc. *Pakistan Journal of Botany*, 24(1): 54–59.

IPGRI, 1996 Pragamme activities, germplsm maintenance and use. In: *Annual Report*, IPGRI. Rome. pp. 56–65.

Iwananga, M., 1994. Role of International organisations in global genetic resource management. In:*Proc. 27th International Symposium on Tropical Agriculture Research*, Japan International Research Centre for Tropical Agricultural Sciences, Ministry of Agriculture, Forestry and Fisheries, Tsukuba, Japan, August, 25–26, 1993, p.1–6.

Johnson, T.S., Ravishanker, G.A and Venkataraman, L.V., 1996. Biotransformation of ferulic acid vanillamine to capsaicin and vanillin in immobilised cell cultures of *Capsicum frutescens PlantCell, Tissue and Organ Culture*, 44(2): 117–123

Jung-Bong Kim, Haeng-Hoon Kim, Hyung-Jin Baek, Eun-Gi Cho, Yong-Hwan Kim and Florent Engelmann, 2005. Changes in sucrose and glycerol content in garlic shoot tips during freezing using pvs3 solution. *CryoLetters*, 26(2): 103–112.

Kartha K K & Engelmann F 1994 Cryopreservation and germplasm storage. In *Plant Cell and Tissue Culture*. Vasil, I. K. and Thorpe, T. A. (eds.), Kluwer Academic Publishers, Dordrecht/Boston/London. pp : 195–230.

Kartha, K.K., 1985. In: *Cryopreservation of Plant Cells and Organs*, (Ed.) Kartha, K.K. CRC Press, Boca Raton, Florida, pp. 243–267.

Keller E. R. J.; Kaczmarczyk A.; Senula A. 2008. Cryopreservation for plant genebanks—a matter between high expectations and cautious reservation. *CryoLetts* 29: 53–62.

Keller, J., 1991. *In vitro* conservation of haploid and diploid germplasm in *Allium cepa* L. *Acta Hortic.*,289: 231–232.

Leigh, E. Towill and Remi Bonnart, 2003. Cracking in a vitrification solution during cooling or warming does not effect growth of cryopreserved mint shoot tips. *CryoLetters*, 24: 341–346.

Lukose, R., Saji, K.V., Venugopal, M.N. and Korikanthimath, V.S., 1993. Comparative field performance of micropropagated plants of cardamom (*Elettaria cardamomum*). *Indian Journal of Agricultural Sciences*, 63(7): 417–418.

Madhusoodanan, K.J., Kuruvilla, K.M., Vadiraj, B.A., Radhakrishnan, V.V. and Thomas, J., 2005. On farm evaluation of tissue culture vanilla plants vis-à-vis vegetative cuttings. *Proceedings of ICAR National Symposium on Biotechnological Interventions for Improvement of Horticultural Crops: Issues and Strategies*, Kerala Agricultural University, Trissur, Kerala, India, p. 89–90.

Mandal, J., Pattnaik, S.and Chand, P.K. 2000. Alginate encapsulation of axillary buds of *Ocimum americanum* L. (hoary basil), *O. Basilicum* L. (sweet basil), *O. Gratissimum* L. (shrubby basil), and *O. Sanctum*. L. (sacred basil) *In Vitro Cellular & Developmental Biology - Plant* 36(4): 287-292.

Maxted N, Brian, F and Hawkes, J G. 1997. Plant Genetic Conservation : the *in situ* approach. Springer. 446 pp

Mc Cown, B.H. and Amos, R., 1979. Initial trials of commercial micropropagation with birch. *Proc.Inter. Plant Pro. Soc.*, 29: 387–393.

Minoo Divakaran and Nirmal Babu, K. 2009.Micropropagation and In Vitro Conservation of Vanilla (*Vanilla planifolia* Andrews) pp 129-138. In SM Jain and PK Saxena (eds) Springer Protocols, Methods in Molecular Biology 547, *Protocols for In Vitro Cultures and Secondary Metabolite Analysis of Aromatic and Medicinal Plants*, The Humana Press, (Springer), USA.

Minoo Divakaran, Nirmal Babu K. and Pete,r K. V. 2011. Cryopreservation of pollen and inter specific hybridization in important orchid species *V. planifolia* and *V. aphylla*. National Consultation for Production and Utilisation of Orchids 19-21 February, 2011 National Research Center for Orchids, Pakyong- 737106, Sikkim. Abstract. pp.98-99.

Minoo Divakaran, Nirmal Babu K and Michel Grisoni 2010. Biotechnological applications In Vanilla, pp. 51-73 In Eric Odoux and Michel Grisoni (eds) Vanilla, CRC Press, Boca Raton, USA.

Minoo, D., 2002. Seedling and somaclonal variation and their characterization in *Vanilla*. *Ph.D. Thesis*, Calicut University, Kerala, India.

Murashige, T. and Skoog, F., 1962. A revised medium for rapid growth and bioassays with tobacco tissue cultures. *Physiol Plant*, 15: 473–493.

Nadagauda, R S., Mascarenhas, A F & Madhusoodanan K J. 1983. Clonal multiplication of *Elettaria cardamomum* Maton. by tissue culture. *J. Plantn Crops* 11 : 60-64.

Nag, K.K. and Street, H.E., 1975. Freeze preservation of cultured plant cells, 11. The freezing and thawing phases. *Physiol. Plantarum*, 340: 254–260

Nair RR and Dutta Gupta 2003 Somatic embryogenesis and plant regeneration in black pepper (*Piper nigrum* L.): Direct somatic embryogenesis from tissues of germinating seeds and ontogeny of somatic embryos. J. Hortic. Sci. Biotechnol, 78 : 416-421

Nirmal Babu, K., Geetha, S.P., Minoo D., Yamuna, G., Praveen, K., Ravindran, P.N. and Peter, K.V. 2007. Conservation of Spices Genetic Resources through *In Vitro* Conservation and Cryo- Preservation. pp. 210-233. In KV Peter and Z Abraham (ed). *Biodiversity in Horticultural Crops* Vol 1, Daya Publishing house, New Delhi

Nirmal Babu, K., Geetha, S.P., Minoo, D., Ravindran, P.N. and Peter, K.V., 1999. *In vitro* conservation of germplasm. In: *Biotechnology and its Application in Horticulture*, (Ed.) S.P. Ghosh. Narosa Publishing House, New Delhi, pp. 106–129.

Nirmal Babu, K., Minoo, D., Geetha, S.P., Ravindran, P.N. and Peter, K.V., 2005. Advances in Biotechnology of Spices and Herbs. *Ind. J. Bot.Res.*, 1(2): 155–214.

Nirmal Babu, K., Ravindran, P.N. and Peter, K.V., 1997. *Protocols for Micropropagation of Spices and Aromatic Crops*. Indian Institute of Spices Research, Calicut, India, p. 35.

Nirmal Babu, K., Ravindran, P.N. and Sasikumar, B., 2003. Field evaluation of tissue cultured plants of spices and assessment of their genetic stability using molecular markers. Final Report submitted to Department of Biotechnology, Government of India, pp. 94.

Nirmal Babu, K., Rema, J., Sree Ranjini, D.P., Samsudeen, K. and Ravindran, P.N., 1996. Micropropagation of an endangered species of Piper, *P. barberi* Gamble and its conservation. *Journal of Plant Genetic Resources*, 9(1): 179–182.

Niwata, E., 1995. Cryopreservation of apical meristems of garlic (*Allium sativum* L.) and high subsequent plant regeneration. *CryoLetters*, 16: 102–107.

Panis B.; Lambardi M. Status of cryopreservation technologies in plants (crops and forest trees). In: Ruane J.; Sonnino A. (eds) 2006.The role of biotechnology in exploring and protecting agricultural genetic resources. FAO, Rome, pp 61–78;

Parani, M., Anand, A. and Parida, A., 1997. Application of RAPD finger printing in selection of micropropagated plants of *Piper longum* for conservation. *Current Science*, 73(1): 81–83.

Peter, K.V., Ravindran, P.N., Nirmal Babu, K., Sasikumar, B., Minoo, D., Geetha, S.P. and Rajalakshmi, K., 2002. *Establishing In vitro Conservatory of Spices Germplasm*. ICAR Project report. Indian Institute of Spices Research, Calicut, Kerala, India, pp. 131.

Philip, V.J, Joseph, D., Triggs, G.S. and Dickinson, N.M., 1992. Micropropagation of black pepper (*Piper nigrum* L.) through shoot tip cultures. *Plant Cell Report*, 12: 41-44.

Phunchindawan, M., Hirata, K., Sakai, A. and Miyamot K. Cryopreservation of encapsulated shoot primordia induced in horseradish (*Armoracia rusticana*) hairy root cultures *Plant Cell Reports* 16(7): 469-473. DOI: 10.1007/BF01092768

Quatrano, R.S., 1968. Freeze preservation of cultured flax cells using DMSO. *Plant*, 43: 2057–2061.

Rajasekharan, P.E. and Ganeshan, S., 2003. Pollen cryopreservation in Capsicum species: A feasibility study. *Capsicum and Eggplant Newsletter*, 22: 87–90.

Ravindran, P.N., Nirmal Babu, K., Saji, K.V., Geetha, S.P., Praveen, K. and Yamuna, G., 2004. Conservation of Spices genetic resources in *in vitro* gene banks. ICAR Project report. Indian Institute of Spices Research, Calicut, Kerala, India pp. 81.

Ravishankar, G A., Sarma K S., Venkataraman L V. and Kalyan A K. 1988. Effect of nutritional stress on capsaicin production in immobilized cell cultures of *Capsicum annuum*. *Curr. Sci.* 57 : 381-383.

Ravishankar, G.A., Sudhakar, J.T. and Venkataraman, L.V., 1993. Biotechnological approach of *in vitro* production of capsaicin. In: *Proceedings of the National Seminar on Post Harvest Technology of Spices*, Trivandrum, p. 75–82.

Reed B. M.; Engelmann F.; Dulloo M. E.; Engels J. M. M. 2004.Technical guidelines for the management of field and in vitro germplasm collections. IPGRI Handbooks for Genebanks No. 7. IPGRI, Rome, Italy.

Rekha Chaudhury and Malik, S.K., 2004. Genetic conservation of plantation crops and spices using cryopreservation. *Indian Journal of Biotechnology*, 3(3): 348–358.

Sajina, A., Minoo, D., Geetha, S.P., Samsudeen, K., Rema, J., Nirmal Babu, K., Ravindran, P.N. and Peter, K.V., 1997. Production of synthetic seeds in a few spice crops. In: *Biotechnology of Spices, Medicinal and Aromatic Plants*, (Eds.) Edison, S., Ramana, K.V., Sasikumar, B., Nirmal Babu, K. and Santhosh, J.E. Indian Society for Spices, Calicut, India, p. 65–69.

Sambrook J, Fritsch E F & Maniatis T 1989 Molecular Cloning – a laboratory manual. Vol.3. Cold Spring Harbor Laboratory Press, New York.

Schenk R U & Hildebrandt A C 1972 Medium and techniques for induction and growth of monocotyledonous and dicotyledonous plant cell cultures. *Can. J. Botany*, 50:199-204.

Sharma, T.R., Singh, B.M. and Chauhan, R.S., 1994. Production of encapsulated buds of *Zingiber officinale* Rosc. *Plant Cell Reports*, 13: 300–302.

Shatnawi, M. A., Johnson K.A. and Torpy F.R. (2004) *In vitro* propagation and cryostorage of *Syzygium francissi (Myrtaceae)* by the encapsulation-dehydration method, *In Vitro Cellular & Developmental Biology - Plant* 40 (4): 403-407.

Baghdadi, S.H, Shibli, R.A, Syouf M.Q., Shatnawi MA., Arabiat, A. and Makhadmeh I.M (2010) Cryopreservation by encapsulation-vitrification of embryogenic callus of wild crocus (*Crocus hyemalis* and C. *moabiticus*) Jordan J. Agricultural Sciences. 6 (3) 436-442.

Umetsu, H., Wake, H., Saitoh, M., Yamaguchi, H. and Shimomura, K., 1995. Characteristics of cold preserved embryogenic suspension cells in fennel *Foeniculum vulgare* Miller. *Journal of Plant Physiology*, 146(3): 337–342.

Venkataraman, L.V., Ravishanker, G.A., Sarma, K.S. and Rajasekaran, T., 1989. *In vitro* metabolite production from saffron and capsicum by plant tissue and cell cultures. In: *Tissue Culture and Biotechnology of Medicinal and Aromatic plants*, (Eds.) Kukreja *et al.* CIMAP, Lucknow, India, p. 147–151.

Williams J G K, Kubelik A R, Livak K J, Rafalski J A & Tingey S V 1990 DNA polymorphisms amplified by arbitrary primers are useful as genetic markers. Nucleic acid research 18:6531-6535

Withers, L.A., 1985. Cryopreservation of cultured cells and meristems. In: *Cell Culture and Somatic Cell Genetics of Plants, Vol. 2: Cell Growth, Nutrition, Cyto-differentiation and Cryopreservation*, (Ed.) I.K. Vasil. Academic Press, Orlando, Florida, pp. 253–316.

Withers, L.A., 1987. Long-term preservation of plant cells, tissues and organs. *Oxford Surveys of Plant Mol. and Cell Biology*, 4: 221–272.

Withers, L.A., 1991. Biotechnology and plant genetic resources Conservation. In: *Plant Genetic Resources Conservation and Management: Concepts and Approaches*, (Eds.) R.S. Paroda and R.K. Arora. IBPGR, New Delhi, pp. 273–297.

Yamuna, G. 2007. *Studies on Cryopreservation of Spices Genetic Resources* Ph. D Thesis, University of Calicut, Kerala, India.

Yamuna, G, Sumath,i V., Geetha, S. P., Praveen, K., Swapna, N. and Nirmal Bab, K. 2007. Cryopreservation of *In Vitro* grown shoot of Ginger (*Zingiber officinale* Rosc). *CryoLetters*.28(4):241-252

7

Cryopreserving Vegetatively Propagated Tropical Crops – The Case of *Dioscorea* Species and *Solenostemon rotundifolius*

Marian D. Quain[1], Patricia Berjak[2],
Elizabeth Acheampong[3] and Marceline Egnin[4]
[1]Council for Scientific and Industrial Research Crops Research Institute
[2]School of Biological and Conservation Sciences,
University of KwaZulu-Natal, Westville Campus, Durban
[3]Tissue Culture Laboratory, Department of Botany, University of Ghana, Legon, Accra
[4]Plant Biotechnology and Genomics Research Laboratory, Tuskegee University
[1,3]Ghana
[2]South Africa
[4]USA

1. Introduction

Root and tuber crops in the Sub-Saharan African region play a major role in daily diet, accounting for over 50% of the total staple. *Dioscorea* spp. and *Solenostemon rotundifolius* are among the tuber staples in West Africa. *Solenostemon rotundifolius* (Poir) J.K. Morton is an edible starchy tuber crop known to have originated in tropical Africa (Schippers, 2000). It occurs in western, central, eastern and southern Africa. In Ghana, it is popular in the northern part of the country and its common name is Frafra potato (Tetteh and Guo, 1993). In South Africa, it occurs mainly in coastal KwaZulu-Natal, eastern Mpumalanga and northwestern Cape and it is commonly known as Zulu round potato (Schippers, 2000). It is used to combat famine as it has high protein content, and has medicinal and social values (Kay, 1973). It flowers profusely, yet has rare seed production and is therefore propagated vegetatively by means of vine cuttings and tuber sprouts. Storage of the tuber in hot climates is a problem. In Ghana, it is stored in dry places or left on the ground under trees where conditions are cool. The tuber is stored buried in the ground to maintain the good quality for about two months. Otherwise the tuber sprouts within a shorter period. However, in South Africa, the tuber stores well through the winter months (Schippers 2000).

The germplasm is endangered because although field and *in vitro* gene banks are being used for conservation, these serve short to medium term purposes, and are expensive. Efforts to conserve the germplasm in the longer-term under slow growth *in vitro* are hampered by the relatively rapid growth of the cultures. Cryostorage which is recognised as the very safe cost effective option for the long-term conservation of genetic resources, especially vegetatively propagated species and crops with recalcitrant seeds (Engelmann & Engels., 2002) therefore provide a viable alternative to the long-term storage, and ensure recovery of stable germplasm (Gonzalez-Arnao *et al.*, 1999).

Dioscorea species, colloquially known as yams, of family Dioscoreaceae are perennial monocotyledonous climbers with underground tubers which, in some species are edible and serve as major staples in sub-Saharan Africa. Propagation is routinely vegetative, using either the tubers or vine cuttings. Farmers ensure the production of true-to-type crops by using clonal planting material, because of the social and staple importance attached to yams in sub-Saharan Africa. Hence the conservation of clonal germplasm of yam is extremely important. *Dioscorea* spp. has about 700 species within the family, nine of which are medicinal plants that accumulate steroid saponins in their rhizomes. Six species of Dioscorea *D. bulbifera, D. cayenensis, D. dumentorum, D. prahensilis, D. alata* and *D. rotundata* contain mealy starch with a good level of vitamin C and other nutritive substances, which serve as major staples in sub-Saharan Africa. *Dioscorea rotundata* is native of West Africa, where it plays important role in the socio cultural life of the people. *Dioscorea alata* is the most widespread worldwide and is most cultivated in Southeast Asia, the Caribbean and West Africa. *Dioscorea rotundata* is now utilised in other parts of the world, and it has become a foreign exchange earner particularly in Ghana.

In vitro slow growth tissue culture methods have been used in conserving the germplasm (Ashun 1996; Ng & Daniel 2000; Ng & Ng 1991). Although this method usefully complements the traditional form of conservation, it serves only short- to medium-term storage purposes. Thus cryopreservation, which imposes a stasis on metabolic and deteriorative processes, is a worthwhile option to be explored.

Explant treatment to attain low water content which is critical for cryopreservation has in some protocols been by exposing tissues to stress, which enhance desiccation and cold tolerance (Withers 1985; Jitsuyama *et al.*, 2002). Such stress has been induced by abscisic acid, sugars, mannitol and sorbitol (Mastumoto *et al.*, 1998; Jitsuyama *et al.*, 2002; Veisseire *et al*, 1993; Panis *et al.*, 2002; Walter *et al.*, 2002). The use of cryoprotectants, which exert osmotic stress and lead to loss of free water from tissues and vitrification when frozen has also been induced by using reagents such as sucrose, glycerol, DMSO, ethylene glycol, proline and many others (Engelmann et al., 1994; Harding & Benson, 1994; Matinez-Montero *et al*, 1998; Plessis *et al.*, 1993; Nishizawa *et al.*, 1993). Desiccation of tissues on activated silica gel (Hatanaka *et al.*, 1994; Cho *et al.*, 2002), in laminar air flow cabinets (Gonzalez-Benito & Pezez, 1994; Thammasiri, 1999) and flash driers (Berjak *et al.*, 1999; Pammenter *et al.*, 1991; Wesley-Smith *et al.*, 1992; Walter *et al.*, 2002; Potts & Lumkin 1997) have all been used to appreciably reduce water content to enhance cryotolerance. Although these treatments have all been reported to be successful in enhancing cryopreservation of some tissues, there are differences in response to known protocols which have been mainly attributed to specie and variety specificity (Gonzalez-Benito *et al.*, 2002; Martinez-Montero *et al.*, 1998; Panis *et al.*, 2002; Gonzalez-Arnao *et al.*, 1999).

The prevention of the formation of lethal ice crystals when tissue is exposed to sub-zero temperatures is essential for successful cryopreservation, of vegetatively propagated germplasm. This chapter looks at the various attempts made to cryopreserve germplasm of *Solenostemon rotundifolious* and possible underlying mechanism that might have led to failure of tissues to respond to all methods utilized. Tissue survival, water contents and ultrastructure are used as parameters for analyzing response to various treatments. Also, response of yam *in vitro*-grown explants (shoot tips and axillary buds) to various desiccation

procedures and their ability to survive after exposure to cryogenic temperatures is
investigated here, with the ultimate aim of developing a simple protocol for long-term
conservation of the germplasm of *Dioscorea* species via cryopreservation. Parameters that
need critical investigation are discussed.

2. Materials and methods

2.1 *Solenostemon rotundifolius*

2.1.1 Source of explant

In vitro cultures of *Solenostemon rotundifolius* accession number UWR 002 was obtained from
the *in vitro* gene bank that had been maintained under slow growth conditions at 18°C. *In
vitro* cultures were multiplied on Murashige and Skoog (MS) medium (Murashige and
Skoog, 1962) supplemented with 2% sucrose and 0.7% agar. Subculturing was carried out at
four-weeks intervals. Cultures were maintained under a 16 h photoperiod (40 µM /m²/s¹) at
25°C ±1°C.

2.1.2 Conditioning donor plant material in culture (pregrowth)

Nodal cuttings were cultured on MS medium supplemented with either 0.058 M (2%) or 0.1
M sucrose or 0.1 M mannitol and 0.8% agar. Cultures were incubated for two to three weeks
after which uniformly developed plantlets were used for various experiments.

2.1.3 Conditioning excised explants in culture (preculture)

Solenostemon rotundifolius nodal cuttings consisting two buds (having lateral buds which are
microscopically globular and covered by leaf primodia as described by Niino *et al.*, 2000)
were obtained from pregrown cultures, positioned on sterile nylon mesh cut side down and
placed on fresh pregrowth media overnight. Explants were then transferred on mesh to
media with higher sucrose concentrations (0.2, 0.3, 0.4, 0.5, 0.6, 0.7, 0.8, 0.9 and 1.0 M)
sucrose for one to seven days. Media supplemented with 0.1 and 0.3 M mannitol were also
used as pregrowth treatment. Incubation was under a 16 h photoperiod (40µM/m²/s⁻¹), at
25°C±1°C.

2.1.4 Silica gel dehydration

Explants were placed on oven-sterilised aluminium foil and dehydrated over approximately
35 g activated silica gel in covered 90 mm glass Petri dishes for 30 min to 16 h under sterile
conditions.

2.1.5 Rehydration

Explants were rehydrated following cryoprotection treatment, silica gel dehydration and
cooling. This was carried out in cryovials containing liquid MS medium supplemented with
0.1 M sucrose, 1mM $MgCl_2.6H_2O$ and 1 µM $CaCl_2.2H_2O$ for 30 minutes. Re-hydrated buds
were cultured on growth medium, incubated under continuous dark conditions till signs of
growth and development were observed (at least one week) before they were transferred to
a dual photoperiod.

2.1.6 Cryoprotection

Liquid medium (MS) supplemented with 0.2, 0.4 M, 0.07 M, 0.14 M, sucrose, 2.0 M, 0.64 M, 1.28 M, 3.23 M glycerol, 2.42 M ethylene glycol, 0.017 M raffinose, 0.64 M, 1.28 M and 1.92 M DMSO, in varied combinations, plant vitrification solution II (PVS2), (Sakai et al., 1990) and half strength PVS2 were used to cryoprotect nodal cutting explants for 5, to 40 minutes. Explants used were obtained from shoots grown on 0.1 M mannitol for 2-3 weeks. Excised explants were cultured on 0.3 M mannitol for 72 hours, and cryoprotected in cryovials, using 1 ml cryoprotectant solution (the 1 ml cryoprotectant was decanted and replace with 0.5 ml during cryoprotection). To enhance explant cryoprotection, dehydration over activated silica gel for 60 minutes either before or after cryoprotection was also investigated. Following cryoprotection, the cryoprotectant solution was decanted and explants were washed three times with rehydration solution (described above). However, explants subjected to cooling, (LN or freezing to -70°C) were immediately on retrieval, rewarmed in water bath at 40°C for two - three minutes. Following rewarming, tissues were then allowed to stay for 30 min in the rehydration solution before blotting dry, and cultured on growth medium. Incubation was in continuous dark conditions till signs of growth and development were observed. Developing cultures were transferred to 16 h photoperiod.

2.1.7 Frafra potato assessments

2.1.7.1 Water content determination

Individually weighed explants were oven-dried at 80°C for 48 hours to determine dry mass. Water content was determined individually for 5-10 explants, and expressed on a dry mass (g H_2O g^{-1} dry mass) basis.

2.1.7.2 Survival

Explant survival was assessed weekly for three weeks after culturing. Generally, 8-10 explants were used per treatment and experiments were replicated three times. Surviving explants were those which showed shoots with buds, leaves, and root development.

2.1.7.3 Transmission electron microscopy

A standard glutaraldehyde-osmium fixation method was used, followed by dehydration through an acetone series embedding in a low viscosity epoxy resin (Spurr, 1969). Sections of the meristematic regions of axillary buds and shoot tips were collected on 200 mesh hexagonal copper 3.05 mm grids. Sections were post-stained with uranyl acetate and lead citrate, washed with distilled water, and viewed and photographed with a JEOL 100-S transmission electron microscope.

2.2 *Dioscorea rotundata*

2.2.1 Source of explant

In vitro cultures of *Dioscorea rotundata* ("Pona"), accession number PS 98 013 were obtained from the *in vitro* gene bank of the Department of Botany, University of Ghana, Legon, where the plants were maintained under long-term slow growth conditions at 18°C. Cultures were multiplied and sub-cultured at six-week intervals on Murashige and Skoog (MS) medium

(Murashige and Skoog 1962) with 2.5 µM kinetin, 20 mg l-1 L-cysteine, 2% (0.056 M) sucrose, 0.7% agar, and maintained under a 16 h photoperiod (40 µ mol m-2 s-1) at 25±1°C.

2.2.2 Conditioning excised explants in culture (preculture)

Yam shoot tips (~1 - 2 mm) were excised from cultures grown on MS pregrowth medium [as above, except containing 3% (0.09 M) sucrose instead of 2% sucrose] for five weeks, placed on sterilised nylon mesh, which was then positioned, explant cut side down, on fresh medium, overnight. Explants were then transferred on the mesh to semi-solid medium with higher sucrose concentrations (0.3, 0.5, 0.7 and 1.0 M, the control material continuing to be exposed to 0.09 M sucrose) in 90 mm Petri dishes for one, three, five or seven days, each followed by transfer either to growth-enhancing medium (MS complete salts with vitamins, 3% sucrose, 5 µM kinetin, 20 mg l-1 L-cysteine, 0.8% agar, 1% filter-sterilised casein hydrolysate at pH 5.7 ± 0.1), or further conditioned for cryopreservation.

2.2.3 Silica gel dehydration and cooling of explants

Yam explants dehydrated using the same methodology as described above for Frafra potato. Dehydrated explants were placed in cryovials, which were plunged into, and maintained in, liquid nitrogen for one hour, or cooled at 1°C min-1 in a Nalgene cryo freezing container (Mr Frosty™), to -70°C, and maintained for at least four hours at this temperature. Rewarming was effected immediately on retrieval from the cryogen, in a water bath at 40°C for two to three minutes for all treatments.

2.2.4 Rehydration

The rehydration solutions consisted of MS complete salts with vitamins, 2.5 µM kinetin, 20 mg l-1 L-cysteine, 1mM $MgCl_2.6H_2O$, 1µM $CaCl_2.2H_2O$, 1% casein hydrolysate (filter-sterilised), and 1 M sucrose, at pH 5.7±0.1, magnesium and calcium chlorides having been shown to enhance explant recovery of date palm somatic and pea zygotic embryos (MyCock 1999). Rehydration was for 30 minutes. Rehydrated buds were blotted dry and cultured on growth medium (as above). Cultures were incubated under continuous dark conditions at 24±1°C until signs of growth and development were observed, before they were transferred to 16 h photoperiod (40 µ mol m-2 s-1) at 25±1°C.

2.2.5 Cryoprotection with modified plant vitrification solution 2(MPVS2)

Explants that had been pregrown and precultured were exposed to 1 ml plant vitrification solution 2 (PVS2) as designed by Sakai and colleagues (Sakai et al., 1990), but modified as follows (MPVS2): basic MS medium, 30% glycerol, 15% ethylene glycol, 15% DMSO (v/v), 0.4 M sucrose (w/v), 0.1 M $CaCl_2.2H_2O$ and 1% D-raffinose at pH 5.7±0.1. Inclusion of calcium chloride and raffinose has been found to be beneficial in promoting recovery after cryopreservation in other species (Mycock, 1999). Explants were treated for 0, 10, 20, 30 or 40 minutes in cryovials. The vitrification solution was decanted and the explants washed three times in 1 ml rehydration solution (as described above) for 30 minutes, then cultured on growth medium and incubated in the dark. Cultures were transferred to the alternating light/dark conditions once signs of growth and development were observed. Prior to being cooled to -70 or -196°C, cryoprotected explants were suspended in fresh 0.5 ml MPVS2 in cryovials. The explants to be cryopreserved were then exposed to cryogenic conditions for specified durations, rewarmed, vitrification solution removed, rehydrated, and incubated as described above.

2.2.6 Yam assessment

2.2.6.1 Water content determination

Water content of yam explants was determined following the same procedure as described for Frafra potato above.

2.2.6.2 Survival assessment

Yam explant survival was determined following the same procedure as described above of Frafra potato.

2.2.6.3 Tetrazolium test for viability

Shoot tips which were pretreated with high sucrose concentrations; pretreated and cryoprotected with MPVS2; pretreated, cryoprotected, and vitrified, were cultured on growth medium for 5-7 d following which they were transferred to a 0.1% aqueous solution of 2,3,5-triphenyl tetrazolium chloride (TTZ) and incubated in the dark overnight. Control material was obtained from cultures under standard growth room conditions. Patchy red staining, located around the meristematic region as a result of respiratory activity in viable cells was scored as the tissue having survived the various treatments.

3. Results and discussion

3.1 Frafra potato

3.1.1 Pregowth

Water contents of Frafra potato nodal cutting explants following pregrowth on medium supplemented with 0.058 M (2%) sucrose was extremely high (22.25 ± 1.7 gH$_2$O g^{-1} dry wt) to enhance successful cryopreservation. As explant size and geometry have a marked effect on the success of freezing hydrated material (Wesley-Smith et al., 1995), nodal cuttings (two buds per explant) to be used for cryopreservation experiments were split into two halves with one bud per explant, this adequately lowered water content of explants (8.78 ± 1.07 gH$_2$O/g dry wt). The water content of the single (well trimmed) bud is similar to that of explants excised from greenhouse established plant (10.16 ± 0.98 g/g dry wt. data shown in Table 2). Since cultures grown on medium supplemented with 2% sucrose were extremely wet for cryopreservation, higher sucrose or mannitol concentration (0.1 M) was employed to enhance desiccation tolerance, which subsequently improves cryotolerance .

Pregrowth of explants on medium supplemented with 0.1 M mannitol lowered the water content of explants from 19.5 under control condition (medium supplemented with 0.058 M sucrose), to 10.4 gg^{-1}, which did not affect survival (Fig. 1). The ultrastructure was as well constituted as that of the control explants (Plate 1a), with ongoing metabolism indicated by abundant cristate mitochondria (Plate 1b), Golgi bodies and profiles of endoplasmic recticulum (insert). Growth on 0.1 M sucrose supplemented medium, also lowered the water

content of explants (11.4 gg⁻¹) as shown in Fig. 1, this observation is in agreement with
response of oil palm explant water content when treated with sucrose (Dumet *et al.*, 1993)
and while this did not reduce survival it resulted in severely distended organelles and
evidence of tonoplast disruption and lobed nuclei (Plate 1d and e). Sucrose has been
extensively used to treat plant tissues prior to cryopreservation (Panis *et al.*, 2002;
Grospietsch *et al.*, 1999; Gonzalez-Benito & Perez, 1994; Santos & Stushnoff 2003), studies
have however, not investigated the structural effect of sucrose on tissue. The damage
revealed by ultrastructure (Plate 1f) could have predisposed explants negatively to
subsequent steps.

3.1.2 Preculture

Culturing individual Frafra potato buds on 0.3 M sucrose for 3 d (Table 1), lowered water
content from 11.4 g g⁻¹ (after growth on 0.1 M sucrose medium) to 7.3 g g⁻¹ and explant
survival was at 100 %. This level of sucrose has been applied in other crops such as carrots
(Dereuddre *et al.*, 1991), wasabi (Mastumoto *et al.*, 1998), and African violet (Shibili *et al.*,
2004). Similarly, explant on medium supplemented with 0.3 M mannitol which were
derived from 0.1 M mannitol supplemented medium, water contents reduced further from
10.47 to 7.42 gg⁻¹ and survival was still at 100 % (Table 1). Mannitol and its isomer, sorbitol
have been used for pre-treatment of plant tissues before cryopreservation (Wang *et al.*, 2001)
as well as in long term storage culture media as osmoticums (Ashun, 1996; Egnin et al,
1998). The growth of explants on regrowth medium following preculture varying sucrose
media is shown in the Plate 2.

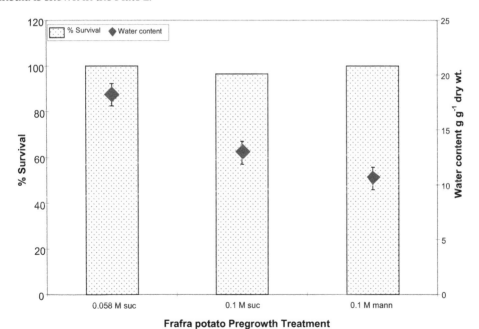

Fig. 1. Survival and water content of Frafra potato cultures on three pregrowth media ± SD.
Survival P ≥0.05, n=30, and WC P≤ 0.05, n=15-30

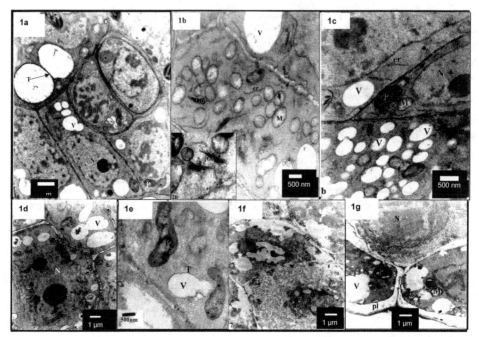

Plate 1. Ultrastructure of *Solenotemon rotundifolius* explant meristematic cells. Legend are as follows: Ch, heterochromatin, V, vacuoles, T, tonoplast, P, plastids, M, mitochondrial, ER, endoplasmic reticulum, N, nucleus Gb, Golgi bodies, pl, plasma membrane.

Plate 1a Control explants cultured on 2% sucrose supplemented medium. Cells show oval nuclei, normally-distributed heterochromatin, vacuoles each with a well-defined tonoplast, small, relatively dense plastids, circular mitochondrial profiles with dense matrices, and profiles of endoplasmic reticulum.

Plates 1b (insert), & c. Explants pregrown on 0.1 M mannitol for three weeks cells (1b). Ongoing metabolic activity indicated by abundance of mitochondria, Golgi bodies and profiles of endoplasmic reticulum (insert), many plasmodesmata are visible. A group of relatively small vacuoles is shown (1c), which appeared typical of mannitol treatment to reduce water content.

Plate 1d After sucrose (0.1 M) pregrowth for three weeks, there was evidence of tonoplast disruption, lobed nuclei with possibility of vacuole fission or fusion. Plate 1(e) shows somewhat distorted plastids and a potentially autolysing cell (lower left) where vacuolar dissolution (tonoplast disruption) appears to have occurred. Such events would have predisposed these explants negatively to subsequent steps.

Plate 1f. Explants pregrown on 0.1 M sucrose, precultured on 0.3 M sucrose for 3 d and then dehydrated over activated silica gel for 120 min. Water content was 0.16 g g-1 while survival of the sample was only 2.5%. Most specimens presented this appearance of advances intracellular deterioration, nuclear remains; and plasma membrane.

Plate 1g. Explants pregrown on 0.1 M mannitol and then preculture on 0.3 M mannitol for 3 d, dehydrated over activated silica gel for 120 min, during which water content was lowered to 0.11 g g-1. This was accompanied by 38.1% survival. Nuclear and cytoplasmic derangement had occurred although some cells had few intact organelles, shown in this illustration of what was probably a non-surviving explant.

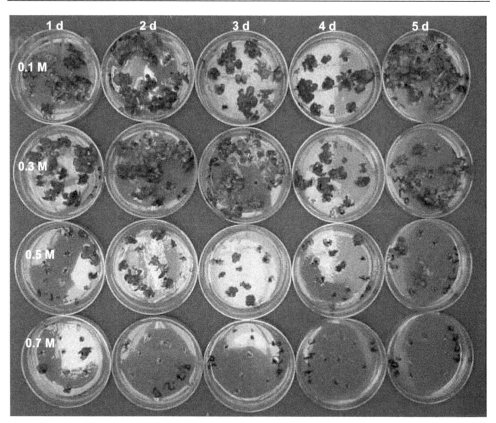

Plate 2. Development of *Solenostemon rotundifolius* explants pregrown and precultured on
medium supplemented with increasing sucrose concentrations 0.1 M, 0.3, 0.5 M, and 0.7 M
for 1 d, 2 d, 3 d, 4 d, and 5 d and grown on growth medium in 90mm Petri plates. Data taken
at 3-weeks.

3.1.3 Dehydration

Dehydrating explants over silica gel, considerably lowered water content (Table 1). In
experiments by other investigators, this technique has been successfully used to dehydrate
and cryopreserve citrus axes, (Santos & Stushnoff, 2003) and encapsulated somatic embryos
of *Coffee canephora* (Hatanaka *et al.*, 1994).

Mannitol treated (0.3 M) explants of Frafra potato used in this particular experiment, when
dehydrated over activated silica gel, the lowest water content (0.11 g g^{-1}) was recorded for
120 min dehydration and survival was 73% (Table 1). However, sucrose treated (0.3 M)
explants indicated lowest water content of 0.16 g g^{-1} which although, is higher than in
mannitol (stated above), survival was as low as 5% (Table 1). It is possible that damage
suffered by tissues as revealed by ultrastructure (Plate 1d&e) during sucrose treatment
predisposed them to further damage on dehydration. It was evident here that mannitol
treated tissue are more desiccation tolerant than sucrose treated tissues although other

Treatment	Water Content g/g dry weight	% Survival
0.058 M sucrose for 1 d	18.22 ± 0.96	100 ± 0
0.058 M sucrose for 3 d	22.37 ± 2.02	86.67 ± 4.7
0.058 M sucrose for 5 d	16.7 ± 0.65	100 ± 0
0.1 M sucrose for 1 d	12.93 ± 0.66	96.67 ± 0.86
0.1 M sucrose for 3 d	14.18 ± 0.03	100
0.1 M sucrose for 5 d	12.74 ± 0.4	100
0.3 M sucrose for 1 d	8.82 ± 0.4	100
0.3 M sucrose for 3 d	7.98 ± 0.73	96.67 ± 0.86
0.3 M sucrose for 5 d	6.17 ± 0.39	100
0.5 M sucrose for 1 d	4.13 ± 0.38	96.67 ± 0.86
0.5 M sucrose for 3 d	4.22 ± 0.41	86.67 ± 2.27
0.5 M sucrose for 5 d	3.37 ± 0.27	96.67 ±0.86
0.1 M sucrose dehydrated for 0 min	11.45 ± 0.50	100
0.1 M sucrose dehydrated for 60 min	0.58 ± 0.15	50 ± 1.8
0.1 M sucrose dehydrated for 90 min	0.56 ± 0.12	30 ± 3.6
0.1 M sucrose dehydrated for 120 min	0.14 ± 0..02	35 ± 10.6
0.3 M sucrose dehydrated for 0 min	6.09 ± 0.16	96.67 ± 0.8
0.3 M sucrose dehydrated for 60 min	0.79 ± 0.17	56.25 ± 1.14
0.3 M sucrose dehydrated for 90 min	0.37 ± 0.07	33.18 ± 0.58
0.3 M sucrose dehydrated for 120 min	0.16 ± 0.61	5 ± 2.4
0.1 M mannitol for 3 d and dehydrated for 0 min	10.51 ± 0.6	80 ± 2.08
0.1 M mannitol for 3 d and dehydrated for 60 min	0.42 ± 0.07	40 ± 0.11
0.1 M mannitol for 3 d and dehydrated for 90 min	0.29 ± 0.05	40 ± 0.11
0.1 M mannitol for 3 d and dehydrated for 120 min	0.08 ±±0.01	20 ± 0.09
0.3 M mannitol for 3 d and dehydrated for 0 min	7.42 ± 0.38	96.67 ±
0.3 M mannitol for 3 d and dehydrated for 60 min	0.31 ± 0.05	71 ± 0.09
0.3 M mannitol for 3 d and dehydrated for 90 min	0.24 ± 0,05	70 ± 0.10
0.3 M mannitol for 3 d and dehydrated for 120 min	0.11 ± 0.4	73 ± 0.10

Table 1. Treating FP 002 with different sucrose concentrations during development of Preculture conditions using nodal cuttings with single buds

reports have successfully used sucrose to induce dehydration tolerance (Dumet et al., 1993; Grospietsch et al., 1999; Santos & Stushnoff, 2003).

Sucrose treated tissues had totally been deranged after 120 minutes (Plate 1f) of dehydration compared with mannitol tissues (Plate 1g) which has some intact nuclei and few organelles present. These must be responsible for the survival recorded (Table 1). It is possible the presence of the intact nuclei and organelles in the mannitol treated cells could be reconstituted for normal plant growth and development to occur.

Crop accession	Water content
FP UER 001	18 + 5.14
FP UER 002	10 + 0.9
FP UWR 003	9.6 + 2.7
FP UER 004	14.6 + 2.1
99/053	8.4 + 1.1
99/1033	11.2 + 0.6
99/016	11.2 + 1.6
99/022	11.35 + 0.6

Table 2. Water content of screenhouse of Frafra potato established in greenhouse for three months

3.1.4 Cryoprotection

Explants treated with 0.1 and then 0.3 M sucrose or mannitol, on exposure to PVS2 indicated only about 20% survival (Table 3). This observation was contrary to report by Niino et al., (2000) that S. rotundifolius innala recorded high survival on treating with PVS2 and subsequently, 85% survival on exposure to liquid nitrogen. Solenostemon rotundifolius used in this study, were extremely sensitive to both the loading solution (0.4 M sucrose + 0.2 M glycerol, data not shown) and PVS2, which in other reported studies, led to successful cryopreservation of other crops including S. rotundiflius (Wang et al., 2003; 2001; Turner et al., 2001; Niino et al., 2000). On screening for appropriate vitrification (cryoprotection) solution, the following cryoprotection solutions listed in Table 3 were tested. It was indicative from results that DMSO and Ethylene Glycol at the concentration (15%) that they occur in PVS2 did not have any lethal effect on the explants. However, sucrose and Glycerol at the concentrations that they occur in PVS2 (0.4 M and 30% respectively) were found to be lethal to the tissues (Table 3). The use of PVS2 at half concentration and a combination of 2.5% Glycerol, 5% sucrose, 7.5% DMSO and 7.5% Ethylene glycol (coded PVSB) resulted in survival and growth of explants. The responses confirm indication that cryoprotectants at full-strength are toxic to plant cells (Rheinhoud et al., 1995). These treated explants, however did not survive on exposure to liquid nitrogen. Combining the cryoprotection treatment with dehydration (data not shown) as has been reported by other investigators as enhancing high cryosurvival (Wang et al., 2003; 2001; Turner et al., 2001), did not result in survival after cryopreservation in this study. Encapsulating explants prior to treatment with PVS2 also did not result in explant survival.

Ultrastructural studies indicated that tissues treated with ½PVS2 (Plate 3) and PVSB (Plate 4) for 15 min, which survived had well constituted cells, however, some tonoplasts were not

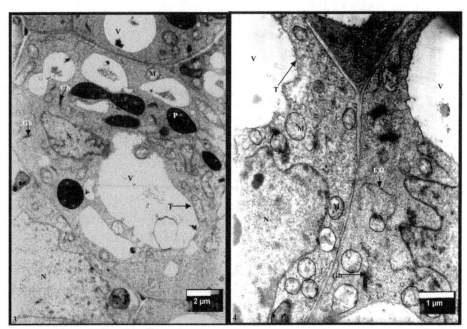

Plate 3 & 4. Ultratructure of surviving Frafra potato after mannitol pregrowth and preculture treatments, followed by subjecting explants for 15 min to ½PVS2 (Plate 3) and PVSB (Plate 4). (3) Although cells were well organised, some extent of autophagy was concluded to have occurred, in terms of intravacuolar inclusions in the ½ PVS2-treated material. (4) Cells appeared exceptionally active but showed distinctly lobed nuclei (N). Other organelles that can be recognized are mitochondria (M), endoplasmic reticulum (ER), Golgi bodies (Gb), plastids (P), and vacuoles (V) with tonoplast (T) intact.

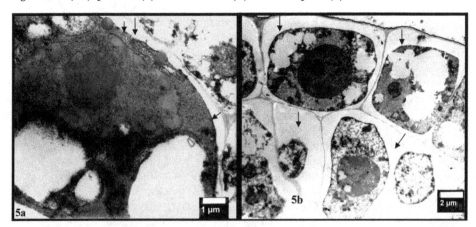

Plate 5a & b. Ultratructure of non-surviving Frafra potato following mannitol pregrowth and preculture treatments, then subjecting explants to ½PVS2 for 15 (Plate 5a) and 40 min (Plate 5b) Cells were highly plasmolysed (arrows) and damaged to the extent that organelles could not be easily recognized.

distinct and some extent of autophagy was observed in ½ PVS2, treated tissues (Plate 3).
Slight deposition of starch was observed in plastids of tissues treated with PVSB. Tissues,
which did not survive after the 15min treatments (Plate 5a&b), showed highly plasmolysed
cells and totally deranged cells, exposing the lethal effect of vitrification solution.

Increasing glycerol and sucrose concentration as well as long exposure duration led to
reduced survival and regeneration of encapsulated nodal cuttings treated with the plant
vitrification solutions as indicated in (Table 4). This observation is contrary to other crops
where encapsulation vitrification has been highly recommended for successful
cryopreservation (Charoensub et al., 1999; Wang et al, 2003) however confirms that
cryoprotectants can be damaging to plant tissues and the extent of effect varies according to
type and concentration of cryoprotectant as well as plant species (Berjak et al., 1996).

Treatment	% Survival
3.23 M Glycerol + 0.4 M Sucrose + 0.017 Raffinose	33.64 ± 1.39
3.23 M Glycerol + 0.4 M Sucrose	22.71 ± 2.77
PVS2 + 0.017 M raffinose	8.28 ± 0.20
PVS2	20 ± 0
0.07 M S + 0.3 M G 15 min	100
0.07 M S + 0.3 M G+0.14 M S +0.6 M G, 15 min	100
0.14 M S + 0.6 M G, 15 min	100
0.64 M DMSO for 15 min	90±0
1.28 M DMSO for 15 min	100± 0
1.92 M DMSO for 15 min	100 ± 0
2.42 M EG + 0.017 M Raffinose	84.24±1.46
2.42 M EG	77.28±0.97
½ PVS2 + 1% Raffinose	80.63±1.82
0.3 M G + 0.2 M S + 0.96 M DMSO + 2.42 M EG + 0.017 M Raffinose (PSVB)	69.17±2.26

S: sucrose, G: Glycerol, EG: Ethylene Glycol

Table 3. Survival (± SD) of Frafra potato explants, pregrown (0.1 M mannitol) and
precultured (0.3 M mannitol) prior to cryoprotection treatments. n=30 – 40, P<0.05

Frafra potato Variety	Treatment	Water Content g/g dry wt	% Survival
	PVSB for 15 min	4.13 ± 0.16	55 ± 2.04
	PVSB for 40 min	4.45 ± 0.25	55 ± 2.04
FP 002	½ PVS2 for 15 min	3.36 ± 0.06	38.75 ± 0.3
	½ PVS2 for 40 min	3.26 ± 0.17	0
	PVS2 for 15 min	2.00 ± 0.10	20
	PVSB for 15 min	2.66 ± 0.16	59.01 ± 1.5
FP 003	½ PVS2 for 15 min	2.58 ± 0.23	53.76 ± 2.3
	Control	5.63 ± 0.31	100
	PVSB for 15 min	2.91 ± 0.28	54.56 ± 1.4
FP 004	½ PVS2 for 15 min	2.7 ± 0.14	66.62 ± 3.47
	Control	6.59 ± 0.62	100

½PVS2 and PVSB were always supplemented with 0.017 M Raffinose

Table 4. FP explants treated with Plant Vitrification Solution

3.1.5 Cryopreservation

Although explants treated with mannitol and dehydrated over activated silica gel for 90 minutes had water content 0.24 gg^{-1} and survival was 70 % (Table 1), when exposed to liquid nitrogen, survival was nil. Ultrastructure indicated extensively degraded cells with withdrawn and broken plasmalemma, cytoplasm and nucleoplasm were all damaged (Plate not shown). Explants from all dehydration treatments did not survive on exposure to liquid nitrogen as well as ultra-cold liquid nitrogen (slash), although, it has been reported that rapid cooling enhance cryosurvival (Wesley-Smith *et al.*, 1992). Having dehydrated explant to water content of 0.11gg-1, it is obvious from ultrastructure (not shown) that the prolonged stress exerted decreased explant ability to withstand freezing since there is a level below which dehydration stress is increasingly apparent (Wesley-Smith *et al.*, 1992). Unlike the loss of viability in *S. rotundifolius*, explant at higher water contents have been reported to survive on exposure to liquid nitrogen (Berjak *et al.*, 1995; Kioko *et al.*, 1998 and 2000)

During cryopreservation all metabolic processes cease, it is possible that mannitol treated explants were too active metabolically judging from the high number of mitochondria occurring in the cytoplasm (Plate 1b). Hence bringing the systems to a halt caused a breakdown in all the plant metabolic systems causing cytoplasm to lose its viability since following dehydration, only few organelles could be observed in cytoplasm (Plate 1f). It is also possible that with the occurrence of high number of small vacuoles in mannitol treated explant (Plate 1c) which is a characteristic whereby, large vacuoles volumes are reduced by redistributing them into smaller vesicles on exposure to mannitol (Gnanapragasam & Vasil, 1992), being an advantage for survival since water contents are relatively low (Reinhoud *et al.*, 1995). However, it is probably that, the water present in the vacuoles did not have high viscosity, which would prevent the formation of ice crystals during cooling and thawing hence causing degeneration of plant cell integrity.

Sucrose treated explant, ultrastructure indicated cytoplasmic breakdown at all stages of treatment. Although the plant cell were not in a high metabolic state prior to exposure to liquid nitrogen, cellular degeneration had already set in and may have had a major role to play, leading to loss of viability on exposure to liquid nitrogen.

The above and all associated factors need to be investigated further. These will help optimise plant cell structure prior to cryopreservation. Based on the ultrastructural studies carried out, it is obvious that the use of mannitol for pregrowth and preculture treatment, the plant tissues develop capability to tolerate other stress (desiccation). *S. rotundifolius* tissues besides yielding high explant survival, results in stable ultrastructure for further plant growth and development. However the treatment does not necessarily result in survival on exposure to cryopreservation. The use of higher concentration of mannitol may enhance cryotolerance. Other critical factors that have to be investigated include maturation of explant supported by constitution of ultrastructure and related water content which play crucial rôle in cryopreservation (Chandal *et al.*, 1994; Berjak *et al.*, 1993). However, the extremely high water content (18.7 – 9.64 g/g dry wt) of plant even in the greenhouse (graph not shown), may still make it difficult to cryopreserved tissues of local accessions of *S. rotundifolius*. Several attempts were made to adequately harden Frafra potato (Table 5) prior to subjecting explants to various treatments and then cooling however, none of them resulted in explants survival after cooling.

Treatment	Water content
Six months in culture	8.1± 1.2
Shoot grown from tuber under sterile conditions (8 weeks)	19.5 ± 1.6
Vitrified shoot grown from tuber under sterile conditions (8 weeks)	24.7 ± 3.9
Normal shoots transferred to vented vessels (3 weeks)	8.2 ± 1.3
Vitrified shoots transferred to vented vessels (3 weeks)	12.7 ± 2.5
Normal shoots transferred to dry air-line (3 weeks)	7.2 ± 1.0
Normal shoots transferred to humid air-line (3 weeks)	18.4 ± 2.2
Cultures transferred to RITA vessels	13.0 ± 2.2

Table 5. Other attempts to acclimatize the new Frafra potato accession 99/053 to lower water
content that might enhance cryosurvival.

3.2 Dioscorea rotundata

Comparatively, yam explants cultured on medium supplemented with 0.3 M sucrose for 3-5
d considerably reduced tissue water content from about 12.2 g g-1 dry mass to between 4.8
and 5.5 g g-1 dry mass before cryoprotection with modified PVS2 (MPVS2) or silica gel
dehydration. Following cryoprotection with MPVS2 the Plate (6) below indicated the
growth of nodal explants.

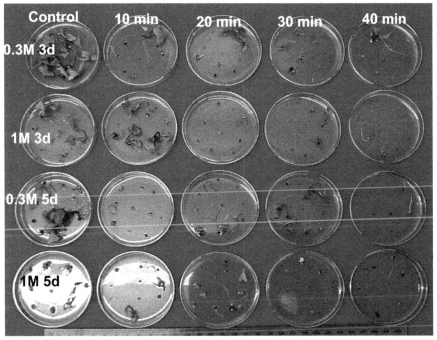

Plate. 6. Growing cultures of yam explant subjected to pregrowth on 0.09 sucrose
supplemented medium for 5 weeks, precultured on medium containing 0.3 m sucrose for 3
or 5 d, treated with MPVS2 for varied duration and unloaded with rehydration solution
containing 0.3 or 1 M sucrose and cultured on regeneration medium for six week

Ultrastructural studies indicated that cells had deposits of starch in plastids following sucrose treatments. Survival for *D. rotundata* shoot tips treated with MPVS2 vitrification solution, and cooled to -70°C, was 16% for 15 min treatment and 44% for 40 min. Explant rehydration was in 1.0M sucrose supplemented solution. After the 40 min MPVS2 treatment the TTZ test indicated 88% viability retention of explants cooled to -70°C, and 44% at -196°C. Plantlet development was obtained for -70°C-cooled shoot tips, whereas only callus development occurred from tissues exposed to liquid nitrogen. Explant regeneration was not obtained with silica gel dehydration techniques. It was concluded that vitrification-solution based cryopreservation presently offers the best option for conservation of this *Dioscorea* species.

4. Conclusions

Undoubtedly, cryopreservation has high potential for the long-term storage of vegetative explants. It is, however, vital to achieve appropriate tissue water content and the quality of the initial material. This study has shown that vitrification-based cryopreservation protocol is useful for yam explants. However, Frafra potato is extremely sensitive to the vitrification based protocol while explants of the latter easily become hyperhydric and are impossible to dehydrate sufficiently for cryopreservation

The findings are relevant for cryopreservation of a range of yam germplasm (Quain et al., 2009) and also provide a sound basis for further attempts to cryopreserve Frafra potato genetic resources. The technique represents developed simple, cost-effective and potentially reliable methodology that does not require sophisticated equipment. Such procedures should be adapted for germplasm conservation of other species, using limited resources in laboratories in sub-Saharan Africa.

Findings in this present study suggest that in order to achieve an optimal recovery of cryopreserved explants, the donor plants should be adequately conditioned and the recovery media enriched while testing the different cryogenic procedures. Although encapsulation vitrification and encapsulation dehydration procedures have been used in cryopreservation procedure, for the yams used in this study, the use of non-encapsulated explant proved to be a better option. The encapsulation of explants after cryopreservation in the production of e.g. synthetic seeds (Naidoo, 2006; Perán et al., 2006), is, however, worth exploring.

The key to successful cryoprotection has been suggested as inducing tolerance to vitrification solutions and the ability of explants to tolerate dehydration treatment by cryoprotectants has been hypothesised by several researchers as the determining factor for successful cryosurvival (Langis & Steponkus, 1990; Reinhoud et al., 1995). It is still not conclusive whether having been hardened, explants being used in procedures would survive cryopreservation. However, Frafra potato explants were not amenable to conditioning by any of these pretreatments. The requirements for successful cryopreservation differ for different species. There is the possibility that the optimum developmental stage of the explant for successful cryopreservation varies from species to species. Therefore there is the need to ascertain and test many parameters on the basis of each species. However, culture conditions especially those that will obviate hyperhydricity, are of paramount importance, as presently indicated for Frafra potato.

It can be concluded from the experiments that:

- Successful cryopreservation of *Dioscorea rotundata* is possible using a simple vitrification protocol.
- The procedure incorporates
  - pregrowth of the donor plant on 0.09 M sucrose-supplemented medium for five weeks,
  - preculture on 0.3 M sucrose supplemented medium for 5 d
  - MPVS2 solution for 40 min,
  - Rapid cooling in liquid nitrogen or slow cooling to -70°C.
- For the first time successful cryopreservation of *Dioscorea rotundata* accession 'Pona' which is an elite variety in Ghana has been achieved.
- The technique represents developed simple, cost-effective and potentially reliable methodology that does not require sophisticated equipment.
- Procedures can be adapted for germplasm conservation of other species, using limited resources in laboratories in sub-Saharan Africa.
- To achieve an optimal recovery of cryopreserved explants the donor plants should be adequately conditioned.
- Frafra potato is extremely sensitive to the vitrification based protocol.

Frafra potato explants easily becomes hyperhydric, and are impossible to dehydrate sufficiently for cryopreservation, this provide a sound basis for further attempts to cryopreserve Frafra potato genetic resources. These observations therefore make available information for further investigation towards development of cryopreservation protocol.

5. Acknowledgements

The authors wish to acknowledge financial support received from the UNU/INRA and the TWOWS. They also wish to thank Mrs B. Asante (University of Ghana Legon) and Mrs P, Maartens (University of Kwa-Zulu Natal) for technical assistance.

6. References

[1] Ashun MD (1996). *In vitro studies on micropropagation of various yam species (Dioscorea* species) M.Phil. Thesis submitted to University of Ghana – Legon.
[2] Berjak, P., Mycock, DJ, Walker, M, Kioko, JI, Pammenter NW, and Wesley-Smith J (1999). Conservation of Genetic Resources Naturally occurring as recalcitrant seeds. In M. B., K.J. Bradford and J Vazquez-Ramus (eds) *Seed Biology Advances in Applications* pp. 223-228.
[3] Berjak P, Mycock DJ, Wesley-Smith J, Dumet D. & Watt MP (1996) Strategies for *in vitro* conservation of hydrated germplasm. M. N. Normah *et al.*, (eds). In vitro *conservation of plant genetic resources* 19-52.
[4] Berjak, P, Mycock DJ, Watt P, Wesley-Smith J and Hope B (1995). Cryopreservation of Pea (Pisum sativum L.). In Y.P.S. Bajaj (ed) Biotechnology in Agriculture and Forestry (32) Springer: 293-307.
[5] Berjak, P, Vertucci CW and Pammenter N.W. (1993). Effects of developmental status and dehydration rate on the characteristics of water and desiccation-sensitivity in recalcitrant seeds of *Camellia sinensis*. *Seed Science Research* 3, 155-166.

[6] Chandel, KPS, Chaudhury R & Radhamani J (1994), Biological mechanisms determining the recalcitrance in seeds of tea, cocoa and jackfruit. IBPGR-NBPGR Report.

[7] Charoensub R, Phansiri S, Sakai A & Yongmanotchai W (1999). Cryopreservation of cassava *in vitro*-grown shoot tips cooled to -196°C by vitrification. *CryoLetters* 20, pp. 89-94.

[8] Cho EG, Noor NM, Kim HH, Rao VR & Engelmann F (2002) Cryopreservation of *Citrus aurantifolia* seeds and embryonic axes using a desiccation protocol. *CryoLetters* 23, 309-316.

[9] Dereuddre J, Bland S. & Hassen N (1991). Resistance of alginate-coated somatic embryos of carrot (*Daucus carota* L.) to desiccation and freezing in Liquid nitrogen: 1. Effect of Preculture. Cryoletters 12 pp.125-134.

[10] Dumet D, Engelmann F, Chardrillange N, Duval Y, & Dereuddre J (1993). Importance of sucrose for the acquisition of tolerance to desiccation and cryopreservation of oil palm somatic embryos. *Cryo-Letters* 14: pp. 243-250.

[11] Egnin M, Mora A & Prakash, C.S. (1998). Factors Enhancing *Agrobacterium tumefaciens*-Mediated Gene Transfer in Peanut (*Arachis Hypogaea* L.). In Vitro *Cellular and Developmental Biology-Plants* 34, 310-318.

[12] Engelmann F & Engels JMM, (2002), Technologies and strategies for existing conservation. In JMM Engels, VR Rao, AHD BROWN and MT Jackson (eds), Managing plant Genetic Diversity, CAB International Wallingfield/IPGRI, Rome, pp 89-104.

[13] Engelmann F, Dambier D & Ollitraut P (1994), cryopreservation of cell suspension and embryogenic callus of citrus using a simple freezing process. *Cryo-Letters* 15: pp. 53-58.

[14] González-Arnao MT, Urra C, Engelmann F, Ortiz R, & Delafe C (1999). Cryopreservation of encapsulated sugarcane apices - effect of storage-temperature and storage duration. *CryoLetters* 20, pp. 347-352.

[15] Gonzales-Benito ME & Perez C (1994). Cryopreservation of embryonic axes of two cultivars of hazelnut (*Corylus avellana* L.). *Cryoletters*, 15, pp. 41-46

[16] Harding K & Benson EE (1994). A study of growth, flowering, and tuberisation in plants derived from cryopreserved potato shoot-tips: implications for *in vitro* germplasm collections. *Cryo Letters* 15, 59-66.

[17] Hatanaka T, Yasuda T, Yamaguchi T & Sakai A (1994). Direct regrowth of encapsulated somatic embryos of coffee (*Coffee canephora*) after cooling in liquid nitrogen. *Cryoletters* 15, pp. 47-52.

[18] Jitsuyama Y, Suzuki T, Harada T & Fujikawa S (2002). Sucrose incubation increases freezing tolerance of Asparagus (*Asparagus officianalis* L) embryonic cell suspensions. *Cryoletters* 23, pp. 103-112.

[19] Kay DE (1973) TPI Crop and Product Digest No. 2 Root CROPS. London: Tropical Products Inst.

[20] Kioko J, Berjak P, Pammenter NW, Watt MP & Wesley-Smith J, (1998). Desiccation and cryopreservation of embryonic axes of *Trichilia dregeana* SOND. *Cryo-Letters* 19: pp. 15-26

[21] Kioko, J, Berjak P, Pritchard H & Daws M (2000). Seeds of African pepper bark (Wurdurgia *salutaris*) can be cryopreserved after rapid dehydration in silica gel. In F. Engelmann and H. Takagi (eds.) *Cryopreservation of Tropical Germplasm, Current Research Progress and Application.* IPGRI: 371-377

[22] Gnanaprasam, S and Vasil IK (1992). Cryopreservation of immature embryos, embryogenic callus and cell suspension cultures of gramineous species. *Plant Sciences* 83, pp. 205-215.

[23] González-Benito ME, Prieto RM, Herradón E & Martín C (2002) Cryopreservation Of *Quercus Suber* And *Quercus Ilex* Embryonic Axes: In Vitro Culture, Desiccation And Cooling Factors *CryoLetters* 23, 283-290

[24] Grospietsch M, Stadulkora E & Jiri Z (1999) Effect of osmotic stress on the dehydration tolerance and cryopreservation of *Solanum Tuberosum* shoot tips. *Cryoletters* 20, 339-346.

[25] Hatanaka T, Yasuda T, Yamaguchi T & Sakai A (1994). Direct regrowth of encapsulated somatic embryos of coffee (*Coffee canephora*) after cooling in liquid nitrogen. *Cryoletters* 15, pp. 47-52.

[26] Langis P, & Steponkus PL (1990). Cryopreservation of rye protoplast by vitrification. *Plant Physiology* 92, pp. 666-671.

[27] Matinez-Montero ME, Gonzalez-Arnao MT, Borroso-Nordelo C, Puentes-Diez C & Engelmann F (1998) Cryopreservation of sugarcane embryogenic callus using a simple freezing process. *Cryo-Letters* 17, 171-176.

[28] Matsumoto T, Sakai A & Nako Y (1998). A novel preculturing for enhancing the survival of *in vitro* – grown meristems of wasabi (*wasabi japonica*) cooled to -196oC by vitrification. *Cryoletters* 19, pp. 27-36.

[29] Murashige T & Skoog E (1962) A revised medium for rapid growth and bioassays with tobacco tissue cultures. *Physiologia Plantarum* 15, 473 – 497.

[30] Mycock D (1999). Addition of Calcium and Magnesium to a Glycerol and Sucrose cryoprotectant solution improves the quality of plant embryo recovery from cryostorage. Cryoletters 20, pp 77-82.

[31] Naidoo, S, (2006). *Investigations into the Post-harvest behavior and germplasm conservation of the seed of selected* Amaryllid *species*. MSc Thesis University of Kwa-Zulu Natal Durban, South Africa.

[32] Niino T, Hettiarachchi A, Takahashi J &. Samarajeewa PK, (2000). Cryopreservation of lateral buds of *in vitro* grown innala plants (*Solenostemon rotundifolius*) by vitrification. *Cryoletters* 21 pp. 349 – 356.

[33] Nishizawa, S, Sakai A, Amano Y & Matsumoto T (1993), Cryopreservation of Asparagus (*Asparagus officinalis* L.) embryogenic cells and subsequent plant regeneration by simple freezing method. *CryoLetters* 13, 379-388.

[34] Ng SYC & Ng NQ (1991) in *Tissue Culture for Conservation of Plant Genetic Resources,* (ed) JH Dodds, Chapman and Hall, London, pp 11-39.

[35] Ng NQ & Daniel IO (2000) in *Cryopreservation of Tropical Germplasm, Current Research Progress and Application,* (eds) F Engelmann and H Takagi, Japan International Research Centre for Agricultural Sciences, Tsukuba, Japan / International Plant Genetic Resources Institute, Rome, Italy, pp 136-139.

[36] Pammenter NW , Vertucci CW & Berjak P (1991). Homoiohydrous (recalcitrant) seeds: dehydration, the state of water and viability characteristics in *Landolphia kirkii*. *Plant Physioplogy* 96, 1093-1098.

[37] Panis B, Strosse H, Van Den Hende S & Swennen R (2002) Sucrose preculture to simplify cryopreservation of banana meristem cultures. *CryoLetters* 23, 375-384 (2002)

[38] Perán R, Berjak P, Pammenter NW & Kioko JI (2006). Cryopreservation, encapsulation and promotion of shoot production of embryonic axes of a recalcitrant species *Ekerbergia capensis*, Sparrm. *Cryoletters* 27 (1), pp. 5-16.

[39] Plessis P, Leddet C, Collas A & Dereuddre J. (1993). Cryopreservation of *Vitis Vinefera* L. Cv Chardonnay shoots tips by encapsulation-dehydration: effects of pretreatment, cooling and postculture conditions. *Cryoletters* 14, pp. 309-320.

[40] Potts SE & Lumpkin TA (1997). Cryopreservation of Wasabi species seeds. Cryoletters. 18, pp.185-190.

[41] Quain, MD, Berjak, P, Acheampong, E, and Kioko, JI, (2009) Sucrose Treatment And Explant Water Content: Critical Factors to Consider in Development of Successful Cryopreservation Protocols for Shoot Tip Explants of the Tropical Species *Dioscorea rotundata* (Yam). *CryoLetters*, 30 (3), 212-223

[42] Reinhoud PJ, Schrijnemakers WM, van Iren F & Kijne JW (1995) vitrification and heat shock treatment improve cryopreservation of tobacco cell suspensions compared to two-step freezing. *Plant Cell Tissue and Organ Culture* 42: pp. 261-267.

[43] Sakai A, Kobayashi S & Oiyama I (1990) Cryopreservation of nuceller cells of navel orange (*Citrus sinensis* Obs, var. *brasiliensis* Tanaka) by vitrification. *Plant Cell Report* 9, 30-33.

[44] Santos IRI & Stushnoff C (2003) Desiccation and freezing tolerance of embryonic axes from *Citrus sinensis* (L) OSB. pretreated with sucrose. *CryoLetters* 24, 281-292.

[45] Schippers, R.R. (2000). African Indigenous Vegetables. An Overview of the Cultivated Species. Chatham. UK: NRI/ACP-EU Technical

[46] Shibli RA, Moges AD, and Karam NS (2004). Cryopreservation Of African Violet (*Saintpaulia Ionantha* Wendl.) Shoot Tips. In vitro *Cellular and Develoment Biology - plant* 40, 4, pp. 389-395(7)

[47] Spurr A R (1969) A low-viscosity epoxy resin embedding medium for electromicroscopy. *Journal of Ultrastructure Research* 26, 31-43

[48] Tetteh JP & Guo JI (1993) *Problems of Frafra Potato Production in Ghana*. A dissertation. School Centre for Agricultural and Rural Cooperation of Agriculture, University of Cape Coast.

[49] Thammasiri K. (1999). Cryopreservation of embryonic axes of jackfruit. *Cryo-letters* 20, pp. 21-28.

[50] Turner SR, Touchell DH, Senarata T, Bunn E, Tan B & Dixon KW (2001) Effect of Plant Growth Regulators on Survival and Recovery Growth Following Cryopreservation. *Cryoletters* 22, 163-174.

[51] Veisseire P, Guerrier J & Coudret A. (1993). Cryopreservation of embryonic suspension of *Hevea brasiliensis*. *CryoLetters* 14, 295-302.

[52] Wang J-H, Bian R-W, Zhang Y-W & Cheng H-P (2001) The dual effect of antifreeze protein on cryopreservation of rice (*Oryza sativa* L.) embryogenic suspension cells. *CryoLetters* 22, 175-182.

[53] Wang Q, Li P, Batuman Ö, Gafny R & Mawassi M., (2003). Effect of benzyladenine on recovery of cryopreserved shoot tips of grapevine and citrus *in vitro*. *CryoLetters* 24, 293-302.

[54] Walter C, DH. Touchell, P Power, J Wesley-Smith and MF. Antolin (2002), A Cryopreservation protocol for embryos of the endangered species *Zizania texana*. *CryoLetters* 23, 291-298.

[55] Wesley-Smith J, Vertucci CW, Berjak P, Pammenter NW & Crane J (1992) Cryopreservation of desiccation-sensitive axes of *Camellia sinensis* in relation to dehydration, freezing rate and the thermal properties of tissue water. *Journal of Plant Physiology* 140, 596-604.

[56] Withers, LA (1985). Cryopreservation of cultured cells and protoplasts. In: K.K. Kartha (ed) *Cryopreservation of plant cells and organs*. CRC press Inc., Boca Raton. pp 243-267.

Part 4

Equipment and Assays

8

Technologies for Cryopreservation: Overview and Innovation

Edoardo Lopez, Katiuscia Cipri and Vincenzo Naso
"Sapienza" University of Rome, Rome
Italy

1. Introduction

The proposed chapter investigates methods, devices and technologies for cryopreservation, explaining the most used cooling processes, as well as conventional and innovative technologies adopted. Main processes used for cryopreservation of oocytes, embryos and sperms can be reassumed in three categories:

1. slow freezing
2. vitrification
3. ultra-rapid freezing

Research is not intended to be exhaustive, but is aimed at covering most of relevant topics.

Slow freezing involves step-wise programmed decrease in temperature. The procedure is lengthy and requires the use of expensive instrumentation. The process does not exclude ice crystal formation, which can have extremely deleterious effects (Pegg, 2005).

In the vitrification process, the use of CryoProtectant Agents and the increasing of cooling rate (from 2,500 °/min to 130,000 °/min) avoid the ice crystal formation, increasing the embryos and oocytes survival. Unfortunately, common cryoprotectants are toxic and the immersion of solution directly in liquid nitrogen can be cause of contamination of embryos and oocytes with bacterium, mushroom and virus.

Ultra-rapid freezing can be considered a midway technique between slow freezing and vitrification, but its application has demonstrated lower performances than the other two processes.

2. Slow freezing

Necessary condition for slow freezing is freezing cells with a cooling rate equal or lower than 1°/min, before storaging them at -130° or lower (De Santis & Coticchio, 2011). If cell is cooled down very slowly, it will be exposed to growing concentrations of cellular solutes due to ice formation inside the solution, with a PH variation and cellular dehydration. If it is cooled down too fast, crystal nucleuses will form in the solution and inside the cell, with the destruction of cell membrane. Usually at temperature below -60°, the samples can be immersed directly in liquid nitrogen or transferred to freezer of maintenance without further loss of viability. Slow freezing generally lasts one or two hours. However, a greater amount of cells can be frozen at a time (Ha et al., 2005), and lower quantity of CPA are used than in vitrification.

2.1 Programmable freezers

Currently, programmable freezers are the most common technology for slow freezing process. Programmable freezers are based on liquid nitrogen technology, but their use is denied in areas without availability of nitrogen or during long transport. Cooling rate is controlled by a heater (Asymptote EF600, Cryologic CL8800) or by the synchronous use of two valves.

Main characteristics of the most common programmable freezers are shown in Tab. 1. [1]

	Kryo 360	Kryo 560M	Cryo-Logic 8800 + Fast CryoChamber	Thermo Scientific Forma 94741
Producer	Planer plc	Planer plc	CryoLogic	Fisher Scientific
Control range [°]	+40 to -180	+30 to -180	+40 to -120	+50 to -180
Cooling rates [°/min]	-0.01 to -50	-0.01 to -50	-0.04 to - 10 (at -40ąC)	-0.1 to -50
Heating rates [°/min]	0.01 to 10	0.01 to 10	—	0.1 to 10
Capacity [l]	1.7 or 3.3	16	11.5	17 or 48

Table 1. Programmable freezers main characteristics

2.2 Stirling engine cryocooler

The **Asymptote EF600** is the first commercially available programmable freezer which does not require liquid-nitrogen. The absence of liquid-nitrogen reduces drastically risk of contamination, and allows to freeze cells where nitrogen is not available (i.e. during transport or in other borderline applications).

Fig. 1. Asymptote EF600 (http://www.asymptote.co.uk/)

---

[1] Research is not intended to be exhaustive

The Asympotote EF600's cooling source is a Stirling Engine, a closed cycle machine in which the refrigerant working fluid is contained inside the machine, and only a source of mechanical or electrical energy is required[2] in order to rich temperature below - 100°.

Studies on human spermatozoa (linear cooling at -2°/min until nucleation followed by linear cooling at -10°/min to -100°), embryonic stem cells (linear cooling at -2°/min until nucleation followed by linear cooling at -1°/min to -45°) mouse embryos (linear cooling at -2°/min until nucleation followed by linear cooling at -0.3°/min to -35° and at -10°/min to -100°) and horse semen (linear cooling at -2°/min until nucleation followed by linear cooling at -4°/min to -80°) were carried on, considering survival rate as a parameter for assessing the performances of the proposed system (Faszer et al., 2006; Morris et al., 2006).

Results show that Stirling Engine cryocooler can established the desired time-temperature profile inside the test tubes and the viability after thawing data confirm that the system can be used for slow freezing applications.

However, Stirling Engines are affected by vibrations, as stated by (Hughes et al., 2000) and (Suárez et al., 2003). Vibrations might damage cells; furthermore, manual nucleation cannot be performed at a desired temperature, since vibrations generally start the nucleation process (as reported by (De Santis et al., 2007; Edgar, 2009; Rosendahl et al., 2011)).

2.3 Pulse tube cryocooler

In order to overcome to problems connected with vibrations of Stirling Engines, a programmable freezer based on a Pulse Tube cryocooler is being developed in "Sapienza" University of Rome Laboratory of Mechanical Engineering, in collaboration with *MES - Microconsulting Energia & Software S.c.a.r.l.*[3] and *LABOR S.r.l.*[4]. Alike the Stirling Engine, the Pulse Tube machine is a closed cycle system and it does not require liquid-nitrogen. The Pulse Tube cryocooler is able to rich temperatures below -150° making the refrigerator fluid (that is generally helium or nitrogen) move oscillatory. The fluid motion is obtained using a compressor and a rotative valve. The Pulse Tube offers low vibrations, as discussed by (Ikushima et al., 2008; Riabzev et al., 2009; Suzuki et al., 2006; Wang & Hartnett, 2010).

Next to the *cold head* (the cooling part of the Pulse Tube), the refrigerator fluid absorbs heat from the test tube, cooling it. The Pulse Tube cryocooler is characterized by a higher cooling rate than the ideal one for cell freezing (0.1°/min ÷ 10°/min) in the temperature range used for cryopreservation (+30° ÷ -60°). The cooling rate is reduced in the proposed solution through a control system that can supply heat to the cryorefrigerator.

A heater is placed by the test tube older (Fig. 2). The power dissipated through the heater for Joule effect varies according to two different control systems proposed:

1. *On-Off* **regulation**. A threshold control system has been implemented: the heater is activated when the real temperature is more than 1° below the desired temperature, and it is turned off when the real temperature is more than 1° over the desired temperature. Using this control system, oscillations of ±6° around the desired temperature were obtained, as it is illustrated in Fig. 3 and Fig. 4.

---

[2] The Asymptote EF600 can be connected to a conventional 240V electricity supply or to a car battery
[3] Via A. Panzini, 3 - 00137 Roma, Italy
[4] Tecnopolo Tiburtino, Via G. Peroni 386 - 00131 Roma, Italy

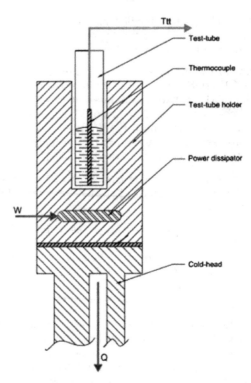

Fig. 2. Representation of the experimental apparatus. T_{tt} represents the temperature inside the test tube, measured by a thermocouple, Q is the heat absorbed by the refrigerating fluid and W is the dissipated power.

However, the oscillation might be reduced optimizing the threshold parameters. The *On-Off* regulation can be easily implemented, and it does not require the regulation of the power dissipated through Joule effect.

2. **Predictive model regulation**: the cooling slow-down is achived by providing an amount of heat, variable with the time, that will be able to raise the temperature of the PT cold head to the desired value (Cipri et al., 2010). The amount of heat is calculated using a predictive and adaptive model. Using this regulation modality, oscillation can be removed. However, it requires the regulation of the power dissipated through Joule effects, increasing the cost of the hardware. Moreover, more computational power is required in order to calculate the amount of heat which has to be dissipated.[5]

Results are shown in Fig. 5 and Fig. 6.

In the determination of the *Predictive model* a lot of simplifying assumptions were made (Cipri et al., 2010), and we believed that the system should have better results if the model was set in more accurate way. Further researches are fostering investigation at Sapienza Laboratory.

At this very moment, the system is not yet commercially viable.

---

[5] An *On-Off* regulation is still used before the transition phase, marked by the abrupt rise of temperature typical of the subcooling

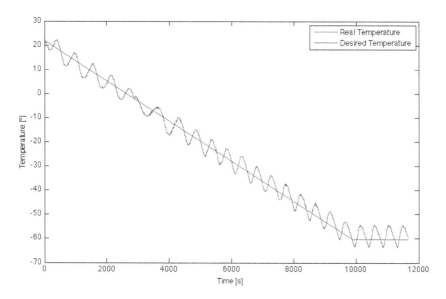

Fig. 3. *On-Off* regulation - Temperature inside the test tube vs time. A desired cooling rate of -0.5°/min was selected.

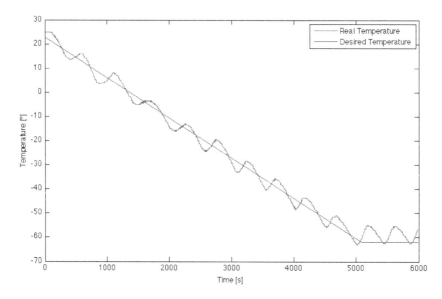

Fig. 4. *On-Off* regulation - Temperature inside the test tube vs time. A desired cooling rate of -1°/min was selected.

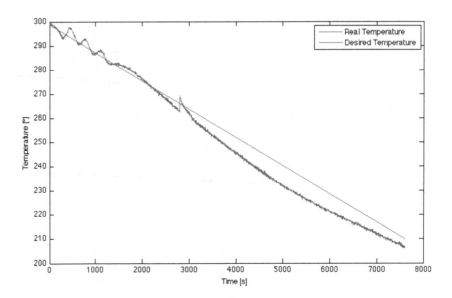

Fig. 5. *Predictive model* regulation - Temperature inside the test tube vs time. A desired cooling rate of -0.7°/min was selected.

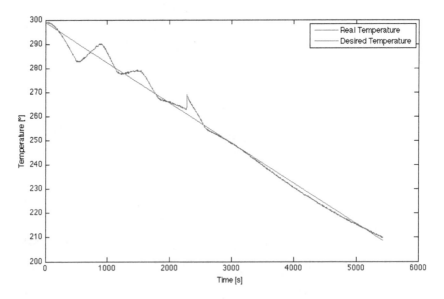

Fig. 6. *Predictive model* regulation - Temperature inside the test tube vs time. A desired cooling rate of -1°/min was selected.

3. Vitrification

A criticality of common cryopreservation methods consists in the formation of ice crystals that drastically reduces the survival of treated embryos and oocytes. Vitrification process produces a glasslike solidification of living cells which completely avoids ice crystal formation. The process is based on the principle that water, characterized by high cellular viscosity increased by the adding of CryoProtectant Agent (CPA), and frozen using a high cooling rate, is not capable of forming ice. The main limits of Vitrification process are represented by: use of potentially toxic cryoprotectant; risk of contamination of embryos and oocytes with bacterium, mushroom and virus when directly immersed in liquid nitrogen or during the storage phase. Studies have demonstrated that reduced quantity of CPA can be used if the cooling rate is increased.

A freezing rate of 2,500°/min and CPA concentration of 5-7 M is reached with the immersion of embryos and oocytes in micro-capillary straws, while in the pulled straws the cooling rate is about 20,000 °/min (Kuleshova & Lopata, 2002). Theoretically, the reaching of a cooling rate of 10^{7}°/sec should allow to vitrify also in pure water, but this rating is not practicable at the moment. Several studies are also oriented to formulate nontoxic and more efficient Vitrification solutions, also combining different cryoprotectants such as sugars and polymers or establishing modern solutions that include non-penetrating additives.

Moreover, the implementation of *Minimum Volume methods* has allowed to reduce the concentration of cryoprotectant. EG (ethylene glycol), characterized by low toxicity, is an important component of vitrification solution, commonly combined with DMSO or PROH (propanediol). In particular, non-permeable cryoprotectans (such sucrose or PVP) can be added in the solution on order to reduce the concentration of permeable cryoprotectans and facilitate dehydration and vitrification. Researches oriented to improve the characteristics of cryoprotectans have been carrying on in order to reduce toxicity. An EG and sucrose (non-permeable cryoprotectans) solution has been tested for cryopreservation of all preimplantation stages of *in vivo* generated mouse and day-6 sheep embryos. Experiments have not shown a loss of viability in vitro or in vivo. The same solution has been proved for vitrification of human oocytes, attaining high surviveal rates using conventional straws.

Another solution used to reduce toxicity is to equilibrate the cryoprotectant using a two-step method: the pretreatment solution, named *equilibration solution*, contains 20-50% concentrations of permeating cryoprotectants. The lower concentration of permeating cryoprotectans in the equilibration solution is much less toxic than the vitrification solution. The permeating cryoprotectant enters into the cells and facilitytes the intracellular vitrification. The cells pretreatment with equilibration solution is used in oocytes vitrification: this method has been demonstrated to increase the survival rate after thawing.

Main devices, commonly use in vitrification, are *Open Supports*: *Pulled Straws, CryoLoop, CryoEM, Cryoleaf* and *CryoTop*. The risk of contamination, due to the use of *Open Supports* for vitrification, limits the use of this process for human cells and tissues, according to the European regulations. In order to reduce contamination risks, *Close Supports* have been introduced: unfortunately their use decreases the cooling rate with consequently need to improve the quantity of CPA for guaranteeing the same survival rate. Vitrification process has demonstrated high performance in term of survival after thawing, comparable to slow cooling and it has become a promising alternative in cryopreservation of mammalian embryos and especially oocytes, through application of slow-rate freezing process.

3.1 Open supports

3.1.1 Open Pulled Straw (OPS)

Open Pulled Straw (OPS) have been designed to guarantee a ultra rapid freezing without ice crystals formation. The system, ideated by G. Vajta in 1998, is based on the hypothesis that decreasing the standard straw diameter, the volume of solution to vitrify is reduced too, raising the cooling rate. This method is so characterized by a very high cooling and warming rates (over 20,000°/min) and a short contact with concentrated cryoprotective additives (less than 30 sec over -180°). This approach reduces the possibility of chilling injury and toxic and osmotic damage. Several OPS have been developed reducing the diameter of standard straws of a half, increasing the cooling rate by 10 times and reducing by 30% the concentration of CPA and the time of exposition. Common OPS are standard 0.25 mL straws with one extremity pulled and thinned by heating. This solution increases the superficies/volume rate and hastens the cooling rate of the 2 μL drop set to contain the embryo. The Open Pulled Straw produced by *MTG* are made of PVC: with a length of 93 mm, straws can have an inner tip diameter approximately of 0.65 mm for Standard OPS and of 0.3 mm for super fine OPS. Before plunging the thin straw into liquid nitrogen, embryos are treated with highly concentrated cryoprotectant (CPA) solutions of ethylene glycol (EG) and dimethyl-sulfoxide (DMSO), in variable percentage.

3.1.2 Cryoloop

Cryoloop is generally applied to investigate the contribution given by cortical areas to network interactions and cerebral functions.

The Cryoloop is manufactured from straight 23 gauge hypodermic stainless steel tubing, having external and internal diameters respectively of 0.635 mm and 0.33 mm. Methanol, drawn from a external reservoir, is pumped in a Teflon tube directly in the Cryoloop that is in contact with the brain. Before reaching the Cryoloop, tubes containing Methanol are coiled and immersed in a bath of methanol and dry-ice pellets. The mixture cools the flowing methanol at a temperature of -75°. A microthemocouple, connected to a digital thermometer allows to monitor the temperature of the Cryoloop.

The use of Cryoloop device in human oocytes vitrification is under investigation. Experiments are now focused on animal oocytes and blastocytes cryopreservation. Cryoloops used for vitrification consist of a nylon loop of 10 or 20 micron diameters mounted on a stainless steel pipe inserted into the lid of a cryovial (Fig. 7). One of the main producers is the Hampton Research Corporation.

For vitrification, blastocytes are placed on a cryoloop that has been coated with a thin film of cryoprotectant solution. Blastocytes on the cryoloop are placed into the cryovial, which is submerged and filled with liquid nitrogen and the vial is sealed. Studies demonstrate that both mouse and human blastocystes can be successfully vitrified by suspension on a nylon loop and immersing directly into nitrogen. Mouse oocytes cryopreservation has provided successful results, but this method has not been applied to human oocytes. Tests on rabbit oocytes showed a good survival rate approximately of 80% for four different protocols.

3.1.3 Cryo-electron microscopic (CryoEM)

The Cryo-electron microscopic technique involves freezing biological samples in order to view the samples with the lowest distortion and the fewest possible artifacts.

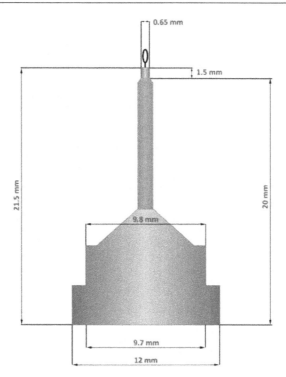

Fig. 7. CryoLoop Properly Installed in a CrystalCap Copper Magnetic - *Hampton Research Corporation*

Biological material is spread on an electron microscopy grid and is preserved in a frozen-hydrated state by rapid freezing (about $3,000°/\text{min}$), usually in ethane slush close to liquid nitrogen temperature. Specimens, maintained at liquid nitrogen temperature or colder, are contained into the high-vacuum of the electron microscope column. The frozen sample grid is then kept at liquid nitrogen temperature in the electron microscope and digital micrographs are collected with a camera. Images obtained from the cryo-electron microscopy are usually very noisy and have very low contrast. It is necessary to smooth the noise as well as enhance the contrast.

3.1.4 Cryoleaf

Cryoleaf is an open device for embryos and oocytes vitrification and storage (Fig. 8). Developed by Dr. Chian and Prof. Tan at McGill University, Montreal, the system uses PROH, EG and sucrose as cryo-protectants in the cooling phase, while in the warming procedure media contains sucrose. The recommended maximum load of the McGill Cryoleaf is 2-3 oocytes or embryos.

Oocytes or embryos are prepared for vitrification according to laboratory protocols. The outer cover of the McGill Cryoleaf is plunged into the liquid nitrogen bath, allowing the air to come out. Vitrified oocytes or embryos are quickly loaded into the McGill Cryoleaf using a suitable pipette. The excess of media, that must be less then 1 μl, has to be removed. The McGill

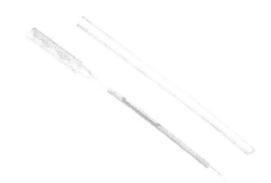

Fig. 8. McGill Cryoleaf Open System for Vitrification

Cryoleaf is inserted, with oocytes or embryos, directly into liquid nitrogen. Then, the Cryoleaf is blocked, sliding the protective sleeve over the tip.

3.1.5 Cryotop

Cryotop method, developed by Kuwayama in the Advanced Medical Research Institute of Kato Ladies Clinic, is based on the assumption that minimizing the volume[6] of the vitrification solution, increases both cooling and warming rates, also decreasing the chance of ice crystal nucleation/formation. Moreover the high-rate cooling decreases CPA concentration, also reducing chilling injury occurring between +15° (in human GV oocytes even +25°) and -5°, which can be minimized by passing embryos or oocytes rapidly through this temperature zone. Finally, studies have demonstrated that the use of small devices eliminates embryo fracture damages, especially in open systems.

The Cryotop tool consists of a narrow, thin film strip (0.4 mm wide, 20 mm long 0.1 mm thick) attached to a hard plastic handle for a minimum volume cooling. To protect oocytes and embryo on strip from mechanical damage and virus contamination during storage, a 3 cm long plastic tube cap is attached to cover the film part (Fig. 9).

The tool and the solutions for Vitrification and warming are market by Kitazato Co., Fujinomiya, Japan.

After a two-step equilibration in a vitrification solution containing EG, DMSO and sucrose, oocytes and embryo are loaded with a narrow glass capillary onto the top of the film strip in a volume of <0.1 ml. After loading, almost all the solution is removed so as to leave only a thin layer covering the oocytes or embryos, and the sample is quickly immersed into liquid nitrogen. Subsequently, the plastic cap is pulled over the film part of the Cryotop, and the sample is stored under liquid nitrogen (Kuwayama, 2007).

The minimal volume increases the cooling and warming rates up to 40,000°/min, contributing positively to the embryos or oocytes survival.

Cryotop vitrification method is applied successfully in various areas of animal technology and now it is indicated as the process which guarantees the highest number of babies born

---

[6] According to common use, for Minimum Volume is intended less than 1 ml for direct dropping of samples into liquid nitrogen or the open pulled straw (OPS) method.

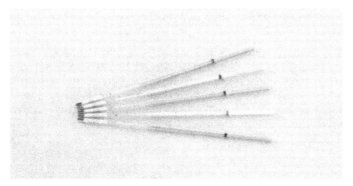

Fig. 9. Kitazato Cryotop - Kitazato Industries

after vitrification of human embryos and after cryopreservation of human oocytes worldwide. Clinical results are shown in Tab. 2

Reference	Cell	n	% Survival	% Pregnancy
Teramoto 2004	Blastocysts	197	100	57.7
Kuwayama 2005	Oocytes	64	91	41.3
Kuwayama 2005	PN Embryos D14	5881	100	-
Kuwayama 2005	Embryos D3	897	98	27
Kuwayama 2005	Blastocysts	6328	90	53
Lucena 2006	Oocytes	159	97	56.5
Antinori 2007	Oocytes	330	91	32.5
Cobo 2008	Oocytes	243	97	65.2
Cobo 2008	Oocytes	797	96	63.2

Table 2. Results achieved with Cryotop vitrification in human

3.1.6 Direct Cover Vitrification - DCV

The Direct Cover Vitrification - DCV is a new cooling method base on the minimum use of concentrated cryoprotectans and direct application of liquid nitrogen to the ovarian tissue. This way, the toxicity derived by cryoprotectants is reduced and the ice crystal injury is prevented. The ovary is immersed in a vitrification solution (0.8 ml) consisting of 15% EG, 15% DMSO and 0.5 M sucrose for 2 min.

The ovary is put in a 1.8-ml plastic standard cryovial, placed on a piece of gauze to remove the surrounding vitrification medium. Liquid nitrogen is directly applied onto the ovary for vitrification. The cap of the cryovial is closed. The lid does not have a hole. The vial is then placed into a liquid nitrogen tank.

DCV cryopreservation method, explored on mousse ovarian, has demonstrated to be highly efficient at increasing morphologically normal and viable follicles from cryopreserved ovarian tissue, compared with slow freezing and conventional vitrification.

3.1.7 Solid Surface Vitrification - SSV

The Solid Surface Vitrification - SSV has been developed at the Department of Animal Science, University of Connecticut. The method aims at defining an effective protocol to cryopreserve bovine oocytes for research and practice of parthenogenetic activation, in vitro fertilization and nuclear transfer.

Bovine oocytes matured in vitro are transferred to a vitrification solution (35% EG, 5% polyvinyl-pyrrolidone, 0.4 M trehalose inTCM 199 and 20% FBS). A metal cube covered with aluminum foil is partially submersed into liquid nitrogen (Fig. 10): the surface reaches the temperature of -150°. Microdrops of vitrification solution, containing the oocytes, are dropped onto the cold upper surface of the metal cube and are instantaneously vitrified. The vitrified microdrops are then stored in liquid nitrogen (Dinnyés et al., 2000).

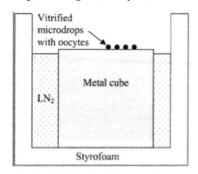

Fig. 10. The solid surface vitrification (SSV) device

3.2 Closed supports

3.2.1 Cryotip

CryoTip consists of a plastic straw with a thin part (250 μm inner diameter, 20 μm wall thickness and 3 cm length) connected to a thick part (2000 μm inner diameter and 150 μm wall thickness, 4.5 cm length) and equipped with a movable protective metal sleeve (Fig. 11) (Kuwayama, Vajta, Ieda & Kato, 2005).

Fig. 11. The CryoTip is a finely pulled straw designed for holding gametes or embryos

Embryos are loaded in approximately 1 μl solution into the narrow part of the CryoTips without any air bubbles by aspiration of medium. Subsequently, the straw is heat-sealed at both ends, the protective sleeve is pulled over the narrow part and the device is plunged into liquid nitrogen. The time required for loading, sealing, adjustment of the sleeve and plunging does not exceed 90 s. The use of the closed CryoTip system eliminates potential embryo's contamination during cryopreservation and storage without compromising survival and developmental rates in vitro and in vivo (Kuwayama, Vajta, Ieda & Kato, 2005).

3.2.2 Isachenko Method

In the *Isachenko Method* (Isachenko et al., 2005), embryos are located inside a open-pulled straws (OPS). The OPS is placed inside a sterile insemination straw (indicative size 90-mm), manufactured from standard 0.5-mL insemination straws. One end of sterile insemination straw is previously sealed using a hand-held sealer. The open end is hermetically closed by a metal ball and this container (OPS and sterile insemination straw) is plunged into liquid nitrogen ("straw in straw" vitrification). The *Isachenko Method*, applied to biopsied mouse pronuclear embryos is resulted efficient as conventional vitrification, guaranteeing a complete isolation of embryos from liquid nitrogen and avoiding potential contamination by pathogenic microorganisms.

3.3 Innovative vitrification devices

A new solution to increase the cooling rate reducing the use of cryoprotectants consists in the physical reduction of liquid nitrogen temperature, as happens in the **Vit-Master**, a new device developed at IMT, Israel. In order to avoid the vaporization of N2, the temperature of liquid nitrogen is reduced until - 210° (boiling point of nitrogen), applying a negative pressure (Arav et al., 2002). The evaporative cooling causes the nitrogen to partially solidify, thus creating a nitrogen slush. Samples immersed in nitrogen slush cool more rapidly because they come into contact with liquid nitrogen sooner than those immersed in normal liquid nitrogen (Cai et al., 2005). The *Vit Master* vitrification machine can provide a very high cooling rate (up to 135,000°/min). The cooling rate is especially enhanced in the first stage of cooling (from 20 to -10°), when it is six, four or two times higher with 0.25-ml straws, open pulled straws (OPS) or electron-microscope (EM) grids, respectively. Between -10 and -150°, the cooling rate is only about doubled by use of the *Vit Master*, but that was found to be enough to reduce the chances of devitrification and recrystallization during warming.

Research about using a **Pulse Tube** for Vitrification is ongoing at "Sapienza" - University of Rome.

4. Ultra-rapid freezing

Ultra-rapid freezing can be considered a midway technique between slow freezing and Vitrification. It is quicker than the slow-freezing technique, does not involve the use of programmable machines and requires lower concentrations of cryoprotectant agents (CPA) than those used in vitrification.

Experimental results demonstrate that this technique has lower performances than slow freezing's and vitrification's ones (AbdelHafez et al., 2010).

5. Comparison between vitrification and slow freezing

Vitrification is an attractive freezing technique: supports required are cost effective and experimental data show an high survival rate after thawing. For example, a survival rate of 99% was quoted in (*KITAZATO BioPharma Co., Ltd. - http://www.kitazato-biopharma.com/*, n.d.) using a *Cryotop* support.

However, vitrification exposes cells to a high risk of contamination, since cells are generally plunged directly into liquid-nitrogen. Risk of contamination is reported in (Bielanski et al., 2000), where cells frozen using vitrification were exposed to the bovine immunodeficiency virus (BIV), that can be considered a model for retrovirus like the human immunodeficiency virus (HIV). Risk of Hepatitis B contamination is analyzed in (Tedder et al., 1995).

Moreover, vitrification requires a greater amount of CPA (CryoProtectant Agent) than Slow Freezing does, increasing the toxicity of the environment.

In order to reduce the risk of contamination, *closed supports* for vitrification were developed (*Cryotip* and *Isachenko Method*). However, a lower survival rate is obtained than using an *open support*.

A lower survival rate after thawing using Slow Freezing instead of Vitrification was claimed by many authors. (Fadini et al., 2009) reports a survival rate of human oocytes of 78.9% using Vitrification, while it is reduced to 57.9 % using Slow Freezing (*p-value* lower than 0.0001); similar results are shown in (Vutyavanich et al., 2010) (where survival rate of human spermatozoa is measured equal to 64.8% using Vitrification and equal to 50.4% using Slow Freezing, *p-value* equal to 0.0036). However, many authors believe that a better understanding of slow freezing principles will improve its performances (Bianchi et al., 2007; De Santis et al., 2007; Edgar, 2009; Fadini et al., 2009; Mcgrath, 2009).

Both Vitrification and Programmable Freezers (the most common machines use for Slow Freezing) require a supply of liquid-nitrogen, that is a limiting factor in many situations of inefficient or absent nitrogen distribution network, such as small industries, isolated places and during transport of cells. In order to overcome to this limitation, two alternative systems for Slow Freezing (*Asymptote EF600* and *Pulse Tube Cryocooler*) have been developed.

Stirling Engine[7] and Pulse Tube Cryocooler are closed-cycle machines, reducing risk of contamination and toxicity. A cells freezing system based upon closed-cycle machines is a viable commercial solution, especially for those markets where liquid nitrogen supply is difficult or excessively expensive, or during transport.

However, Stirling Engine exhibits high vibration, thus the nucleation process can not be inducted manually. Moreover, vibrations might damage cells. Those problems are avoided using a Pulse Tube cryocooler.

The application of a Pulse Tube Cryocooler for cells cryopreservation is under developing at "Sapienza" - University of Rome Laboratory of Mechanical Engineering. A validation of the proposed system with the assessment of cells survival rate after thawing is envisaged as next step. Future work will also focus on the development of a cost effective control system which allows the operator to set a desired cooling rate.

6. Acknowledgment

We would really like to thank *MES - Microconsulting Energia & Software S.c.a.r.l.* and *LABOR S.r.l.* for their contribution to the realization of this work.

7. References

AbdelHafez, F. F., Desai, N., Abou-Setta, A. M., Falcone, T. & Goldfarb, J. (2010). Slow freezing, vitrification and ultra-rapid freezing of human embryos: a systematic review and meta-analysis., *Reproductive biomedicine online* 20(2): 209–22.
URL: *http://www.ncbi.nlm.nih.gov/pubmed/20113959*

---

[7] The Stirling Engine is used in the Asymptote EF600

Angle, M. (2007). Survival and re-expansion of mouse blastocysts following vitrification in two fda-approved closed devices with and without assisted shrinkage, *Fertility and Sterility* 88(September): S90–S91.
URL: *http://www.fertstert.org/article/S0015-0282(07)01955-3/abstract*

Arav, A., Yavi, S., Zeron, Y., Natan, D., Dekel, I. & Gacitua, H. (2002). New trends in gamete's cryopreservation, *Molecular and Cellular Endocrinology* 187(1-2): 77–81.
URL: *http://linkinghub.elsevier.com/retrieve/pii/S0303720701007006*

Asymptote Ltd - *http://www.asymptote.co.uk/* (n.d.).

Bianchi, V., Coticchio, G., Distratis, V., Di Giusto, N., Flamigni, C. & Borini, a. (2007). Differential sucrose concentration during dehydration (0.2 mol/l) and rehydration (0.3 mol/l) increases the implantation rate of frozen human oocytes, *Reproductive BioMedicine Online* 14(1): 64–71.
URL: *http://linkinghub.elsevier.com/retrieve/pii/S1472648310607651*

Bielanski, A., Nadin-Davis, S., Sapp, T. & Lutze-Wallace, C. (2000). Viral contamination of embryos cryopreserved in liquid nitrogen., *Cryobiology* 40(2): 110–6.
URL: *http://www.ncbi.nlm.nih.gov/pubmed/10788310*

Blayney, M. (2005). *Cryopreservation of embryos and spermatozoa*, Taylor & Francis, chapter 17, p. 322.

Cai, X. Y., Chen, G. a., Lian, Y., Zheng, X. Y. & Peng, H. M. (2005). Cryoloop vitrification of rabbit oocytes., *Human reproduction (Oxford, England)* 20(7): 1969–74.
URL: *http://www.ncbi.nlm.nih.gov/pubmed/15932910*

Canemco and Marivac - *http://www.canemco.com/* (n.d.).

Chen, S.-U., Chien, C.-L., Wu, M.-Y., Chen, T.-H., Lai, S.-M., Lin, C.-W. & Yang, Y.-S. (2006). Novel direct cover vitrification for cryopreservation of ovarian tissues increases follicle viability and pregnancy capability in mice., *Human reproduction (Oxford, England)* 21(11): 2794–800.
URL: *http://www.ncbi.nlm.nih.gov/pubmed/16982660*

Chen, S. U., Lien, Y. R., Chen, H. F., Chao, K. H., Ho, H. N. & Yang, Y. S. (2000). Open pulled straws for vitrification of mature mouse oocytes preserve patterns of meiotic spindles and chromosomes better than conventional straws., *Human reproduction (Oxford, England)* 15(12): 2598–603.
URL: *http://www.ncbi.nlm.nih.gov/pubmed/11098033*

Chen, S. U., Lien, Y. R., Cheng, Y. Y., Chen, H. F., Ho, H. N. & Yang, Y. S. (2001). Vitrification of mouse oocytes using closed pulled straws (CPS) achieves a high survival and preserves good patterns of meiotic spindles, compared with conventional straws, open pulled straws (OPS) and grids., *Human reproduction (Oxford, England)* 16(11): 2350–6.
URL: *http://www.ncbi.nlm.nih.gov/pubmed/11679519*

Chen, S.-U. & Yang, Y.-S. (2009). Slow freezing or vitrification of oocytes: their effects on survival and meiotic spindles, and the time schedule for clinical practice., *Taiwanese journal of obstetrics & gynecology* 48(1): 15–22.
URL: *http://www.ncbi.nlm.nih.gov/pubmed/19346187*

Chian, R., Son, W., Huang, J., Cui, S., Buckett, W. & Tan, S. (2005). High survival rates and pregnancies of human oocytes following vitrification: preliminary report, *Fertility and Sterility* 84(September): S36–S36.
URL: *http://www.fertstert.org/article/S0015-0282(05)01544-X/abstract*

Cipri, K., Lopez, E. & Naso, V. (2010). Investigation of the use of Pulse Tube in cell cryopreservation systems, *Cryobiology* 61(2): 225–30.
URL: *http://www.ncbi.nlm.nih.gov/pubmed/20691677*

CryoLogic - *http://www.cryologic.com/* (n.d.).

De Santis, L., Cino, I., Rabellotti, E., Papaleo, E., Calzi, F., Fusi, F., Brigante, C. & Ferrari, a. (2007). Oocyte cryopreservation: clinical outcome of slow-cooling protocols differing in sucrose concentration, *Reproductive BioMedicine Online* 14(1): 57–63.
    URL: *http://linkinghub.elsevier.com/retrieve/pii/S147264831060764X*

De Santis, L. & Coticchio, G. (2011). Theoretical and experimental basis of slow freezing., *Reproductive BioMedicine Online* 22(2): 125–32.
    URL: *http://www.ncbi.nlm.nih.gov/pubmed/21237713*

Dinnyés, a., Dai, Y., Jiang, S. & Yang, X. (2000). High developmental rates of vitrified bovine oocytes following parthenogenetic activation, in vitro fertilization, and somatic cell nuclear transfer., *Biology of reproduction* 63(2): 513–8.
    URL: *http://www.ncbi.nlm.nih.gov/pubmed/10906058*

Edgar, D. (2009). Increasing dehydration of human cleavage-stage embryos prior to slow cooling significantly increases cryosurvival, *Reproductive BioMedicine Online* 19(4): 521–525.
    URL: *http://linkinghub.elsevier.com/retrieve/pii/S1472648309000169*

Fadini, R., Brambillasca, F., Renzini, M. M., Merola, M., Comi, R., De Ponti, E. & Dal Canto, M. (2009). Human oocyte cryopreservation: comparison between slow and ultrarapid methods, *Reproductive BioMedicine Online* 19(2): 171–180.
    URL: *http://linkinghub.elsevier.com/retrieve/pii/S1472648310600697*

Faszer, K., Draper, D., Green, J. E., Morris, G. J. & Grout, B. W. W. (2006). Cryopreservation of horse semen under laboratory and field conditions using a Stirling Cycle freezer., *Cryo letters* 27(3): 179–86.
    URL: *http://www.ncbi.nlm.nih.gov/pubmed/16892166*

Fisher Scientific Italia - http://www.it.fishersci.com/ (n.d.).

Grant Instruments (Cambridge) Ltd - http://www.grant.co.uk/ (n.d.).

Grossman, G., Bradley, P. E., Lewis, M. a. & Radebaugh, R. (2011). Model for transient behavior of pulse tube cryocooler, *Cryogenics* 51(3): 124–131.
    URL: *http://linkinghub.elsevier.com/retrieve/pii/S0011227510002560*

Ha, S. Y., Jee, B. C., Suh, C. S., Kim, H. S., Oh, S. K., Kim, S. H. & Moon, S. Y. (2005). Cryopreservation of human embryonic stem cells without the use of a programmable freezer., *Human reproduction* 20(7): 1779–85.
    URL: *http://www.ncbi.nlm.nih.gov/pubmed/15760949*

Hampton Research - http://hamptonresearch.com/ (n.d.).

Hu, J., Dai, W., Luo, E., Wang, X. & Huang, Y. (2010). Development of high efficiency Stirling-type pulse tube cryocoolers, *Cryogenics* 50(9): 603–607.
    URL: *http://linkinghub.elsevier.com/retrieve/pii/S0011227510000548*

Hughes, W., McNelis, M., Goodnight, T. & Center, N. G. R. (2000). Vibration Testing of an Operating Stirling Convertor, *Seventh International Congress on Sound and Vibration*, number November, National Aeronautics and Space Administration, Glenn Research Center.
    URL: *http://gltrs.grc.nasa.gov/reports/2000/TM-2000-210526.pdf*

Ikushima, Y., Li, R., Tomaru, T., Sato, N., Suzuki, T., Haruyama, T., Shintomi, T. & Yamamoto, A. (2008). Ultra-low-vibration pulse-tube cryocooler system – cooling capacity and vibration, *Cryogenics* 48(9-10): 406–412.
    URL: *http://linkinghub.elsevier.com/retrieve/pii/S0011227508000490*

Imura, J., Shinoki, S., Sato, T., Iwata, N., Yamamoto, H., Yasohama, K., Ohashi, Y., Nomachi, H., Okumura, N., Nagaya, S., Tamada, T. & Hirano, N. (2007). Development of high capacity Stirling type pulse tube cryocooler, *Physica C: Superconductivity* 463-465: 1369–1371.
    URL: *http://linkinghub.elsevier.com/retrieve/pii/S0921453407010404*

Irvine Scientific - http://www.irvinesci.com/ (n.d.).

Isachenko, V., Montag, M., Isachenko, E. & van der Ven, H. (2005). Vitrification of mouse pronuclear embryos after polar body biopsy without direct contact with liquid nitrogen., *Fertility and sterility* 84(4): 1011–6.
URL: *http://www.ncbi.nlm.nih.gov/pubmed/16213857*

KITAZATO BioPharma Co., Ltd. - http://www.kitazato-biopharma.com/ (n.d.).

Koh, D. Y., Hong, Y. J., Park, S. J., Kim, H. B. & Lee, K. S. (2002). A study on the linear compressor characteristics of the Stirling cryocooler, *Cryogenics* 42(6-7): 427–432.
URL: *http://linkinghub.elsevier.com/retrieve/pii/S0011227502000644*

Kongtragool, B. & Wongwises, S. (2003). A review of solar-powered Stirling engines and low temperature differential Stirling engines, *Renewable and Sustainable Energy Reviews* 7(2): 131–154.
URL: *http://linkinghub.elsevier.com/retrieve/pii/S1364032102000539*

Kongtragool, B. & Wongwises, S. (2007). Performance of low-temperature differential Stirling engines, *Renewable Energy* 32(4): 547–566.
URL: *http://linkinghub.elsevier.com/retrieve/pii/S0960148106000772*

Kuleshova, L. L. & Lopata, A. (2002). Vitrification can be more favorable than slow cooling., *Fertility and sterility* 78(3): 449–54.
URL: *http://www.ncbi.nlm.nih.gov/pubmed/12215314*

Kuleshova, L. L. & Shaw, J. M. (2000). A strategy for rapid cooling of mouse embryos within a double straw to eliminate the risk of contamination during storage in liquid nitrogen., *Human reproduction (Oxford, England)* 15(12): 2604–9.
URL: *http://www.ncbi.nlm.nih.gov/pubmed/11098034*

Kuwayama, M. (2007). Highly efficient vitrification for cryopreservation of human oocytes and embryos: the Cryotop method., *Theriogenology* 67(1): 73–80.
URL: *http://www.ncbi.nlm.nih.gov/pubmed/17055564*

Kuwayama, M., Vajta, G., Ieda, S. & Kato, O. (2005). Comparison of open and closed methods for vitrification of human embryos and the elimination of potential contamination, *Reproductive BioMedicine Online* 11(5): 608–614.
URL: *http://linkinghub.elsevier.com/retrieve/pii/S1472648310611698*

Kuwayama, M., Vajta, G., Kato, O. & Leibo, S. P. (2005). Highly efficient vitrification method for cryopreservation of human oocytes, *Reproductive BioMedicine Online* 11(3): 300–308.
URL: *http://linkinghub.elsevier.com/retrieve/pii/S1472648310608371*

Lane, M., Schoolcraft, W. B. & Gardner, D. K. (1999). Vitrification of mouse and human blastocysts using a novel cryoloop container-less technique., *Fertility and sterility* 72(6): 1073–8.
URL: *http://www.ncbi.nlm.nih.gov/pubmed/10593384*

Liebermann, J. (2002). Potential Importance of Vitrification in Reproductive Medicine, *Biology of Reproduction* 67(6): 1671–1680.
URL: *http://www.biolreprod.org/cgi/doi/10.1095/biolreprod.102.006833*

Liebermann, J. & Tucker, M. J. (2002). Effect of carrier system on the yield of human oocytes and embryos as assessed by survival and developmental potential after vitrification., *Reproduction (Cambridge, England)* 124(4): 483–9.
URL: *http://www.ncbi.nlm.nih.gov/pubmed/12361466*

Lomber, S. G., Payne, B. R. & Horel, J. a. (1999). The cryoloop: an adaptable reversible cooling deactivation method for behavioral or electrophysiological assessment of neural function., *Journal of neuroscience methods* 86(2): 179–94.
URL: *http://www.ncbi.nlm.nih.gov/pubmed/10065985*

Lucena, E., Bernal, D. P., Lucena, C., Rojas, A., Moran, A. & Lucena, A. (2006). Successful ongoing pregnancies after vitrification of oocytes., *Fertility and sterility* 85(1): 108–11. URL: *http://www.ncbi.nlm.nih.gov/pubmed/16412739*

Martino, a., Songsasen, N. & Leibo, S. P. (1996). Development into blastocysts of bovine oocytes cryopreserved by ultra-rapid cooling., *Biology of reproduction* 54(5): 1059–69. URL: *http://www.ncbi.nlm.nih.gov/pubmed/8722627*

Matson, P., Kappelle, W. & Webb, S. (2008). Maximum rates of cooling by three programmable freezers, and the potential relevance to sperm cryopreservation., *Reproductive biology* 8(1): 69–73. URL: *http://www.ncbi.nlm.nih.gov/pubmed/18432308*

Matsumoto, H., Jiang, J. Y., Tanaka, T., Sasada, H. & Sato, E. (2001). Vitrification of large quantities of immature bovine oocytes using nylon mesh., *Cryobiology* 42(2): 139–44. URL: *http://www.ncbi.nlm.nih.gov/pubmed/11448116*

Mcgrath, J. (2009). *Predictive Models for the Development of Improved Cryopreservation Protocols for Human Oocytes,* number 7, Informa Helatcare, pp. 62–82.

Morris, G., Acton, E., Faszer, K., Franklin, A., Yin, H., Bodine, R., Pareja, J., Zaninovic, N. & Gosden, R. (2006). Cryopreservation of murine embryos, human spermatozoa and embryonic stem cells using a liquid nitrogen-free, controlled rate freezer, *Reproductive BioMedicine Online* 13(3): 421–426. URL: *http://linkinghub.elsevier.com/retrieve/pii/S1472648310614484*

Mukaida, T. (2003). Vitrification of human blastocysts using cryoloops: clinical outcome of 223 cycles, *Human Reproduction* 18(2): 384–391. URL: *http://www.humrep.oupjournals.org/cgi/doi/10.1093/humrep/deg047*

Mukaida, T., Wada, S., Takahashi, K., Pedro, P. B., An, T. Z. & Kasai, M. (1998). Vitrification of human embryos based on the assessment of suitable conditions for 8-cell mouse embryos., *Human reproduction (Oxford, England)* 13(1O): 2874–9. URL: *http://www.ncbi.nlm.nih.gov/pubmed/9804248*

Nottola, S., Coticchio, G., Sciajno, R., Gambardella, a., Maione, M., Scaravelli, G., Bianchi, S., Macchiarelli, G. & Borini, a. (2009). Ultrastructural markers of quality in human mature oocytes vitrified using cryoleaf and cryoloop, *Reproductive BioMedicine Online* 19(October): 17–27. URL: *http://linkinghub.elsevier.com/retrieve/pii/S1472648310602805*

Nowshari, M. a. & Brem, G. (2001). Effect of freezing rate and exposure time to cryoprotectant on the development of mouse pronuclear stage embryos., *Human reproduction (Oxford, England)* 16(11): 2368–73. URL: *http://www.ncbi.nlm.nih.gov/pubmed/11679522*

Oberstein, N., O'Donovan, M. K., Bruemmer, J. E., Seidel, G. E. J., Camevale, E. M. & Squires, E. S. (2001). Cryopreservation of Equine Embryos by Open Pulled Straw, Cryoloop, or Conventional Slow Cooling Methods, *Theriogenology* 55: 607–613. URL: *http://www.ncbi.nlm.nih.gov/pubmed/21235366*

Orief, Y., Schultze-mosgau, A., Dafopoulos, K. & Al-hasani, S. (2005). Vitrification : will it replace the conventional gamete cryopreservation techniques ?, *Middle East Fertility Society Journal* 10(3): 171–184.

ORIGIO - http://www.origio.com/ (n.d.).

Papis, K., Shimizu, M. & Izaike, Y. (2000). Factors affecting the survivability of bovine oocytes vitrified in droplets, *Theriogenology* 54(5): 651–658. URL: *http://linkinghub.elsevier.com/retrieve/pii/S0093691X00003800*

Pegg, D. E. (2005). The role of vitrification techniques of cryopreservation in reproductive medicine, *Human Fertility* 8: 231–239.

Planer Controlled Rate Freezer - http://www.planer.co.uk/ (n.d.).

Popescu, G., Radcenco, V., Gargalian, E. & Ramany Bala, P. (2001). A critical review of pulse tube cryogenerator, *International Journal of Refrigeration* 24: 230–237.

Reubinoff, B. E., Pera, M. F., Vajta, G. & Trounson, a. O. (2001). Effective cryopreservation of human embryonic stem cells by the open pulled straw vitrification method., *Human reproduction (Oxford, England)* 16(10): 2187–94.
URL: *http://www.ncbi.nlm.nih.gov/pubmed/11574514*

Rezazadeh Valojerdi, M., Eftekhari-Yazdi, P., Karimian, L., Hassani, F. & Movaghar, B. (2009). Vitrification versus slow freezing gives excellent survival, post warming embryo morphology and pregnancy outcomes for human cleaved embryos., *Journal of assisted reproduction and genetics* 26(6): 347–54.
URL: *http://www.pubmedcentral.nih.gov/articlerender.fcgi?artid=2729856&tool=pmcentrez &rendertype=abstract*

Riabzev, S., Veprik, A., Vilenchik, H. & Pundak, N. (2009). Vibration generation in a pulse tube refrigerator, *Cryogenics* 49(1): 1–6.
URL: *http://linkinghub.elsevier.com/retrieve/pii/S001122750800115X*

Rosendahl, M., Schmidt, K. T., Ernst, E., Rasmussen, P. E., Loft, A., Byskov, A. G., Andersen, A. N. & Andersen, C. Y. (2011). Cryopreservation of ovarian tissue for a decade in Denmark: a view of the technique., *Reproductive biomedicine online* 22(2): 162–71.
URL: *http://www.ncbi.nlm.nih.gov/pubmed/21239230*

Roth, T. L., Bush, L. M., Wildt, D. E. & Weiss, R. B. (1999). Scimitar-horned oryx (Oryx dammah) spermatozoa are functionally competent in a heterologous bovine in vitro fertilization system after cryopreservation on dry ice, in a dry shipper, or over liquid nitrogen vapor., *Biology of reproduction* 60(2): 493–8.
URL: *http://www.ncbi.nlm.nih.gov/pubmed/9916019*

Scollo, L., Valdez, P. & Baron, J. (2008). Design and construction of a Stirling engine prototype, *International Journal of Hydrogen Energy* 33(13): 3506–3510.
URL: *http://linkinghub.elsevier.com/retrieve/pii/S0360319908000098*

Sheehan, C. B., Lane, M. & Gardner, D. K. (2006). The CryoLoop facilitates re-vitrification of embryos at four successive stages of development without impairing embryo growth., *Human reproduction (Oxford, England)* 21(11): 2978–84.
URL: *http://www.ncbi.nlm.nih.gov/pubmed/16950825*

Smith, G. D., Serafini, P. C., Fioravanti, J., Yadid, I., Coslovsky, M., Hassun, P., Alegretti, J. R. & Motta, E. L. (2010). Prospective randomized comparison of human oocyte cryopreservation with slow-rate freezing or vitrification., *Fertility and sterility* 94(6): 2088–95.
URL: *http://www.ncbi.nlm.nih.gov/pubmed/20171613*

Stanic, P., Tandara, M., Sonicki, Z., Simunic, V., Radakovic, B. & Suchanek, E. (2000). Comparison of protective media and freezing techniques for cryopreservation of human semen., *European journal of obstetrics, gynecology, and reproductive biology* 91(1): 65–70.
URL: *http://www.ncbi.nlm.nih.gov/pubmed/10817881*

Stănescu Pascal, M. & Birțoiu, A. I. (2010). Comparative Studies of Canine Semen Freezing Protocols, *Veterinary Medicine* 67(2): 209–215.

Suárez, V. J., Goodnight, T. W. & Hughes, W. O. (2003). Vibration Modal Characterization of a Stirling Convertor via Base-Shake Excitation, *the Proceedings of the International Energy Conversion Engineering Conference, Portsmouth, Virginia*, number November.
URL: *http://gltrs.grc.nasa.gov/reports/2003/TM-2003-212479.pdf*

Suzuki, T., Tomaru, T., Haruyama, T., Sato, N., Yamamoto, A., Shintomi, T., Ikushima, Y. & Li, R. (2006). Pulse tube cryocooler with self-cancellation of cold stage vibration, *Cryo*

Prague 2006.
URL: *http://arxiv.org/abs/physics/0611031*
Tedder, R., Zuckerman, M., Brink, N., Goldstone, A., Fielding, A., Blair, S., Patterson, K., Hawkins, A., Gormon, A., Heptonstall, J. & Others (1995). Hepatitis tank B transmission from contaminated cryopreservation, *The Lancet* 346(8968): 137–140.
URL: *http://linkinghub.elsevier.com/retrieve/pii/S014067369591207X*
Vajta, G. & Kuwayama, M. (2006). Improving cryopreservation systems., *Theriogenology* 65(1): 236–44.
URL: *http://www.ncbi.nlm.nih.gov/pubmed/16289262*
Valbuena, D., Sánchez-Luengo, S., Galán, A., Sánchez, E., Gómez, E., Poo, M. E., Ruiz, V., Genbacev, O., Krtolica, A. & Pellicer, A. (2008). Efficient method for slow cryopreservation of human embryonic stem cells in xeno-free conditions, *Reproductive BioMedicine Online* 17(1): 127–135.
URL: *http://linkinghub.elsevier.com/retrieve/pii/S1472648310603021*
van den Abbeel, E., van der Elst, J., van der Linden, M. & van Steirteghem, a. C. (1997). High survival rate of one-cell mouse embryos cooled rapidly to -196 degrees C after exposure to a propylene glycol-dimethylsulfoxide-sucrose solution., *Cryobiology* 34(1): 1–12.
URL: *http://www.ncbi.nlm.nih.gov/pubmed/9028912*
Varghese, A. C., Nagy, Z. P. & Agarwal, A. (2009). Current trends, biological foundations and future prospects of oocyte and embryo cryopreservation, *Reproductive BioMedicine Online* 19(1): 126–140.
URL: *http://linkinghub.elsevier.com/retrieve/pii/S1472648310600569*
Vutyavanich, T., Piromlertamorn, W. & Nunta, S. (2010). Rapid freezing versus slow programmable freezing of human spermatozoa., *Fertility and sterility* 93(6): 1921–8.
URL: *http://www.ncbi.nlm.nih.gov/pubmed/19243759*
Vutyavanich, T., Sreshthaputra, O., Piromlertamorn, W. & Nunta, S. (2009). Closed-system solid surface vitrification versus slow programmable freezing of mouse 2-cell embryos., *Journal of assisted reproduction and genetics* 26(5): 285–90.
URL: *http://www.pubmedcentral.nih.gov/articlerender.fcgi?artid=2719071&tool=pmcentrez &rendertype=abstract*
Wang, C. & Hartnett, J. G. (2010). A vibration free cryostat using pulse tube cryocooler, *Cryogenics* 50(5): 336–341.
URL: *http://linkinghub.elsevier.com/retrieve/pii/S0011227510000044*
Ware, C. B., Nelson, A. M. & Blau, C. A. (2005). Controlled-rate freezing of human ES cells., *BioTechniques* 38(6): 879–80, 882–3.
URL: *http://www.ncbi.nlm.nih.gov/pubmed/16018548*
Wilding, M. G., Capobianco, C., Montanaro, N., Kabili, G., Di Matteo, L., Fusco, E. & Dale, B. (2010). Human cleavage-stage embryo vitrification is comparable to slow-rate cryopreservation in cycles of assisted reproduction., *Journal of assisted reproduction and genetics* 27(9-10): 549–54.
URL: *http://www.pubmedcentral.nih.gov/articlerender.fcgi?artid=2965340&tool=pmcentrez &rendertype=abstract*
Yaqi, L., Yaling, H. & Weiwei, W. (2011). Optimization of solar-powered Stirling heat engine with finite-time thermodynamics, *Renewable Energy* 36(1): 421–427.
URL: *http://linkinghub.elsevier.com/retrieve/pii/S0960148110003101*

9

Precision in Cryopreservation – Equipment and Control

Stephen Butler and David Pegg
Planer plc, University of York
UK

1. Introduction

1.1 Different samples may have differing cryopreservation requirements

For any cryopreservation protocol there are five key questions that govern the methodology and logistics of the freezing and storing process.

- What is to be stored?
- How many batches are to be stored?
- What is the expected duration of storage?
- What properties are the retrieved samples required to possess?
- Are there packaging requirements in addition to those dictated by the cryopreservation process?

Reduction of temperature results in the retardation of metabolic processes and this can, in some circumstances, provide sufficient stability for the required period of storage. However, at temperatures below 0 °C the biological effects of cooling are dominated by the crystallization of ice: typically, water constitutes around 80 % of tissue mass. Freezing is the conversion of liquid water to crystalline ice but the term is commonly misused in circumstances where samples are cooled below their expected freezing point but without the formation of ice, for example by supercooling or by vitrification. The result of the freezing of water in a complex solution is that the concentration of the solutes in the remaining liquid phase increases and some solutes may precipitate if their concentration exceeds their solubility limits. This realisation provides two potential mechanisms of damage: direct mechanical effects of the formation of ice, and the rise in concentration of dissolved solutes.

In 1948 a method was discovered that permitted the freezing of many types of animal cells with good post-thaw recovery of living cells: Polge, Smith, and Parkes (1948) showed in a landmark paper that adding 10-20 % of glycerol enabled avian spermatozoa to survive freezing at -80 °C. Theories of freezing injury that were current at the time envisaged ice crystals damaging the cells and intracellular structures, and because glycerol increased the total solute concentration in the system, the amount of ice that formed was reduced. A little later, in the 1950s, Lovelock (1952) showed that the increase in concentration of salts as the volume of the suspending solution decreased was in fact the dominant damaging mechanism: salt concentration, rather than ice formation, was a major cause of freezing injury to cells. Subsequently other cryoprotectant solutes were explored along with different rates of cooling, resulting in solidification of the stored samples but with a range of mixtures of ice and vitrified solid in the stored samples.

The physical nature of the sample dictates the thermal transfer characteristics of the cooling process for that specific sample and either the physical size or cell-type will affect the appropriate cooling rate and other parameters of the cryopreservation protocol. Similarly, the physical type and ultimate intended use of the sample (for example dose requirement in the case of future therapeutic use) will determine the size of the individual packaging. An additional layer of packaging may be necessary to prevent microbiological contamination – so-called 'double bagging'. Likewise the ultimate destination of the sample will also dictate the care required during the freezing process and the conditions necessary for long-term storage. Some tissues and most larger biological samples are currently difficult or impossible to cryopreserve successfully and new techniques, such as Liquidus Tracking (discussed later in this chapter) may address some of the problems associated with cryopreservation of these types of sample.

It is sometimes the case that the ultimate use of the samples stored is not known at the time of the initial collection and storage and sometimes the significance of particular samples may change with time. However, in many cases, the potential of the stored samples is fixed or limited at the time of selecting the cryopreservation and storage methods. The importance of these choices will be covered later; however it is pertinent to note here that the storage process may have an important impact on the value of samples when they are recovered from storage; changes in the properties of the recovered samples may be irreversible and this is therefore a key to maximising the sample's potential.

The term "viability" is frequently used in the context of cell and tissue banking. Strictly speaking it means the potential to exhibit the signs of life at some future stage, whereas it is often misused to mean the extent to which a sample demonstrates attributes of life at the present time. But that is "vitality" not "viability." However, it is also the case that not all the attributes of life are exhibited by all living things, and the possession of one attribute does not imply the presence of them all. In fact, few of the properties that characterise "life" can be measured quantitatively. The term is best avoided; functional measurements should be named to describe what they actually measure: membrane integrity; a specified metabolic function; ability to reproduce. In addition there are obvious cases where the tissue does not have to be alive in order to function; in bone for example. But equally, in many cases fully functional survival is paramount; the haemopoietic stem cells in cord blood will not graft in the recipient if the cell concentration is lower than a threshold value. In such cases a low total recovery of living cells in the thawed sample will limit the use of the thawed sample. Another common situation is where samples are stored in order to ensure that a supply of identical cells will be available throughout a long-term study. Although it is possible to regrow new cell batches from recovered samples, repeating this process can lead to progressive degradation due to mutations.

1.2 The physics of freezing

The process of freezing is ultimately simple; it is merely the application of an environment that removes energy from the sample over a period of time and changes the physical state of water in the sample from liquid to crystalline. Crystalline water (ice) excludes the solutes previously dissolved in the water, resulting in two potentially dangerous mechanisms – direct effects of ice and secondary effects in the solute composition. At a sufficiently low temperature all biological activity is prevented and the physical state of the sample is preserved. In simple cases, where the only requirement is to preserve the physical state or where cellular structure is absent (viruses, DNA etc.), that is the end of the story; physical deterioration can be prevented at relatively high temperatures, and in many institutions worldwide this task is completed in the banks of -80 °C refrigeration units that proliferate in medical and biological research establishments.

The preservation of living cells and tissues and the post-thaw ability of cells to proliferate and thrive are determined by a number of factors: the laboratory techniques and the thermodynamic processes that a sample experiences during processing and freezing; the environment in which it resides between freezing and the ultimate use post-thaw. The potential of many samples is severely limited at this stage by the choices made by, or enforced upon, the technician regarding the freezing protocol. It may be that some stored samples lose significant value due simply to the omission of a few simple additional steps.

The cryopreservation process has two main aims. The first is to reduce the temperature of the sample to a point where biological stability is achieved. The application of an external cryogenic environment will remove energy from the sample and create a very low-energy solid state within which biological and chemical activity are limited or prevented altogether. The second is that during the freezing process it is necessary to prevent the formation of intracellular ice crystals: such crystals damage the cellular structure and can lead to limited post-thaw recovery and post-thaw failure of the cell sample to function as required. Additionally, the protocol must take into account the stresses to which the cells are exposed during the freezing process (dehydration, hypothermia, chemical toxicity, and solute concentration) and the potential for an apoptotic response post-thaw.

The objective, therefore, is to create an environment in which, as the sample is cooled, the chemical composition inside the cell, is managed in such a way as to create an intracellular composition with a lower freezing point than the applied environment, whilst maintaining an external suspending composition that is able to solidify at the same temperature. The balance between the internal and external environment is managed chemically via the solutes in the micro-environment and thermodynamically via the application of an energy reducing (cooling) macro-environment. It is the combined action of these two factors that determines the success or otherwise of a cryopreservation protocol for the conservation of vitality.

The appropriate solute composition is created by including cryoprotective agents (CPAs) in the medium. These operate in one of two ways: either they modify the extracellular composition or alternatively they also replace some of the intracellular water. The first mechanism involves the addition of non-penetrating CPAs such as trehalose, polyethylene glycol (PEG) or Polyvinyl-pyrrolidone (PVP), to the medium. The second mechanism requires the addition of penetrating solutes that can traverse the cell membrane, such as glycerol, ethylene glycol and dimethyl sulphoxide (DMSO). Since water does not retain solutes when it freezes, a solution at equilibrium with ice will vary in osmotic potential as it freezes and because of this, the micro-environment of a cell will require either the cell to lose water to the environment or exchange water for CPA molecules, thereby maintaining osmotic balance. The concentration of intracellular material lowers the effective freezing point of intracellular material and, provided the external temperature is correctly managed, prevents the formation of intracellular ice. As such, the creation of ice crystals within the cell is avoided. At temperatures below -130 °C (close to the glass transition temperature of the medium) the residual liquid has too little energy to orientate into long range molecular matrices and will form short range semi-solid structures; i.e. an amorphous solid or glass. At this point there is no possibility for significant chemical transport; biological activity, and hence deterioration, effectively ceases.

The options for control of this process are the chosen CPA and its concentration, and the cooling rate. Water and solute permeability are temperature dependent and nominally the higher the concentration of extracellular CPA, the less ice will form during cooling. With a very high applied concentration of CPA, very rapid cooling without the formation of ice may be possible – a process that is known as vitrification. At the other extreme, lower CPA concentrations that allow ice to form, require more precisely managed cooling rates which can be provided by programmable controlled rate freezers. The issue here is the toxicity of the applied CPA since high concentrations, even for short periods, can lead to excessive dehydration and high cell stress, whereas lower concentrations may involve prolonged cellular exposure to essentially toxic material. DMSO, for example, is an organic solvent and has been linked to cellular mutation. The choices made for the preparation and subsequent freezing of cells is a complex balance between thermodynamic and biochemical variables, the choice and management of which can have a profound effect on the post-thaw recovery of living cells and hence the value of the sample.

1.3 Long term cell survival and contamination

All biological materials will, without intervention, naturally deteriorate, and if they are to be preserved it is necessary to utilize a method that will preserve both morphology and functionality while preventing any alteration of the fundamental nature of the material. The most common methodology available for this is cryopreservation. Biological materials, however, have widely different properties and in order to create a truly effective cryopreservation protocol, it is necessary to consider these properties as they affect the preservation of vital characteristics both during the freezing process and the subsequent environment in which the samples are to be stored long term.

Regarding the minimum storage temperature, no temperature is too cold. Once a sample is frozen and the residual liquid phase has vitrified, further cooling simply reduces molecular energy and vibration. It is possible for short-range structural changes to occur at a molecular level, but they do not affect post-thaw biological properties. It is worth noting that because the cell micro-environment within a frozen sample is chemically different from the majority of the frozen material, biological activity may continue, albeit slowly, at temperatures several degrees below the freezing point of the material.

The minimum melting point of the multi-dimensional phase diagram for typical cryopreservation media occurs at around -80 °C but the cell contents do not finally solidify to an amorphous state until around -120 °C. It is not sufficient simply to keep the samples frozen because, at a micro-environmental level, if the material retains the ability to diffuse it may also degrade, albeit at a much reduced rate. The glass transition temperature is therefore regarded as the "critical" temperature if truly long term storage is required

Best practice dictates that freezers should maintain sample temperatures as far as possible below this critical temperature. By storing well below the critical temperature, transitory warming events above that temperature can be avoided during sample handling, retrieval, storage and in the event of any disruption to the availability of cryogen or power. Freezing a sample in such a way as to maintain maximum biological potential is not a trivial task, and the same care applied to this process should be brought to bear when designing and building storage environments.

The key considerations when looking at a cryopreservation process were listed at the beginning of this chapter. Clearly, the process should be able to maximise the potential for use after processing and storage. Because the future use may be unclear, the preservation and storage procedures should be designed to provide the best possible opportunity for future exploitation. The storage of cells without either adequate care during the initial cryopreservation process or at too high a temperature during subsequent long-term storage are key problems that should be avoided and when the purpose of storage is to maintain biological potential, it is vital that the mechanism of freezing injury be considered.

As the liquid in which the cells are suspended begins to freeze, any solutes in the unfrozen solution become more concentrated and this results in a depression of the freezing point of the remaining solution. The result of this, when the temperature is reduced, is that the cells are exposed to a solution of progressively higher concentration. The increasing concentration increases the osmotic gradient across the cell membrane which results in water leaving the cell in order to maintain balance. Hence, controlling the cooling rate provides a mechanism for controlled dehydration of the cells. Eventually the aqueous phase is so viscous that there is insufficient energy available for the water molecules to form a crystalline solid and the solution becomes an amorphous solid or glass. The temperature at which this condition is reached is known as the "glass transition" temperature (T_g). Once the sample is below this temperature, diffusion within and without the cell stops and the sample is biologically inert. At temperatures below T_g the sample can be maintained indefinitely; other physical interactions, such as background radiation, may have an impact on extremely long-term storage but such effects are probably without any significance in practice (Glenister et. al., 1984).

Unlike freezing, the glass transition is not based upon a thermodynamically defined phase change but rather on the observed dramatic change in viscosity that occurs in cryoprotectant solutions typically at around -120 °C. It is important to ensure that samples are maintained below this temperature throughout the storage term. A temperature of -150 °C is typically stated to be the critical storage temperature for cell products since this temperature provides a reasonable safety margin to ensure that that samples remain below the glass transition temperature during transitory events such as handling, but in practice, storage in liquid nitrogen at -196 °C is a convenient and reliable way to meet this requirement; moreover the additional safety margin provides even greater sample security.

However, storage in liquid nitrogen is not without its disadvantages which include the risk of explosion during warming should liquid nitrogen have entered the vials. Microbiological cross-contamination is another hazard of storage in liquid nitrogen (Byers, 1999) and may lead to the application of a secondary enclosure ('double bagging'). Storage in the gas phase has been advocated to avoid these problems. In the past, the temperature gradient in the vapour phase of liquid nitrogen refrigerators has been a problem, and there may have been increased vulnerability to inadequate amounts of liquid nitrogen between refills. Modern high-efficiency liquid nitrogen cooled vessels now allow storage in the vapour phase without these problems. These vessels are vacuum insulated and the surface area that is not insulated by the vacuum is minimised ensuring that the evaporation rate of liquid nitrogen is kept low. Restricting the amount of energy entering the vessel ensures that the temperature in the vapour phase is maintained close to the liquid nitrogen temperature. The upper region of the refrigerator, close to its access point where the temperature would otherwise be higher, can be efficiently cooled if it is ensured that the heat exchange surface extends right from the bottom to the top of the refrigerator: gas phase temperatures of around -190 °C can be achieved.

The weak point in the process of maintaining safe low temperatures for samples focuses on the time in transport to and from its storage. Small samples of low thermal mass, such as vitrified straws, can warm at the rate of thousands of degrees Celsius per minute and regulatory inspections requiring the removal of samples for identification can be another weak point.

1.4 Traceability

Under most regulatory environments, a rigorous sample tracking system is a key and mandatory component of compliance. It is vital that the individual location of any sample is recorded accurately, and that the sample is labelled with a unique identifier such that the identity of a sample at any location can be verified. For many research and therapy provision operations it is also necessary to have all processing, analytical and, if relevant, patient data linked in a central database.

Labelling can be a challenge as sample containers can be small and the cryogenic environment hostile; however commercially available cryogenic-proof labels and label printing systems are readily available. RFID tags are also a promising solution.

Sample location databases should be organized hierarchically, such that the location of any individual sample can be readily identified; for example: Room / Freezer / Shelf; or Segment / Rack / Position of Box in Rack / position of Vial in Box; or Room / Freezer / Canister / Cane / Goblet / Straw position in Goblet.

Most regulatory environments require the label to include both machine and human readable identifiers (bar code plus text) and where a sample is stored in a secondary container (such as a blood bag in a cassette) it is vital that both the external container and the primary sample container be correctly labelled; see for example the European Directives 2004/23/EC and 2006/86/EC

Concomitant with good identification procedures are good location and retrieval methods and there are a number of commercial software systems available with varying degrees of sophistication to accommodate larger or smaller numbers of stored samples. However an often overlooked part of the storage process is the logging, monitoring and associated alarms. Recording the parameters of storage is sometimes seen only as a regulatory obligation but liquid nitrogen levels or temperatures and the performance of mechanical freezers is of front-line importance. Alarms that work in practice rather than in theory are vital additions to a comprehensive storage environment.

2. Types of technique

There are various options to consider when choosing the methodology and equipment for the cooling process. In conventional cryopreservation, where the intention is to control the rate of formation of ice in the material, it is necessary either to vary the rate of application of a cryogen when working against a constant warm environment, or to provide insulation or energy while maintaining a constant external cold environment. If insulation is used, the cooling rate at any point is approximately proportional to the difference in temperature between the sample and the environment as modified by the insulation and the change in specific heat of the sample as cooling proceeds. Thus, during the process, the cooling rate asymptotically approaches zero as thermal equilibrium is achieved. Applying variable energy to a sample in a cold environment allows the rate of cooling to be modified during the process. The aim is to maintain a composition within the cells that varies as cooling proceeds such that its freezing point remains below the applied environmental temperature. Alternatively, if the concentration of cryoprotectant is high enough, it may be possible to cool the sample sufficiently rapidly that ice cannot form – an approach called vitrification. The required cooling rate will depend on the cryoprotectant and its concentration, the latter being dependant on the concentration that the cells will tolerate. In general, very rapid but uncontrolled cooling is used. The new technique of liquidus tracking allows slow cooling and vitrification.

2.1 Freezing in mechanical freezers

Passive cooling uses insulation to moderate the cooling rate of samples that have been equilibrated with low concentrations of cryoprotectant and then placed inside traditional electromechanical refrigerators at -20 °C, -40 °C, -80 °C or even at lower temperatures. The cells are dessicated slowly during the cooling process. This method can be used for most robust cells but even under the best circumstances the post-thaw recovery rates may not be ideal. In addition, in most cases no instrumentation monitoring or recording of the process is provided. The variation of temperature within mechanical refrigerators is well known with one study reporting values of -43.5 °C to -90 °C in upright freezers (Su et. al., 1996). Since there is no active control during the process, it follows that the poor repeatability of the process can affect the cooling rate and hence the efficiency of the whole procedure. Variability might be improved if the local environment were more stable and protected from instantaneous variation due to external factors such as door openings etc. It is generally preferable to use a liquid nitrogen gas phase freezer for this approach since the internal temperature variation is small and the environment disturbed less frequently.

This approach to cooling and freezing is increasingly being used for material provision in pharmaceutical drug screening programmes as they move from supply by continuous culture towards a "cells-as-reagents" concept. In this approach, the cells are insulated in polystyrene containers as they are cooled initially to -80 °C and then transferred for cryogenic storage into liquid or vapour phase nitrogen. The need for rapid use of the cells for drug assays following cryopreservation, places an increased importance on the post-thaw quality of the cells. In recent work carried out at LGC (Teddington,UK) this has been shown to be compromised by this type of freezing where the cooling rates are not actively controlled but rely on the passive characteristics of the system. In particular, temperature fluctuations within the polystyrene container and the storage time at -80 °C can significantly impact the post-thaw recovery of the cells and their biological function.

2.2 Controlled freezing, protocols and seeding

Liquid nitrogen may be applied via a pressurised supply and cryogenic valve to create a very accurate cooling profile of temperature over time. This methodology offers the most options for optimization since the cooling rate can be varied at multiple stages in the process. As freezing proceeds the concentration of solutes in the medium increases causing cell dehydration in the sample.

As described in the opening section, cooling protocols are designed to manage the intracellular solute concentration. The key point is the nucleation temperature of the suspending medium - that is, the temperature at which ice starts to form. The ice is extracellular, resulting in an increase in the extracellular solute concentration and and hence an osmotic pressure difference between the intracellular and the extracellular solutions that leads to the withdrawal of water from the cells. It is important to recognize that under normal circumstances, solutions do not freeze at their freezing point; they freeze at their nucleation temperature, which is variable and depends on the availability of nucleation centres in the sample. The nucleation temperature is normally several degrees below the nominal freezing point.

Once the extracellular fluid begins to freeze, two major events occur. First, as explained above, the concentration of CPA increases in the fraction of the extracellular fluid that has not at this point frozen, and this causes the cells to dehydrate. Secondly, the temperature of the suspension where freezing has commenced rises towards the nominal freezing temperature and remains at or close to this temperature until the freezing process is complete. This is followed by a drop in temperature as the sample catches up with the temperature of the surrounding medium, but if the cooling rate is too rapid the intracellular CPA concentration may be insufficient to prevent intracellular freezing – with severe consequences for the cells

In order to avoid this hazard, the control program may be designed to allow equilibration of the sample and its suspending medium at a temperature marginally below the calculated freezing point and at this temperature the sample forced to begin to freeze by applying either a physical nucleation point via a cold instrument placed on the external wall of the sample container, or via a sudden, short-lived introduction of cryogen into the environment. This causes the sample to commence freezing. As the sample was originally held only marginally below the nominal fusion temperature, the cell experiences a much more moderate reduction in temperature when the fusion is complete and the temperatures re-equilibrate. After this, the cooling processes is started and continues with a temperature program that is designed to effect the necessary concentration changes to maintain the intracellular composition in the liquid region of the phase diagram. This process is called "seeding".

2.3 Vitrification

The process of vitrification usually uses the highest concentration of cryoprotectant that the cells in the tissue will tolerate and follows this by very rapid but uncontrolled cooling, usually by plunging the sample into liquid nitrogen. The crucial element is exposure of the cells or tissue to potentially toxic levels of CPA: too low a level or too short a time and ice will form killing the cells. If the levels are too high or the process time is too long, the chemicals employed will prove toxic to the cell and post-thaw viability will be limited. Since this process depends on rapid cooling, vitrification has only ever proven applicable for samples with very small volumes, ideally those with very high surface area-to-volume ratios; for example cryogenic straws can fit this description. It is important to be aware of the Leidenfrost effect where a sheath of vapour will surround a warm sample when plunged into liquid cryogen, essentially insulating the sample for a short period of time. For many vitrification protocols however, even this short additional time period before the sample is vitrified has proven fatal to the cells due to increased toxic exposure to the CPA and decreased cooling rates. In conventional vitrification, very high cooling rates are achieved by exposing small samples directly to the liquid nitrogen. The sample is surrounded with as little physical material as possible to achieve the maximum cooling rates. With large samples, however, such high cooling rates are impracticable.

Although vitrification is normally associated with cooling rates in the tens of thousands of degrees Celsius per minute, slower techniques have been reported such as the S3 vitrification technique for blastocysts (Stachecki & Chen, 2008); this uses rates <200 °C/minute. But in fact vitrification does not necessarily require rapid cooling at all. It all depends on the dependence of the critical cooling rate required to prevent freezing on the concentration of the cryoprotectant. (Sutton, 1991). As the following section describes. vitrification can be produced at really low cooling rates.

2.4 Warming and thawing

It is usual to thaw cryopreserved or vitrified samples rapidly – typically by plunging them in a 37 °C water bath. The warming rate does have an effect on the recovery of living cells but this is not as great an influence as cooling rate is during cooling. In fact, optimum cooling rates have usually been determined using rapid warming so it is hardly surprising that rapid warming then gives the highest recovery! However, there are circumstances when the warming rate is of importance in its own right. The first is when the sample has been vitrified but is nucleated without a significant amount of ice being present. This is an unstable situation and in such circumstances the warming must be rapid to avoid intracellular freezing during warming. This consideration argues for rapid warming. The other situation occurs when the frozen material contains a significant amount of vitrified material, as is always the case in conventional cryopreservation. Glasses are brittle and the hazard here is that rapid warming will generate thermal stresses and cause the vitreous material to fracture. This will not matter greatly with cell suspensions where a fracture running through the sample is unlikely to traverse many cells but it is very important when the extracellular matrix must be intact – as it must, for example, in grafted blood vessels and heart valves. The solution here is to warm through the vitreous zone, that is from -196 °C to -123 °C, relatively slowly: once above the Tg there is no hazard from fractures and the sample can be warmed as rapidly as you like. A convenient way to do this is to allow the sample to warm slowly in a -80 °C refrigerator or packed in solid CO_2 until its temperature is at above -100 °C. Alternatively the sample can be surrounded by a layer of insulation during the initial stage of warming in room air. A warming rate of around 50 °C/minute up to -100 °C was 'slow' enough to prevent fractures in cryopreserved rabbit carotid arteries (Pegg et al., 1997).

2.5 Liquidus tracking – A new method

The controlled-rate freezing process achieves its results by preventing the formation of intracellular ice. In some samples, however, even extracellular ice can be severely damaging.

An example of this is articular cartilage. Isolated chondrocytes can be cryopreserved using conventional techniques (Pegg et. al., 2006a) but results when attempting to cryopreserve chondrocytes in situ have proven to be very disappointing. It was found that traditional cryopreservation results in the formation of ice crystals within the chondrons and not just in the acellular matrix (Pegg et. al., 2006b) which might have been expected from experience with conventional cryopreservation. In articular cartilage it is important to prevent both intracellular and extracellular ice. With this requirement in mind, the most appropriate cryopreservation approach would appear to be vitrification; that is the prevention of any ice formation at all. However, it will be clear that conventional vitrification is out of the question because of the heat transfer problems with bulky samples. Liquidus tracking (LT) provides a new approach to this problem.

During conventional cryopreservation, with a moderate concentration of CPA (say 10 %w/w) and relatively slow cooling (say 1 °C/minute), the cells are exposed to gradually increasing concentrations of cryoprotectant as progressively more extracellular ice is formed. The instantaneous CPA concentration is determined by the temperature according to the phase diagram of that specific system. The idea of LT is to control the instantaneous concentration of CPA throughout the cooling process so that the CPA concentration follows the liquidus line by external control rather than by progressive freezing of the medium. In this way the medium remains just above its freezing point at all times and no ice is formed. It is important to note that the cells are exposed only to the concentrations of CPA that they would experience during conventional cryopreservation. And we know that isolated chondrocytes in suspension can be cryopreserved by standard methods. In effect, the LT process takes advantage of the decrease in cytotoxicity of cryoprotectants as the temperature is decreased: hence, rather than starting with a very high concentration of cryoprotectant, the LT approach controls the concentration dynamically throughout the cooling process. In this way, vitrification can be achieved without using the extremely high concentrations of cryprotectant at the start of the process and without the need for rapid cooling. Of course, allowance has to be made for the time that diffusion of CPA into the tissue takes and this can be very considerable. On the other hand, if an organ can be perfused with the cryoprotectant solution, via the vascular system during cooling, then the diffusion distances will be very short and mass transport delays much less significant. In practice, when designing an LT process for a particular tissue, it is crucial to determine the concentration that is actually achieved in the tissue as the process continues and to adjust the concentration/ temperature/ time program to achieve the desired tissue concentration at all stages of the process. This necessitates slow cooling, commonly of the order of 0.1 to 0.3 °C per minute. The cooling of the samples can be achieved in a conventional controlled rate cooler and the solution composition can be controlled by standard peristaltic pumps, the whole system being under computer control – see section 3.7 below for a discussion of the methods that are now available for research use.

3. Types of equipment

Due to the many different types of samples that can be cryopreserved and their differing sensitivities, a number of different techniques and types of equipment are used.

3.1 Minus 80 °C and lower: Mechanical freezers

Mechanical refrigeration always applies the same methods no matter the degree of cooling desired: a gas is passed through a compression system and liquefied. The energy which is released during this liquefaction is dissipated to the environment via heat exchanger coils. The liquid is then passed through cooling coils within the freezer chamber and absorbs energy from the chamber as it vaporizes. It is the vaporization process that creates the cooling effect. As lower temperatures are required, lower liquid point gases must be employed. In order to liquefy these gases, higher pressures are required and often the liquefaction cannot be completed in a single process; this results in larger, multiple compressors being employed.

The most commonly used freezers for cryogenic purposes are upright, front-opening freezers with a cold point at a nominal -80 °C. It should be noted that there is no biological significance for this temperature, merely a physical significance since it approximates to the sublimation temperature of dry ice (solid carbon dioxide, -79 °C). This type of freezer can be employed to store biomaterial in which living cells are not a prime concern, or when it is to be stored for only a short period of time. This type of equipment is intended to be for transactional storage – holding material required daily and which will either be consumed or transferred to more appropriate conditions within a short time - 6 to 12 weeks typically.

The front opening design, while adding considerable convenience, creates a significant issue with temperature stability and variability. Because cold air is significantly heavier than warm air, opening the door causes massive air exchanges and temperature rises in the sample area in a short period of time. In addition, because the compressor systems run on a very high cycle time, there is little spare capacity to effect a cooling after the temperature has risen and it can take some time to return to equilibrium after a warming event. This property is similarly exhibited when the freezer is in normal operation and as has been previously noted, there can be significant temperature variations. The use of deep drawers within the refrigerator for the storage of samples is helpful in reducing the loss of cold air when the door is opened.

Because of the high cycle times, compressor failures are quite common and expensive to repair. It should also be noted that as the energy removed from the sample area is 100 % dissipated into the room in which the freezer is located, the term cost of operating a unit such as this should take into account not only the electricity consumption required for the compressor system, but also the significant air conditioning costs associated with the expelled heat from the freezers. If this energy is not removed by air-conditioning, the freezers become less efficient as room temperature rises, compressors are required to cycle even longer, power usage rises and compressors fail more quickly. Environmental management at a macro as well as micro level is therefore important.

3.2 Alcohol bath freezers

These commonly used laboratory units are essentially refrigerated circulators. A reservoir of cooling medium (normally an alcohol) is passed through a cooling system and re-enters a reservoir, reducing the temperature. The degree of refrigeration applied and the flow rate through the cooling coils determine the derived temperature of the reservoir. The relatively large volume of cooling liquid creates two noticeable effects: temperatures are very stable due to the large heat capacity of the available fluid and cooling rates can be controlled very accurately for a similar reason. The corollary to this however is that the rates achievable are very low and so rapid (> 1 °C/minute) rates are very hard to achieve. In addition, alcohol bath freezers are normally limited to temperatures above -80 °C.

3.3 Liquid nitrogen vessels: Liquid and vapour

Storage of important biomaterial in liquid nitrogen at -196 °C is widely practised. This method allows for a 70 °C plus safety zone when considering the -120 °C threshold for long-term storage; the significance of -120 °C, the glass transition temperature, has been previously discussed. Liquid nitrogen storage does provide the greatest safety zone. However, it also presents a number of problems, including personal safety and potential microbiological cross-contamination via the liquid nitrogen.

Storage in the vapour stage is felt to address these issues but it does come with its own set of problems. The vapour is not as cold as the liquid nitrogen itself and as such the 70 °C safety margin is diminished. However, modern vapour storage vessels use carefully designed vacuum insulation to minimise the heat leakage from the environment into the vessel. This allows the vessel to maintain a vapour temperature at around -190 °C resulting in samples still being maintained at a safe distance from the glass transition temperature. Efficient designs also result in very low liquid nitrogen usage and temperatures can be maintained for up to a month without additional filling; temperatures are even maintained with the lid removed for short periods.

3.4 Controlled rate liquid nitrogen freezers

As described previously, up to present times, the controlled rate freezer offers the widest control options for a freezing protocol. With a truly variable application system for cryogen, most sample sizes can be easily accommodated and rates from the very slow (< 0.1 °C/minute) to in excess of 50 °C/minute are both achievable and controllable. Sample size, container dimension, cell volume, membrane permeability etc. are all variable factors. As the controlled-rate freezer allows complex, fully controlled temperature versus time profiles to be created, protocols can be designed that are appropriate to the cell type and cryoprotectant concentration. Additional steps such as pauses for manual 'seeding' or rapid plunges to initiated freezing can be added to the profile. Transition to different rates can be triggered from the chamber temperature or representative sample; triggering the transition from the sample temperature can help remove variability introduced by different sample loads..

From an instrumentation standpoint, the programming and record-keeping intrinsic within the system meet most external compliance standards and optional software packages are generally available to enhance this aspect beyond the current requirements of any legislative authority. The fact that it is possible to optimise processes for every unique cell type together with the compliance aspect lend great versatility to this type of instrument in most application areas.

3.5 Equipment for conventional (high cooling rate) vitrification

Such protocols call for extremely rapid solidification of the sample, typically by plunging it directly, and in a somewhat uncontrolled manner, into liquid nitrogen. Intracellular ice formation is avoided by the application of very high concentrations of CPA. Equipment such as the VitMaster (IMT ltd.) can be used to increase the cooling rate. This uses negative pressure to depress the freezing point of liquid nitrogen to below -205 °C thereby increasing the cooling rate. Several open techniques have been developed to minimise the sample volume and achieve high cooling rates; for example the Cryotop method which uses a thin film strip to hold the sample. These open systems typically expose the sample directly to the liquid nitrogen which assists in achieving the very high cooling rates. Of course exposing the sample directly to liquid nitrogen in this manner raises questions of potential contamination from the cryogen. Other approaches, such as the Cryologic Vitrification Method, still use an open device at the stage of vitrification but cool the sample by touching on a liquid nitrogen cooled aluminium block. This means that the sample is not directly exposed to the liquid nitrogen and the block avoids the Leidenfrost effect. Alternative approaches use closed straws. These avoid the contamination issues but at the expense of the cooling rate. By definition, the vitrification stage of the process is difficult to measure, monitor or document, so validation and on-going quality control are qualitative exercises only.

3.6 Stirling engines

Originally conceived in 1816 by the Reverend Stirling, the Stirling engine converts heat energy into mechanical work. The principal also works the other way round to convert mechanical energy to heat, when the Stirling engine forms a heat pump able to move heat 'uphill' from a cold place to a warmer one. This gives the Stirling engine an application as a refrigeration unit.

Most refrigerators operate on the Rankine cycle which depends on refrigerants existing with appropriate boiling points. Triple stage Rankine machines are at the limit of the technology and achieve roughly -140 °C. Although the Stirling cycle is less efficient than Rankine cycle machines, it is capable of cooling to lower temperatures and therefore comes into its own below -140 °C; miniature cryo-coolers based on Stirling engines are now quite common. Due to relative inefficiency, these Stirling based cryo-coolers can normally freeze only quite small samples of a few tens of grams maximum and cannot compete with liquid nitrogen powered machines for cooling capacity. On the other hand, they excel in clean rooms where it is not possible to obtain a supply of liquid nitrogen and it is only desired to freeze very small samples.

3.7 Liquidus tracking equipment

Because liquidus tracking is a relatively novel technique, there is little choice of equipment to assist with research into its use. Planer plc do manufacture a Liquidus Tracking controller that can be used for research into this approach. The equipment comprises a conventional slow-rate chamber coupled with a liquidus tracking controller and two peristaltic pumps.

The controller cools the sample in a similar manner to the conventional slow-rate freezing process. The cooling profile is typically a simple linear ramp. During the cooling of the sample, the controller monitors the current chamber temperature and adjusts the speeds of the two pumps to dynamically alter the concentration of cryoprotectant surrounding the samples. In the ideal process the concentration of cryoprotectant is maintained just above the liquidus curve. As the temperature decreases, the concentration of cryoprotectant is therefore increased. However, as the temperature of the sample decreases, the toxicity of the cryoprotectant decreases and this allows the sample to tolerate the ever increasing concentrations; see figure 1.

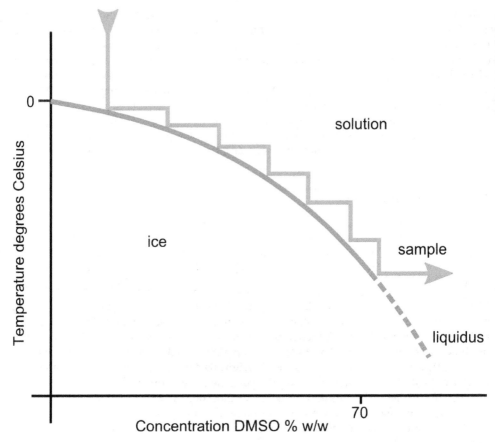

Fig. 1. Concentration of sample tracking liquidus

Two specific requirements of this process are the rather large volumes of cryoprotectant required and the need to ensure good mixing around the sample. The Planer Liquidus Tracker supports two modes of operation each with its own advantages and disadvantages; these are the single solution and dual solution modes.

In the dual solution mode, shown in figure 2, a solution containing a high concentration of CPA (typically 72 % w/w DMSO plus isotonic salts) and a solution containing only isotonic salts (nominal 0 % solution) are used. Each solution is pumped through a mixing junction and a heat exchanger into the sample container and thence into a waste collection container. The relative speeds of the pumps are continuously adjusted via a computer program to deliver the correct concentration to the sample. The pump speeds are adjusted to maintain a constant flow rate through the sample container. The dual solution system requires a small volume surrounding the sample so that the incoming, premixed solution is able to displace the existing solution completely as it flows through the container. The total volume of cryoprotectant can be quite large; for example, a run from 0 °C to -70 °C at 0.3 °C/minute requires a total volume of solution equal to 233 times the sample container volume. This method is suitable for use with small sample containers and has been used for discs of ovine articular cartilage (see Wang et al., 2007).

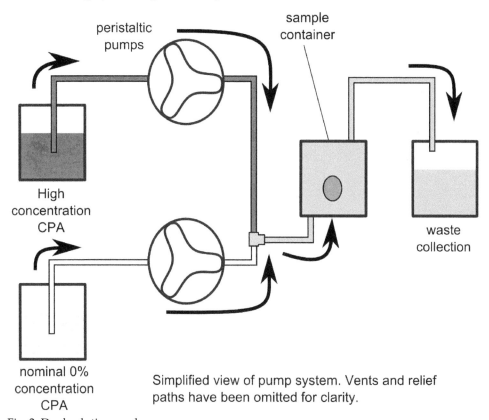

Fig. 2. Dual solution mode.

The single solution mode is more suitable for use with larger samples. It is illustrated in figure 3. Here a highly concentrated solution is cooled and delivered to the sample container. This increases the concentration of the CPA solution surrounding the sample. To maintain a constant volume within the container, the second pump extracts the excess solution from the container. This technique is suitable for larger samples as it reduces the total volume of cryoprotectant required. For a sample container volume of Vs, a cryoprotectant concentration of Ksol and a target concentration of Kt, the volume of cryoprotectant V can be calculated from this equation.

$$V = Vs.\ln(Ksol / (Ksol - kt)) \tag{1}$$

For a sample container volume of 50 ml, depending on the actual values of Ksol and Kt this approach could require less than 100 ml of concentrated DMSO solution. Because the incoming solution has to be thoroughly mixed within the container, additional stirring equipment running at cryogenic temperatures is required. This results in a mechanically more complex arrangement than the dual solution approach.

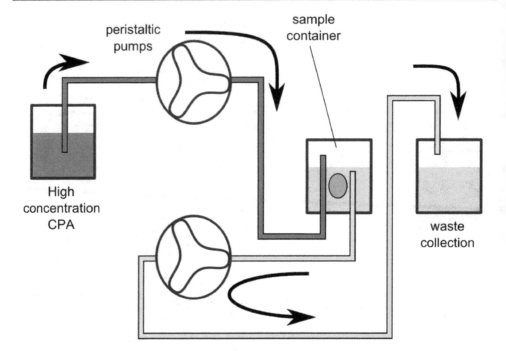

Simplified view of pump system. Vents and relief paths have been omitted for clarity.

Fig. 3. Single solution mode.

3.8 Examples of equipment used

The BioCool Controlled Rate Freezer from FTS Systems/SP Scientific is a mechanically refrigerated bath with temperature control to -40 °C or -80 °C. The fluid in the 2 litre bath provides temperature stability and dispersion of the heat of fusion without a concomitant temperature rise. Courtesy SP Scientific and Gary Gold Photography.

BioCool™ Controlled Rate Freezer

The Asymptote EF600, a unique liquid nitrogen free, controlled-rate freezer, is electrically powered by a Stirling Cycle Cryocooler rather than liquid nitrogen. This allows the freezer to be used where liquid nitrogen is in short supply, where extra high air quality is needed, or where there is a risk of LN2 contamination to samples.

Asymptote liquid nitrogen free controlled-rate freezer

The Planer Kryo 360 cell freezer controls down to a -180 °C end temperature to ensure sample integrity during transfer to storage. Fully programmable, it allows the use of protocols associated with the most advanced cryopreservation techniques and is widely used in laboratories around the world.

Planer Kryo 360 programmable cell freezer

The Crysalys controlled rate freezer: programmes, time and temperatures may be entered via a touch screen with up to 100,000 cycles held on the onboard SD card; the data can be retrieved by any PC or Mac computer. A battery back up operates the system for 3 hours; its portability and 3.2 kg weight make it especially suited to veterinary purposes.

Crysalys controlled rate freezer

The Gemini Tinytag View 2 data logger, when used with a specially designed probe, is used for temperature monitoring in cryogenic environments down to -200 °C.

Tinytag View 2 data logger

The CoolCell is an alcohol-free cell freezing container which provides a reproducible cooling rate of 1 °C/minute when placed in a -80 °C freezer. No alcohol is required to control the freeze rate as the design and materials of the CoolCell ensure precise and uniform heat removal from cryovials.

CoolCell®, an alcohol-free cell freezing container

The Planer ShipsLog is a datalogger specifically designed for vapour shippers, which maintains a downloadable temperature history of samples during transit.

ShipsLog datalogger for vapour shippers

The Liquidus Tracker is a new controlled vitrifier for cryopreservation of samples using the liquidus tracking technique. This approach may have uses in vitrifying larger samples and those which are currently difficult to cryopreserve.

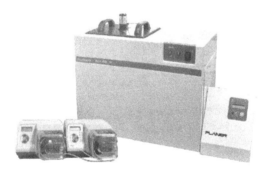

Liquidus Tracker controlled vitrifier for cryopreservation

4. Acknowledgement

We would like to thank Dr. Damian Marshall, Principal Scientist, LGC Ltd, Teddington, UK and Mr Ian Pope, Coreus Consulting, Faribault, MN, USA, for their valuable assistance with this chapter.

Additional material can be found in 'Cryopreservation and Freeze-Drying Protocols'; Day, J.; Stacey, G. (Eds.); 2nd ed., 2007 Humana Press Cryopreservation of Animal and Human Cell Lines; See

Chapter 3; Pegg,D.E. Principles of Cryopreservation.
Chapter 16; Morris, C; Cryopreservation of Hematopoietic Stem/Progenitor Cells;
Chapter 17; Watt S., Austin E., Armitage S.,

5. References

Byers, K. (1999) Risks associated with liquid nitrogen cryogenic storage systems. Journal of the American Biological Safety Association, Vol 4, No 3. pp. 143-146.

Glenister, P., Whittingham, Lyon, M. (1984) Further studies on the effect of radiation during the storage of frozen 8-cell mouse embryos at -196 °C. Journal Reproductive Fertility, Vol 70, No. 1, (Jan 1984), pp. 229-34

Lovelock J E (1953) The haemolysis of human red blood cells by freezing and thawing Biochim Biophys Acta 10, 414-426

Pegg, D., Wusteman, M. & Wang, L. (2006a) Cryopreservation of articular cartilage. Part 1: Conventional cryopreservation methods. Cryobiology, Vol 52, No. 3, (June 2006), pp. 335-346

Pegg, D., Wang, L., Vaughan, D., (2006b) Cryopreservation of articular cartilage 3: Vitrification methods. Cryobiology, Vol 52, No. 3, (June 2006), pp.360-368

Pegg D.E., Wusteman C. and Boylan S. (1997) Fractures in cryopreserved elastic arteries. Cryobiology, 34, 183-192.

Polge C, Smith A U, and Parkes A S. (1949) Revival of spermatozoa after vitrification and dehydration at low temperatures Nature, 164, 666.

Wang L. Pegg D.E., Lorrison J., Vaughan D. and Rooney P. Further work on the cryopreservation of articular cartilage with particular reference to the liquidus-tracking method Cryobiology. 55, 138-147 (2007).

Saragusty, J., Arav, A. (2011) Current progress in oocyte and embryo cryopreservation by slow freezing an vitrification. Reproduction, Vol 141, (Jan 2011), pp. 1-19

Stachecki, J., Cohen, J. (2008) S3 Vitrification system: a novel approach to blastocyst freezing. The journal of clinical embryology, Vol 11. No. 4, (2008), pp. 5-14

Su, S., Garbers, S., Reiper, T. (1996) Temperature variations in upright mechanical freezers. Cancer, epidemiology, biomarkers and prevention, Vol 5, (1996), pp. 139-40

Sutton R.L. (1991) Critical cooling rates to avoid ice crystallization in solutions of cryoprotective agents. J. Chem. Soc. Faraday Trans., 87(1), 101-105.

Wang L. Pegg D.E., Lorrison J., Vaughan D. and Rooney P. Further work on the cryopreservation of articular cartilage with particular reference to the liquidus-tracking method Cryobiology. 55, 138-147 (2007).

10

Methods of Assessment of Cryopreserved Semen

Agnieszka Partyka, Wojciech Niżański and Małgorzata Ochota
Wroclaw University of Environmental and Life Sciences
Poland

1. Introduction

The numerous effects that cryopreservation can induce in spermatozoa, ranging from lethal injuries to those which merely impair their subsequent function. In the last few years, the considerable increase in our understanding of both, the cell physiology of spermatozoa, and the stress of cryopreservation, have contributed to a renewed interest in improving the performance of cryopreserved semen.

Despite the significant progress, the post-thaw viability and fertility of the cryopreserved sperm are still reduced, as a consequence of accumulated cellular injuries that arise throughout the cryopreservation process. Many laboratory tests have already been carried out to verify these detrimental effects and their origin. Their is needed to well understand the whole process of cryopreservation and its influence on sperm function. As a consequence, it would lead to a subsequent improvement of sperm viability by means of reformulated protocols and approaches helping to minimize the detrimental effect of cryopreservation.

Here, we present an overview of the cryopreserved semen assessment methods in the light of sperm physiology, in order to relate these factors to altered functions of cryopreserved sperm and to determine the fertilizing potential of the frozen-thawed semen.

2. Conventional methods of semen assessment

Light microscopy is the most often used to analyze the quality and predict the fertility of the cryopreserved semen in the conventional way. Visual assessment requires such equipment as microscope, heated stage and slides, as well as an experienced evaluator, however the assessment is subjected to the evaluator bias.

2.1 Sperm motility

Motility is one of the most important features of a fertile spermatozoa. It was the first, and continues to be the most widely used indicator of sperm function. Sperm motility is an important attribute, because it is readily identifiable and reflects several structural, and functional competence, as well as essential aspects of spermatozoa metabolism. Sperm motility is expressed as the percentage of total motile or progressively motile spermatozoa. This parameter is usually assessed by the subjective visual examination under a phase contrast microscope at 37°C using low objectives (10 or 20x). Light microscopic evaluation of motile spermatozoa does not require expensive equipment, is a simple and rapid method for assessment of sperm quality, however, it is a highly subjective and not reliable assay for the prediction of fertility (Peña Martínez, 2004).

2.2 Sperm morphology

On account of the fact that freezing and thawing process provokes morphological or biochemical cryogenic damage resulting in sperm dysfunction and changes in cell's membrane, the sperm morphology evaluation is an essential component of any semen analysis and provides the clinical information about the potential fertility of semen sample.

Despite, there are many different and also new methods, as described below, used in semen analysis, semen smears are still employed for routine light microscopic morphological evaluation. However, this assessment is subjective and results are largely dependent on the proficiency and experience of the evaluator. Vital dye in combination with different stains for acrosome evaluation are commonly utilised to assess the spermatozoa morphology and the viability together. For this purpose, India ink, William's, Karras, Spermac, Diff-Quick, Papanicolaou, Fuelgen or combination: Trypan blue and Giemsa, Trypan blue, Bismarck Brown and Rose Bengal, and finally eosin-nigrosin (described in 2.3 section) have been used in birds and mammals including human (Brito et al., 2003; Brito et al., 2011; Didion et al., 1989; Freneau et al., 2010; Łukaszewicz et al., 2008; Partyka et al., 2007; Rodriguez-Gil et al., 1994; Sprecher & Coe, 1996; Talbot & Chacon, 1981). In spite of that, Freneau et al. (2010) have shown that differential interference phase contrast microscopy of wet-mounted semen is the superior method for bulls sperm morphology assessment. For cats sperm morphology, the best differentiation of sperm structures, especially acrosome, with lower artifacts, fast green FCF-rose Bengal staining or Hancock and Glendhill solution staining and phase-contrast microscope are encouraged (Zambelli & Cunto, 2006). However, when frozen-thawed semen is analyzed these stains are negatively affected by egg yolk and glycerol, causing egg yolk agglutination and lack of sperm structures differentiation. Therefore, sperm washing is recommended to prevent these interferences (Zambelli & Cunto, 2006).

Many reports have shown the common classification system for the morphology of spermatozoa from different species. However, classification categories are different for the various species and the adoption of uniform system within each species is needed. Mammalian spermatozoa abnormalities can be divided into primary and secondary abnormalities (Blom, 1950), or in some classification systems into major and minor abnormalities (Blom, 1968, 1983). Primary sperm defects are assumed to have occurred during spermatogenesis, and secondary defects are assumed to have occurred during maturation in the epididymis and the transit through the ductal system and specimen preparation. Second system classifies sperm defects according to the perceived effects on fertility. The most common sperm abnormalities (Fig. 1) are related to abnormal acrosomal regions/heads, detached head, proximal droplets, distal droplets, abnormal midpieces, bent/coiled tails. Acrosome defects include knobbed, roughed, and detached acrosomes. Head defects include microcephalic, macrocephalic, pyriform, tapered, other shape defects, nuclear vacuoles, and multiple heads. Midpiece and principal piece (tail) abnormalities enclose simple bent, folded, fractured, thickened, swollen, roughed, Dag-like, disrupted sheet, duplicated, coiled. Various defects are typical for each species.

For each slide, at least 100-300 spermatozoa should be counted at 400-1000x magnification, which allows for accurate calculation of the percentage of different sperm defects.

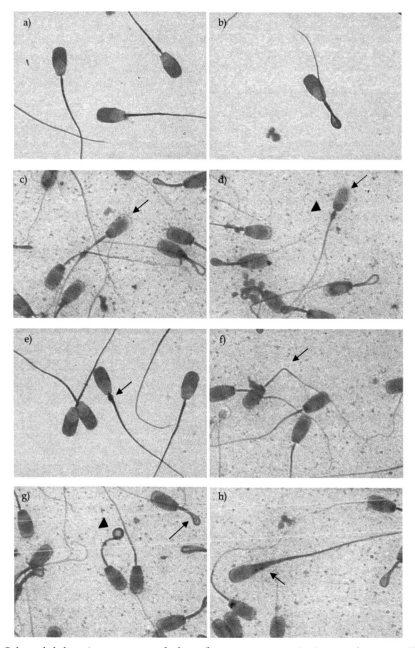

Fig. 1. Selected defects in sperm morphology (boar spermatozoa): a) normal sperm cells; b) looped tail; c) acrosome detachment; d) loss of acrosomal contents (back arrow), proximal cytoplasmatic droplet (arrowhead); e) proximal cytoplasmatic droplet; f) kinked midpiece; g) looped tail (black arrow), coiled tail (arrowhead); h) thickened midpiece.

2.3 Sperm membrane integrity

Live-dead staining. The traditional method for assessing whether the sperm membrane is intact or disrupted involves examining a percentage of viable sperm by a stain exclusion assay. For the determination of cell viability live-dead stains as aniline-eosin, eosin-nigrosin or eosin-fast green are widely used. Integrity of the plasma membrane is shown by the ability of a viable cell to exclude the dye, whereas the dye will diffuse passively into sperm cells with damaged plasma membranes. When stained smears are viewed under the oil immersion objective of light microscope, the percentage of viable, live, properly formed spermatozoa, nonviable and also partially-damaged spermatozoa can be determined. In eosin-nigrosin stain under the microscope, live spermatozoa appear white, unstained against the purple background of nigrosin (Fig. 2a). Dead and damaged spermatozoa which have a permeable plasma membrane are pink (Fig. 2b). The evaluation of the percentage of live and dead spermatozoa and the percentage of morphology defects may be performed on the same nigrosin-eosin stained slides.

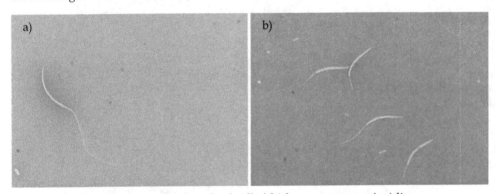

Fig. 2. Eosin-nigrosin staining for live-dead cells (chicken spermatozoa): a) live spermatozoon, b) dead spermatozoa.

The hypoosmotic swelling test (HOS) is a method of investigating membrane integrity in sperm and, as such, is an alternative to supra-vital staining. In fact, the HOS test is thought to have the advantage of indicating not only whether the membrane is intact, but also whether it is osmotically active. Sperm with an intact, functional membrane when are exposed to an hypoosmotic solution incubated for 30 minutes at 37°C, swell to achieve an osmotic equilibrium. An expression of this is a typical swelling of the sperm tail (Fig. 3) (Neild et al., 1999). The HOS test is a simple, inexpensive and easily applicable technique, which has been adapted to assess spermatozoa of several species (Corea & Zavos, 1994; Kumi-Diaka, 1993 Neild et al., 1999; Pérez-Llano et al., 2001; Santiago-Moreno et al., 2009). It has been suggested that this test may supplement the information provided by the conventional parameters of semen analysis, and is useful for fertilizing ability assessment (Brito et al., 2003; Vazquez et al., 1997). This test correlates highly with other predictive tests, such as hamster oocyte penetration (Jeyendran et al., 1992), in-vitro fertilization (IVF) results in human (van der Venn et al., 1986), and with pregnancy rates in pigs (Pérez-Llano et al., 2001). The HOS test seems to be more appropriate for predicting the fertilizing capacity of frozen-thawed than fresh semen, because membrane damage is here a more important limiting factor than in the former (Colenbrander et al., 2003).

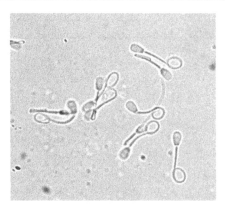

Fig. 3. HOS test (canine spermatozoa).

3. Advanced methods of semen assessment

3.1 Computer assisted sperm analysis (CASA)

Recently, computer assisted sperm analysis has been introduced to veterinary andrology, same as it has been used in reproductive technologies in human andrology (Rijsselaere et al., 2003; Verstegen et al., 2002). This technique assures objective semen assessment, whereas the main disadvantage of conventional semen evaluation is variability of obtained results. Subjectivity of traditional semen analysis is associated mainly with experience and skill of the observer, the method of specimen preparation, staining technique and number of cells evaluated. Variations in the results of conventional evaluation of the same semen samples by different observers and laboratories may achieve up to 30-60% (Coetzee et al., 1999; Davis & Katz, 1992). Subsequently, correlations between spermatozoa characteristics and fertility trials in females are relatively low. Computer assisted sperm analysers allow for calculation of several motility parameters, which characterize movement of individual sperm cells. They include VAP-average path velocity, VSL-straight line velocity, VCL-cell velocity, ALH-amplitude of lateral head displacement, BCF-beat cross frequency (Fig. 4), STR-straightness of cell track, LIN-linearity of cell track, subpopulation of rapid, medium and slow cells (Niżański et al., 2009). Selected characteristics of spermatozoa motility parameters measured by CASA systems are summarized in table 1.

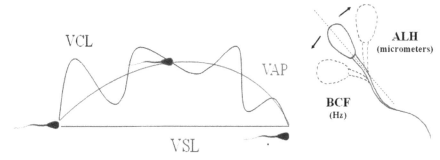

Fig. 4. Scheme of different velocities and parameters of sperm movement measured by CASA systems.

Parameter	Unit	Description
MOT	%	Motility - The population of cells that are moving at or above a minimum speed as determined by values defined under setup.
PMOT	%	Progressive motility- the population of cells that are moving actively forward.
VCL	µm/s	Track speed – Is defined as average velocity measured over the actual point-to-point track followed by the cell.
VAP	µm/s	Path velocity - Is defined as average velocity over smoothed average position of the cell.
VSL	µm/s	Progressive velocity - Is measured in the straight line distance between the beginning and the end of the track.
ALH	µm	Amplitude of Lateral Head Displacement – is the mean width of the head oscillation as the cell moves.
BCF	Hz	Beat Cross Frequency – is the frequency with which the sperm head moves back and forth in its track across the cell path.
STR	%	Straightness - A measure of VCL side to side movement determined by the ratio VSL/VAP.
LIN	%	Linearity - A measure of the departure of the cell track from a straight line. It is the ratio VSL/VCL.
RAP	%	Rapid – subpopulation of rapid cells.
MED	%	Medium – subpopulation of cells with medium velocity.
SLOW	%	Slow – subpopulation of slow cells.
STATIC	%	Static cells.

Table 1. Selected parameters of spermatozoa motility measured by CASA systems.

It was proven in human, that results obtained with CASA systems are better correlated with the outcome of assisted reproductive techniques than results of traditional semen evaluation (Verstegen et al., 2002). Blesbois et al. (2008) showed that some of parameters detected in CASA system are correlated with fertility results obtained with frozen–thawed chicken spermatozoa (PMOT, PROG, VAP, VSL). Most of them were affected by cryopreservation, with the exception of straightness (STR), suggesting that cryopreservation slows down the movement of chicken spermatozoa without changing the shape of trajectories.

The important advantage of computer assisted sperm analysers is the immediate measurement of sperm concentration, total number of sperms in ejaculate and the automated calculation of number of insemination units which could be prepared from one ejaculate. Additionally some machines are equipped with UV excitation module, which gives the opportunity to analyse the percentage of live and dead spermatozoa after staining with vital fluorescent probes, such as Hoechst 33258. Nevertheless, CASA system needs standardization and validation before its use and image settings have been standardized

(Davis & Katz, 1992; Iguer-Ouada & Verstegen, 2002; Rijsselaere et al., 2003; Verstegen et al., 2002). Also other factors as the type and depth of the used chamber, number of fields analysed, temperature during analysis and protocol of semen sample preparation affect results. Optimization and validation of the technical settings would allow to compare intra- and inter-laboratory results, regardless of the instruments that have been used (Agarwal et al., 1992).

Computer assisted sperm analysis allows for a detailed estimation of subtle changes of sperm motion characteristics such as hyperactivation (HA) of spermatozoa associated with capacitation process. Hyperactivation is the process that mammalian spermatozoa exhibit, while they progress through the female oviduct. It is described as vigorous, non-progressive, non-linear sperm motion linked with capacitation. During HA, the pattern of sperm track undergo dramatic changes, characterized by wide-amplitude marked lateral movements of the head and tail of the spermatozoon, with slow or non-progressive 'star-pin' movement (Verstegen et al., 2002). Hyperactivated sperm movement, is assumed to be necessary, for mammalian sperms to penetrate into and pass through, the cumulus cell layer of an oocyte (Meyers et al., 1997; Suarez et al., 1983). To fertilize the oocyte, mammalian spermatozoa must be capacitated, the process that depends on the removal or alteration of substances absorbed on, or integrated in the sperm plasma membrane, resulting in changes in membrane permeability and intracellular ionic composition, with Ca^{2+} movements playing the most critical role (Fraser et al., 1995; Rota et al., 1999). ALH and velocity parameters such as path velocity VAP, progressive velocity VSL are increased in hyperactivated spermatozoa, whereas linearity LIN and straightness STR of movement are lowered. Such changes are characteristic for capacitation induced by specific media (Rota et al., 1999) and for spermatozoa that underwent preservation (cryocapacitation) and are pronounced, especially when media with addition of detergents are used (Niżański et al., 2009). Kawakami et al. (2001) observed that oviduct's epithelium possess the ability to bind hyperactivated spermatozoa, which results in the obvious prolongation of their flagellar movement. On the other hand, the life-span of the free moving non-bound hyperactivated spermatozoa within oviductal lumen, is relatively shorter. It was also found, that Ca influx into the cytoplasm is inhibited in the oviduct-epithelium-binding sperms (Dobrinsky et al., 1997). Active movement of the sperms and Ca influx into cytoplasm negatively affect the maintenance of viability and fertile life of sperm in the lumen of oviduct. Binding to the oviduct epithelium presumably prevents Ca influx, required for sperm capacitation. This phenomenon is available for prolonging viability and fertile life of canine sperms in the oviduct (Kawakami et al., 2001). Considering the obvious lack of such regulatory mechanism, in frozen-thawed semen it is believed, that in vitro post-thaw hyperactivation results in depletion in spermatozoa energy resources, accumulation of metabolites in the extender and cell death, if insemination dose is not deposited into the female's genital tract immediately after thawing.

Nevertheless, the computer assisted sperm analysis of cryopreserved semen should be treated with a dose of criticism. It should be emphasized, that CASA parameters describing kinematic features of frozen-thawed sperm cells may not reflect the real loss of quality of ejaculate after treatment. Absolute CASA parameters (VCL, VSL, VAP, ALH, BCF) should be used with caution, whereas relative CASA parameters (combinations of absolute features-LIN, STR) can not be used directly for estimation of semen quality. Selective death of the most immotile and weakened spermatozoa leads to the situation, where normal CASA parameters show the 'pseudoenhancement' of kinematics. Thus, the mean velocity and linearity parameters may be higher after freezing. This is caused by the fact, that the sub-population of the most resistant cells which survive freezing-thawing may possess higher mean quality parameters, than the larger population of motile sperm cells in fresh semen. In spite of the fact, that only half or one third of population of sperm cells may survive the cryopreservation, their mean velocity may be higher in comparison with velocity parameters of larger population of spermatozoa in fresh semen. Thus, some investigators (Katkov & Lulat, 2000) observed increase in kinematic parameters (KP) of specimen after freezing-thawing, while at the same time substantial losses in post-thaw motility (percentage of motile cells) were observed. The possible explanation of this phenomenon is the selective elimination of the slowest sub-population within the specimens. This "CASA-paradox" is caused by substantial exclusion of slow-moving cells from the motile fraction measured after freezing-thawing.

Therefore, in order to obtain more reliable results of semen assessment after thawing, it was proposed to use Modified Kinematics Parameters (MKP) or Yield of Kinematic Parameters (YKP). MKP can be defined as KP that is average on an entire sample:

$$MKP = KP \times Motility / 100\% \tag{1}$$

YKP is the product of KP and the number of motile cells for which this parameter is average:

$$YKP = Total\ Number\ of\ Motile\ Cells \times KP / 100\% \tag{2}$$

Furthermore, morphology (Assisted Sperm Morphology Assessment-ASMA) of sperm cells can be objectively evaluated, on the basis of morphometric analysis of predefined specific measurements of particular elements in spermatozoa. Usually, on the slides, the head morphometric dimensions of length, width, width/length, area and perimeter of a minimum of 200 sperm are analyzed (Fig. 5). Additionally, parameters of head shape can be evaluated such as ellipticity, circularity, elongation, and regularity (Álvarez et al., 2008). Nevertheless, the accuracy of sperm morphology assessment depends on the careful preparation, fixation and staining of spermatozoa. The analysis of sperm morphology may be done using Diff-Quik stain recommended by World Health Organization (WHO, 2010) or SpermBlue, which has been developed for the evaluation of human and animal sperm morphology (Maree et al., 2010).

Rubio-Guillen et al. (2007) showed that by applying ASMA techniques and multivariate cluster analysis, it is possible to determine three subtle subpopulations of spermatozoa with different morphometric characteristics coexisting in bull ejaculates. The proportion of spermatozoa in each sperm subpopulation showed considerable differences among males and varied significantly throughout the cryopreservation procedure. The cryopreservation of spermatozoa has been found to affect chromatin structure and morphometry of the sperm head (Arruda et al., 2002; Esteso et al., 2003; Gravance et al., 1998; Hidalgo et al., 2006; Rijsselaere et al., 2004). Thus, it is presumed that the adverse effects of cryopreservation on sperm chromatin and head morphology, may be responsible for lowered fertility of spermatozoa, observed after cryopreservation.

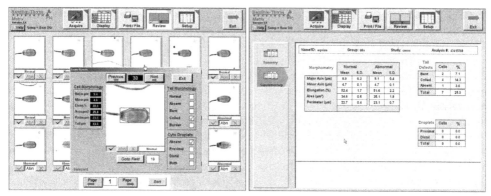

Fig. 5. System for Assisted Sperm Morphology Assessment.

3.2 Flow cytometry and fluorescent probes

During last decades many fluorescent probes have been used for the semen assessment. The fluorescence of these compounds may be estimated using fluorescent microscopy or flow cytometry. Flow cytometry enables the observation of physical characteristics such as cell size, shape, and also any component or function of the spermatozoon that can be detected by a fluorochrome or fluorescently labeled compound. The analysis is objective and accurate. The great number of spermatozoa (>10 000) can be analyzed in a small volume of samples in a short time. This is considerably more than the total of 200 cells generally observed by microscopic analysis. Thus, the analysis of events detected on dot plots gives the accurate and high reliable results (Peña et al., 2001). It is a sensitive method of detection of subtle differences among spermatozoal populations that may not be detected with other techniques.

3.2.1 Sperm membrane integrity

The integrity of sperm membranes is a necessary condition to maintain spermatozoal functions during storage in the female's reproductive tract and penetration of the oocyte (Holt, 2000). When semen is frozen, cells are exposed to a cold shock, ice crystals formation, and cellular dehydratation, which all cause irreversible damage (Amann, 1999; Parks & Graham, 1992). Cellular membranes are one of the primary sites of injury during chilling, freezing and thawing. Damage is caused by alteration of membrane structure and lateral organization (Amann, 1999). The cryopreservation results in temperature-dependent and dehydratation-induced membrane phase changes, which are thought to result in lateral phase separation of membrane components and increased membrane permeability for solutes (Hammerstedt et al., 1990). The disruption of plasma membrane integrity caused by disarrangement of lipids within the membrane during cryopreservation may induce further cellular damage and consequently lead to a sperm death (Watson, 1995).

Membrane integrity of mammalian and avian spermatozoa may be assessed by using many fluorescent probe combinations including: carboxyfluorescein diacetate (CFDA) in combination with propidium iodide (PI), SYBR-14 with PI, carboxy-seminaphthorhodfluor (Carboxy-SNARF) with PI, calcein-AM with ethidium homodimer (EthD-1) and Hoechst

33258 (Christensen et al., 2004; Donoghue et al., 1995; Hewitt & England, 1998; Partyka et al., 2010; Peña et al., 1998; Rijsselaere et al, 2005; Sirivaidyapong et al., 2000). SYBR-14 and CFDA, usually used detectors of live cells, are membrane-permeant and non-fluorescent compounds, which are immediately deacylated and thus rapidly converted into high fluorescent compounds by intracellular esterases. These green fluorochromes are maintained intracellular by intact membranes (Peña et al., 1998; Silva & Gadella, 2006). As plasma membrane deteriorate at cell death, cells lose their ability to resist the influx of red fluorescent PI. PI replaces or quenches green fluorochromes (Garner & Johnson, 1995). Live, viable, intact spermatozoa show a green fluorescence (CFDA, SYBR-14, calcein-AM) while dead stain red (PI) (Fig. 6a). Carboxy-SNARF is pH indicator which stains live spermatozoa orange, while bisbenzimide stain Hoechst 33258 labels dead spermatozoa bright blue (Hewitt & England, 1998). The last one requires flow cytometric analysis with a laser that operates in the ultraviolet light range and is less commonly used in andrology laboratory, however alternatively it may be applied within fluorescent microscope.

SYBR-14/PI fluorochromes have been found to be more sensitive in comparison with conventional method of live-dead cell assessment. The advantage of the use of fluorochromes is the possibility to assess the semen without the interference of fat particles and others material present in the extended semen (Rijsselaere et al., 2005). The detection of the third subpopulation i.e. moribund spermatozoa is the next advantage of this method. Additionally the simultaneous assessment of several functions of spermatozoa may be done in the same specimen by simultaneous staining of sperm cells with fluorescent lectins PNA or PSA for acrosome evaluation with PI for dead cell assessment.

3.2.2 Acrosomal membrane integrity

Acrosome is the acidic secretory organelle filled with hydrolytic enzymes. Assessment of the acrosomal status is a very important part of semen evaluation, in the view of the role of this structure in the maintenance of spermatozoal ability to penetrate the egg's zona pellucida (in mammals), or the egg envelope (in birds) and the ability to fuse with the egg plasma membrane. Cells must retain a normal acrosome to ensure that the acrosome reaction may occur at the suitable time to facilitate fertilization (Esteves et al., 2007). Also the determination of the acrosome status in cryopreserved sperm is of the fundamental importance as cryopreservation directly damages sperm membrane, which could be followed by a loss of the acrosomal matrix contents.

Acrosomal status may be assessed using lectins, such as peanut agglutinin from *Arachis hypogaea* (PNA) or *Pisum Sativum* agglutinin (PSA), conjugated with different fluorescent probes like fluorescein isothiocyanate (FITC), phycoerythryn (PE) or Alexa Fluor®, (Graham et al., 1990; Kawakami et al., 2002; Nagy et al., 2004; Partyka et al., 2010; Peña et al., 2001; Rijsselaere et al., 2005). For human sperm concanavalin A lectins (ConA) is used as well (Holden et al., 1990). The PNA labelling is specific for the outer acrosomal membrane and it binds to β-galactose moieties. Whereas the PSA is labelling α-mannose and α-galactose moieties of the acrosomal matrix. The absence of the fluorescence on the living sperm is indicative for an intact acrosome, and fluorescence is indicative for acrosome disruption or acrosome reaction (Silva & Gadella, 2006). Since PNA agglutinin displays less non-specific binding to other areas of the spermatozoa, it leads some researchers to favour this over PSA (Graham, 2001). Lectins may be also combined with Hoechst 33258, carboxy-SNARF/PI, ethidium homodimer allowing for simultaneous assessment of acrosomal status and membrane integrity (Fig. 6b) (Kawakami et al., 1993; Szász et al., 2000).

3.2.3 Mitochondrial function

The motility of spermatozoa subjected to cryopreservation is reduced by reason of some changes in the active transport and the permeability of the plasma membrane in the tail region (Blesbois et al., 2008; Watson, 1995). A reduction of spermatozoa motility may also be triggered by a change in the availability of energy or an injury of the axonemal elements. Moreover, it has been noted that the alterations in the ultrastructure of mitochondria occurring during cryopreservation are followed by a loss of the internal mitochondrial structure of frozen-thawed spermatozoa (Watson, 1995).

Rhodamine 123 (R123) is the potentiometric membrane dye which is used to selectively stain functional mitochondria. It fluoresces only when the proton gradient over the inner mitochondrial membrane (IMM) is built up and unstained sperm do not contain functional mitochondria (Garner et al., 1997; Gravance et al., 2001). Also the group of Mitotrackers: Mitotracker Deep Red, Red, Orange and Green selectively label the respiring mitochondria. Thus, these probes are suitable to discriminate sperm with deteriorated mitochondria from sperm in which oxidative respiration occurs (Gadella & Harrison, 2002; Garner et al., 1997).

Some of mitotrackers such as 5,5',6,6'-tetrachloro-1,1',3,3'-tetraethylbenzimidazolyl-carbocyanine iodide (JC-1) change their fluorescent properties due to changes in the potential of IMM. JC-1 is a lipophilic cationic fluorescent carbocyanine dye that is internalized by all functioning mitochondria, where it fluoresces green. However, as the concentration of JC-1 inside the mitochondria increases (highly functional mitochondria), the stain forms aggregates which fluoresce orange. Hereby, population of spermatozoa can be divided into high (orange staining), moderate (orange and green) and low (green) mitochondrial potential groups after IMM depolarisation (Fig. 6c) (Garner et al. 1999; Gravance et al., 2000).

3.2.4 Capacitation status

Before fertilizing of the oocyte, mammalian spermatozoa undergo the sequence of membrane alterations associated with accumulation of calcium ion and the increase of tyrosine phosphorylation resulting in sperm hyperactivation (Hewitt & England, 1998; Petrunkina et al., 2003). At the contact with oocyte, capacitated spermatozoa presents the acrosome reaction which enables the zona pellucida penetration. However, in avian spermatozoa it is believed that a period of capacitation within the female's reproductive tract in order to fertilize ova is not required (Howarth, 1971). The hen oocyte is not surrounded by cumulus cells that would require a different way of sperm motility to pass them trough. It may therefore be suggested that there is no need for motility hyperactivation to prepare for the acrosome reaction in the chicken and that this special motility pattern has not been developed in birds (Lemoine et al., 2008).

The capacitation of the mammalian spermatozoa is assessed by using chlorotetracycline assay (CTC), lectins, measurements of CASA motility characteristics and assessment of thyrosine phosphorylation within plasma membrane (Guérin et al., 1999; Hewitt & England,

1998; Petrunkina et al., 2004; Rota et al., 1999). Fluorescent antibiotic CTC is used to assess the destabilization of sperm membrane. Neutral and uncomplexed CTC crosses over the cell membrane, enters intracellular compartments and binds to free calcium ions. During these events, CTC becomes negatively charged and after creating CTC-Ca+2 complexes becoms more fluorescent. Thus CTC can be used as a tool to distinguish capacitated and uncapacitated spermatozoa. Three classes of sperm cells may be assessed: uncapacitated and acrosome intact (F-pattern, an overall staining of the sperm head), capacitated and acrosome intact (B-pattern, a prominent staining of the apical area of the sperm head) and capacitated and acrosome reacted (AR-pattern, loss of staining of the sperm head) (Maxwell & Johnson, 1997). CTC may be combined with Hoechst 33258, to simultaneous assessment of percentage of live cells and capacitation status (Hewitt & England, 1998).

The exposure of spermatozoa to low temperatures shortens their capacitation time, changing the membrane lipid architecture, membrane permeability and the reducing efficiency of enzymes extruding calcium ions. These changes resemble capacitation, and are likely to reduce long-term sperm viability and alter their motility (Watson, 1995). Therefore, the researchers have introduced the term "cryocapacitation" to emphasize the fact that cryopreservation procedures induce capacitation-like changes in spermatozoa (Bailey et al., 2000; Cormier & Bailey, 2003; Watson, 1995). These cooling-related capacitation-like changes in spermatozoa, may affect the fertility of cryopreserved semen, by rendering the cells less stable in the reproductive tract, after artificial insemination and therefore relatively short-lived. Such changes cannot easily be distinguished from true capacitation, but Green & Watson (2001) were able to establish that the capacitation-like changes in pig spermatozoa differed from true capacitation in the pattern of tyrosine phosphorylation of proteins.

An increase in both, plasma membrane phospholipid scrambling and phospholipid disorder, during capacitation is associated with enhanced plasma membrane fluidity (Gadella & Harrison, 2002). During freeze–thaw cycle, the sperm membranes undergo lipid phase transition that also leads to an increased disorder of phospholipid packing and membrane fluidity, which causes poor control of intracellular calcium concentration (Bailey & Buhr, 1994; Holt, 2000). Therefore, an alternative stain for assessment of capacitation status of spermatozoa is the hydrophobic probe Merocyanine 540 (M540). This stain detects a decreased packing order of phospholipids in the outer leaflet of the plasma membrane lipid bilayer. Due to the fact that M540 earlier detects changes in the membrane fluidity than CTC, therefore, the hydrophobic probe is believed to be better for evaluating the early events of capacitation (Rathi et al., 2001).

3.2.5 Lipid peroxidation

A content of polyunsaturated fatty acids (PUFAs) in phospholipids of spermatozoa membranes makes them especially susceptible to lipid peroxidation (LPO) (Aitken et al., 1993). LPO is a chain reaction with the formation of lipid peroxides and ultimately the formation of cytotoxic aldehydes (Aitken, 1995). In spermatozoa, peroxidation of lipids has critical consequences. Oxidation reactions in biomembranes lead to amplification of reactive oxygen species (ROS), change in membrane fluidity, loss of compartmentalization and plasma-membrane integrity, disturbance of ion-gradients, impairment of lipid-protein interactions, modification of DNA and proteins (Halliwell & Chiroco, 1993).

Effect of oxidative stress is particularly important during the storage of sperm and its cryopreservation. The analysis of semen of mammalian and avian species, showed that the production of ROS and LPO occurrence is increased during freezing- thawing (Bilodeau et al., 2000; Chatterjee and Gagnon, 2001, Guthrie & Welch, 2007; Neild et al., 2005; Partyka et al., 2011b). The main site of their formation are mitochondria (Brouwers & Gadella, 2003) and sperm cell membranes (Agarwal et al., 2005), which are particularly vulnerable to damage from sudden temperature changes. Although, aerobic cells have substrates and enzymes to prevent or restrict the formation and propagation of ROS, but the antioxidant defence of spermatozoa are relatively weak and these germ cells are very susceptible to oxidative stress (Jones & Mann, 1977).

As an alternative to the colorimetric detection of lipid peroxide formation, a fluorescent membrane probe 4,4-difluoro-5-(4-phenyl-1,3-butadienyl)-4-bora-3a,4a-diaza-s-indacene-3-undecanoic acid (C_{11}-BODIPY[581/591]) has recently been successfully used in human, equine, bovine, porcine, feline and chicken's and goose's spermatozoa (Aitken et al., 2007; Almeida & Ball, 2005; Brouwers & Gadella, 2003; Brouwers et al., 2005; Neild et al., 2005; Partyka et al., 2011a,b; Thuwanut et al., 2009). This is an oxidation-sensitive fluorescent fatty acid analogue, that is easily incorporating into membranes and fluoresces red in the intact state, but turns green after undergoing peroxidation (Drummen et al., 2002). C_{11}-BODIPY[581/591] oxidation is virtually insensitive to environmental changes and the probe does not spontaneously leave the lipid bilayer after oxidation, moreover the extent of peroxidation is correlated with the formation of hydroxyl- and hydroperoxiphosphatidylcholine (Brouwers & Gadella, 2003; Brouwers et al., 2005). The degree of probe peroxidation can be followed in separate sperm subpopulations using flow cytometry, or localized in individual sperm using fluorescence microscopy. Moreover, the use of combination C_{11}-BODIPY[581/591] with PI makes it possible to distinguish the presence of reactive oxygen and nitrogen species in the hydrophobic part of lipid bilayers of live sperm from dead cell membranes (Fig. 6d).

For monitoring the intracellular level of ROS, such as hydrogen peroxide (H_2O_2) in the spermatozoa, the fluorescent dye 5-(and-6)-carboxy-20,70-dichlorodihydrofluorescein diacetate (carboxy-H2DCFDA) can be used. Viable spermatozoa are differentiated from dead cells by a counterstain - propidium iodide and the subpopulations of sperm with a high H_2O_2 level (strong fluorescence) and with low H_2O_2 level (weak fluorescence) can be distinguished.

3.2.6 Apoptotic changes

Apoptosis is a physiological mechanism required for any organism function. In contrast to necrosis, apoptosis is a process, where cells play an active role in their own death. Apoptosis comprising of a complex phenomenon that includes three stages: induction, execution and degradation. The most significant changes related to apoptosis are the externalization of the phosphatidylserine (PS), DNA fragmentation, caspase activation, loss of mitochondrial membrane potential, and increase in sperm membrane permeability (Bratton et al., 1997; Glander & Schaller, 1999; Martin et al., 2004;). Several pathways are reported for mammalian cell apoptosis. These include the intrinsic, extrinsic, and apoptosis-inducing factors. During the early phases of disturbed membrane function, asymmetry of the membrane phospholipids occurs, before the integrity of the plasma membrane is progressively damaged (Martin et al., 1995). When the cell membrane is disturbed, the phospholipid PS is translocated from the inner to the outer leaflet of the plasma membrane (Desagher & Martinou, 2000).

It is widely known that the cryopreservation usually causes sublethal cryodamage to spermatozoa, decreasing post-thaw cell viability. The freezing-thawing of human (Glander & Schaller, 1999), bull (Martin et al., 2004), and boar (Pena et al., 2003), stallion (Ortega Ferrusola et al., 2008), and dog (Kim et al., 2010) spermatozoa induces membrane PS translocation, what demonstrates that cryopreservation leads to apoptosis. Therefore, detecting early phases of membrane dysfunction, or initial phases of apoptosis of viable spermatozoa, would be important when evaluating stressed spermatozoa, such as those subjected to freezing and thawing, and would be useful for controlling freezing procedures in semen.

Annexin V is calcium-dependent phosphatidylserine (PS) binding protein conjugated with fluorochrome – FITC or Alexa Fluor®. The properties of Annexin V allow for detection of externally exposed PS. In ejaculated spermatozoa PS is confined to the cytoplasmatic side of the plasma membrane (Gadella et al., 1999). Different categories of apoptotic, necrotic and viable cells can then be sorted out using AnnexinV with PI, through flow cytometer (Fig. 6e), or visually evaluated using fluorescent microscope.

After induction of apoptosis, mitochondrial pores are being opened, leading to a decrease in mitochondrial membrane potential. Therefore, described above JC-1 dye is used for monitoring of apoptotic changes in spermatozoa, too (Ortega Ferrusola et al., 2009). Mentioned above, the opening of mitochondrial pores causes the release of proapoptotic factors into the cytoplasm, where they are activated. These factors – caspases, are central components in the apoptosis signaling cascade. The detection of activated caspases in living spermatozoa can be performed using fluorescence labeled inhibitors of caspases (FLICA™). It allows investigating caspase activation in semen samples with regard to a single cell. The FLICA™ reagent is comprised of 3 segments—it includes a green (FAM 5 carboxyfluorescein) fluorescent label; an amino acid peptide inhibitor sequence targeted by the active caspase; and a fluoromethylketone group (FMK), which acts as a leaving group and forms a covalent bond with the active enzyme. Fluorescence labeled inhibitors of caspases are cell permeable and noncytotoxic (cited by Grunewald et al., 2009). Martin et al. (2004) showed that cryopreservation of bovine spermatozoa induced the significant increase in the proportion of cells with active caspases, which were mainly detected in the intermediate piece of spermatozoa.

3.2.7 DNA status of spermatozoa

DNA integrity has been considered as an important parameter in the determination of spermatozoa ability to withstand the cryopreservation process. It is suggested that chromatin structure should be studied as an independent complementary parameter for the better assessment of the sperm quality (Evenson et al., 2002). The spermatozoal chromatin is much more compact when compared to somatic and spermatogenic cell types (e.g., spermatognia, spermatocytes and spermatids). It appears that during freezing-thawing procedure the integrity of the nuclear DNA, which is related to fertility, could be negatively affected. Although, spermatozoa with DNA damage may be able to fertilize an oocyte, that could potentially disturb (epi)genetic regulation of the early embryo and block its further development (Lewis & Aitken, 2005).

DNA damage can be evaluated at different levels. One of the usually used methods, developed for detecting changes in the chromatin structure of DNA integrity, is the sperm chromatin structure assay (SCSA) (Chohan et al., 2006). The SCSA is a flow cytometric method for identification of changes in the DNA status. It is based on the assumption that a structurally abnormal sperm chromatin shows a higher susceptibility to acid denaturation (Evenson et al., 2002). The SCSA method utilizes the metachromatic properties of acridine orange (AO). This stain fluoresces in the green band when intercalates into the intact double-stranded DNA helix, and in the red band when associated with single strand denaturated DNA and RNA. After denaturation of chromatin by decreased pH, the spermatozoa with structurally abnormal chromatin fluorescence is detected in the red band (Fig. 6f) (Bochenek et al., 2001). The fertility data have been shown to correlate with the results obtained from the SCSA of human (Evenson et al., 1980), bull (Ballachey et al., 1988; Karabinus et al., 1990), stallion (Love & Kenney, 1998) and boar semen (Evenson et al., 1994). SCSA was also used for dog semen assessment (Garcia-Macias et al, 2006) and for evaluation of freezing-thawing effect on chicken and goose DNA status (Partyka et al. 2010; Partyka et al., 2011b).

Another method to detect DNA defragmentation is TUNEL assay, which allows to incorporate of fluorescent nucleotide analogs by a terminal nucleotide transferase into single stranded DNA areas at the 3-OH termini (Chohan et al., 2006). Ramos & Wetzels (2001) using this method have shown that DNA damage is limited in functional human spermatozoa resulting from a swim-up procedure.

The alternative method for detecting the DNA damage at the level of individual cells is the single-cell DNA gel electrophoresis assay (COMET). Although this method does not use such equipment as flow cytometry, application of fluorescent DNA specific stain is required. In COMET assay spermatozoa are spread on a surface covered with an agarose gel, and treated with a solution that lyses the cell components leaving the DNA immobilized in the agarose. They are then subjected to a DNA denaturation process, followed by electrophoresis, causing DNA fragments to migrate away from the main bulk of nuclear DNA. After staining with propidium iodide or ethidium bromide, cells with DNA strand breaks, display a comet-like shape, with the undamaged DNA located in the head of the comet and the fragmented DNA dispersed through the tail. Image analyses provide information on the extent of strand breaks in the DNA molecule. Several studies, conducted with different techniques, including comet assay, showed a negative relationship between the fertilization potential of spermatozoa and alterations at the level of genetic material. In particular in humans, infertility has been associated with higher levels of DNA damage in sperm compared to fertile subjects (Irvine et al., 2000). Fraser & Strzeżek (2007) have shown that the freezing–thawing process provoked sperm chromatin destabilization rendering the boar spermatozoa more vulnerable to DNA fragmentation. COMET assay has also been recently used for the evaluation of cryopreserved avian semen (Madeddu et al., 2010; Gliozzi et al., 2011).

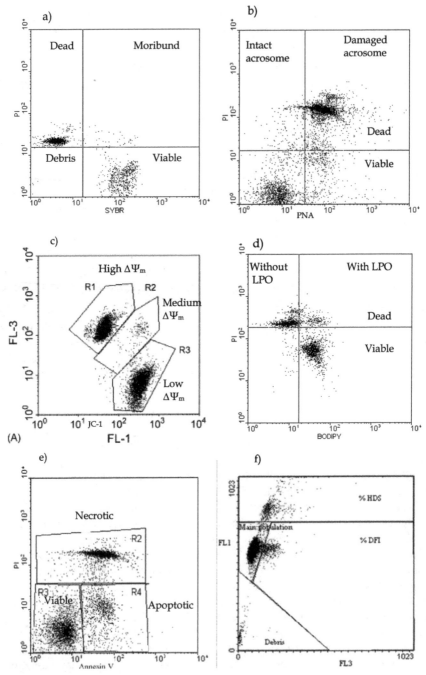

Fig. 6. Examples of flow cytometry analyses of frozen-thawed spermatozoa: a) Dot plot of SYBR-14/PI stain. Four subpopulations can be distinguished: dead sperm (red stained),

moribund sperm (red/green fluorescence), viable sperm (green stained), unstained debris are discarded; b) Dot plot of PNA-AlexaFluor/PI stain. Spermatozoa can be identified as: acrosome - intact/damaged, together with selection - viable/dead, according to their green and red fluorescence; c) Dot plot of JC-1 staining for mitochondrial status analysis. The intensity of orange fluorescence depends on mitochondrial membrane potential ($\Delta\Psi_m$) allowing for differentiation between high, medium and low $\Delta\Psi_m$; d) Dot plot of C_{11} BODIPY$^{581/591}$/PI for assessment of lipid peroxidation (LPO). Spermatozoa can be divided into four subpopulations: dead without LPO, dead with LPO, live without LPO and live with LPO; e) Dot plot of Annexin V/PI stain. Spermatozoa can be identified as: viable, necrotic and apoptotic; f) Dot plot of SCSA using acridine orange. The distribution of spermatozoa is based on green (FL1) and red (FL3) fluorescence. Main population includes sperm without DNA fragmentation, %DFI represents the percentage of sperm with detectable DNA fragmentation and % HDS determines the percentage of immature cells.

3.3 In vitro gamete interaction tests

During fertilization, a sperm initially binds to the oocyte zona pellucida (ZP), undergoes the acrosome reaction (AR), penetrates the ZP, and fuses with the oolemma to form a zygote. Sperm-ZP interactions are carbohydrate-mediated events in various species, including humans (Benoff, 1997). The ZP of mammalian oocytes is a critical site for sperm-oocyte interaction. The ability of sperm to bind to the ZP indicates many functions of spermatozoa, such as viability, motility, morphology, acrosomal status and the ability to penetrate the oocyte (Liu & Baker, 1994), and for that reason this ability is of a diagnostic relevance.

3.3.1 Zona pellucida binding assay

The assessment of the ability of sperm cells to bind the homologous zona pellucida (ZP) is the useful test for prediction of spermatozoal fertilizing ability (Hermansson et al., 2006). It is assumed that it is reliable test to detect sperm damage at a molecular level, which is not visible by microscopic analysis, because binding is receptor-ligand mediated reaction. The test may be done in two ways: by using intact homologous oocytes (ZP-binding assay, ZBA) and by using bisected hemizonae (hemizona binding assay, HZA) (Kawakami et al., 1998; Rijsselaere et al., 2005). In ZBA spermatozoa are coincubated with oocytes obtained from sliced ovaries. The number of spermatozoa that bound to ZP is counted with contrast-phase microscopy. The disadvantage of ZBA is the fact that the attachment of sperm cells to zona depends on the oocyte. This feature was partly overcome in HZA. Bisected by micromanipulation two parts of ZP are coincubated with spermatozoa. As a result the direct comparison of sperm cells from two origins may be done (Ivanova et al., 1999; Mayenco-Aguirre & Pérez Cortés, 1998).

A sublethal damage that occurres during cryopreservation leads to loss of sperm surface proteins, segregation of membrane proteins, inactivation of membrane-bound enzymes and decreased lateral protein diffusion within the membrane (Watson, 1995). Kadirvel et al. (2011) observed significant reduction of the zona binding ability of cryocapacitated buffalo bulls spermatozoa, and further reduction of binding ability of frozen-thawed spermatozoa, after incubation, in either capacitating, or non capacitating medium. Similar results have been obtained in bulls (Fazeli et al., 1997) and humans (Amann et al., 1999) spermatozoa, with significantly reduced binding ability to the zona pellucida after freezing and thawing.

The reduced binding ability of the frozen-thawed spermatozoa might be due to the higher proportion of acrosome reacted and damaged spermatozoa, after cryopreservation and thawing. Moreover, impaired receptor-ligand interaction in frozen-thawed spermatozoa could be caused by cryoelution of an "essential ligand" from the sperm surface that has been described in human (Amann et al., 1999).

3.3.2 Oocyte penetration assay

Oocyte penetration assays (OPT) involve multiple sperm penetrations of each oocyte and permit the observation of pronuclear development (Yanagimachi et al., 1976). The application of the zona-free hamster oocyte assay has been used to assess the fertility of men (Freeman et al., 2001;) and domestic animals (Cormier et al., 1997; De los Reyes et al., 2009; Hewitt & England, 1997; Maxwell et al., 1996;). The OPT is a less time-consuming technique than in vitro fertilization (IVF) test, because oocytes can be immatured, and after evaluation are not further subjected to development. In this assay spermatozoa presented in the pervitelline space and ooplasm of the oocytes are observed under fluorescent microscopy using Hoechst 33258, PI or light microscopy (aceto-orcein) (Hay et al., 1997; Hewitt & England, 1997).

All changes in cryopreserved spermatozoa described in the above sections may affect the final percentage of fertilized oocytes, and also the time course of sperm penetration through the oocyte envelop, as reported previously in frozen-thawed ram and bovine sperm (Cormier et al., 1997; Maxwell et al., 1996). Nevertheless, the previous study has indicated that the major ability of cryopreserved sperm to penetrate oocytes occurs at the 1st hour of co-culture (Cormier et al., 1997; De los Reyes et al., 2009). This finding indicates that these sperm can undergo the events associated with fertilization earlier or faster than fresh sperm in relation with cryocapacitation appearance.

Because the efficiency of oocyte penetration is a result of sperm–oocyte interaction, variation in oocyte properties are likely to produce large diversity in this assay results. However, this can be reduced with the use of a large number of oocytes (Lucas et al., 2003). However, in dogs in vitro maturation (IVM) of oocytes and IVF is difficult to achieve. Nevertheless capacitated dog spermatozoa are able to penetrate immature oocytes, inducing chromatin decondensation and resumption of meiosis (Luvoni et al., 2005; Hay et al., 1994; Sain-Dizier et al., 2001). Thus, in dogs both, immature or mature oocytes, may be use for this test.

4. Conclusions

For many years, scientists have made every endeavour to develop laboratory assays that precisely estimate the fertilizing capacity of semen. Laboratory semen appraisals can be classified in several ways. Nevertheless, an important factor for a laboratory analysis to be useful, it must be objective, repeatable, accurate and as far as possible, rapid. Among others, there can be distinguished one major division into conventional methods and advanced techniques of sperm assessment. It is little questionable, whether the subjective assessment of parameters related to the functional and morphological characteristics of spermatozoa, would increase the predictability of the fertilizing potential of cryopreserved semen. However, conventional methods for sperm evaluation in connection with the more objective computer-assisted sperm analyzers, flow cytometry and in vitro fertilization tests, have enabled researchers to gain accurate information about the morpho-functional status of spermatozoa and mechanisms of sperm cryoinjury.

5. Acknowledgments

This work was supported by the Polish National Science Centre, grant no N N311 530040.

6. References

Agarwal, A.; Ozturk E. & Loughlin, K.R. (1992). Comparison of semen analysis between the two Hamilton-Thorn semen analysers. *Andrologia* 24, 327–329.

Agarwal, A.; Prabakaran, S.A. & Said T.M. (2005). Prevention of oxidative stress injury to sperm. *J Androl* 26, 654-60.

Aitken, R.J. (1995). Free radicals, lipid peroxidation and sperm function. *Reprod Fertil Dev* 7, 659-668.

Aitken, R.J.; Harikss, D. & Buckingham, DW. (1993). Analysis of lipid peroxidation mechanisms in human spermatozoa. *Mol Reprod Dev* 35, 302-315.

Aitken, R.J.; Wingate, J.K.; De Iuliis, G.N. & McLaughlin, E.A. (2007). Analysis of lipid peroxidation in human spermatozoa using BODIPY C11. *Mol Hum Reprod* 13, 203-211.

Almeida, J. & Ball, B.A. (2005). Effect of α-tocopherol and tocopherol succinate on lipid peroxidation in equine spermatozoa. *Anim Reprod Sci* 87, 321-337.

Álvarez, M.; García-Macías, V.; Martínez-Pastor, F.; Martínez, F.; Borragán, S.; Mata, M.; Garde, J.; Anel, L. & De Paz, P. (2008). Effects of cryopreservation on head morphometry and its relation with chromatin status in brown bear (Ursus arctos) spermatozoa. *Theriogenology* 70, 1498-1506.

Amann, R.P. (1999). Cryopreservation of sperm. In: *Encyclopedia of Reproduction*, vol. 1. Academic Press, Burlington, MA, USA, 773-783.

Amann, R.P.; Shabanowitz, R.B.; Huszar, G. & Broder, S.J. (1999). In vitro sperm-binding assay to distinguish differences in populations of human sperm or damage to sperm resulting from cryopreservation. *J Androl* 20, 648–654.

Arruda, R.P.; Ball, B.A.; Gravance, C.G.; Garcia, A.R. & Liu IKM. (2002). Effects of extender and cryoprotectants on stallion sperm head morphometry. *Theriogenology* 58, 253–6.

Bailey, J.L. & Buhr, M.M. (1994). Cryopreservation alters the Ca2+ flux of bovine spermatozoa. *Can J Anim Sci* 74, 45–52.

Bailey, J.L., Bilodeau J.F. & Cormier, N. (2000). Semen cryopreservation in domestic animals: a damaging and capacitating phenomenon. *J Androl* 21, 1-7.

Ballachey, B.E.; Evenson, D.P. & Saacke, R.G. (1988). The sperm chromatin structure assay. Relationship with alternate tests of semen quality and heterospermic performance of bulls. *J Androl* 9, 109-115.

Benoff, S. (1997). Carbohydrates and fertilization: an overview. *Mol Hum Reprod* 3, 599–637.

Bilodeau, J-F., Chatterjee, S.; Sirard, M-A. & Gagnon, C. (2000). Levels of antioxidant defences are decreased in bovine spermatozoa after a cycle of freezing and thawing. *Mol Reprod Dev* 55, 282-288.

Blesbois, E.; Grasseau, I.; Seigneurin, F.; Mignon-Grasteau, S.; Saint Jalme, M. & Mialon-Richard, M.M. (2008). Predictors of success of semen cryopreservation in chickens. *Theriogenology* 69, 252-261.

Blom, E. (1950). Interpretation of spermatic cytology in bulls. *Fertil Steril* 1, 223-38.

Blom, E. (1968). A new sperm defect—"Pseudodroplets"—in the middle piece of the bull sperm. *Nord Vet Med* 20, 279–283.

Blom, E. (1983). Pathological conditions in the genital organs and in the semen as ground for rejection of breeding bulls for import or export to and from Denmark, 1958–1982. *Nord Vet Med* 35, 105–130.

Bochenek, M.; Smorąg, Z. & Pilch, J. (2001). Sperm chromatin structure assay of bulls qualified for artificial insemination. *Theriogenology* 56, 557-567.

Bratton, D.L.; Fadok, V.A.; Richter, D.A.; Kailey, J.M.; Guthrie, L.A. & Henson, P.M. (1997). Appearance of phosphatidylserine on apoptotic cells requires calcium-mediated nonspecific flip-flop and is enhanced by loss of the aminophospholipid translocase. *J Biol Chem* 272, 26159–26165.

Brito, L.F.; Barth, A.D.; Bilodeau-Goeseels, S.; Panich, P.L. & Kastelic, J.P. (2003). Comparison of methods to evaluate the plasmalemma of bovine sperm and their relationship with in vitro fertilization rate. *Theriogenology* 60, 1539-1551.

Brito, L.F.; Greene, L.M.; Kelleman, A.; Knobbe, M. & Turner, R. (2011). Effect of method and clinician on stallion sperm morphology evaluation. *Theriogenology* 76, 745-750.

Brouwers, J.F.H.M. & Gadella, B.M. (2003). In situ detection and localization of lipid peroxidation in individual bovine sperm cells. *Free Radical Biol Med* 35, 1382-1391.

Brouwers, J.F.H.M.; Silva, P.F.N. & Gadella, B.M. (2005). New assay for detection and localization of endogenous lipid peroxidation products in living boar sperm after BTS dilution or after freeze-thawing. *Theriogenology* 63, 458-469.

Chatterjee, S. & Gagnon, C. (2001). Reproduction of reactive oxygen species by spermatozoa undergoing cooling, freezing, and thawing. *Mol Reprod Dev* 59, 451-458.

Chohan, K.R.; Griffin, J.T.; Lafromboise, M. & De Jonge, C.J. (2006). Comparison of chromatin assays for DNA fragmentation evaluation in human sperm. *J Androl* 27, 53-59.

Christensen, P.; Stenvang, J.P. & Godfrey, W.L. (2004). A flow cytometric method for rapid determination of sperm concentration and viability in mammalian and avian semen. *J Androl* 25, 255-264.

Coetzee, K.; Kruger, T.F.; Lombard, C.J.; Shaughnessy, D.; Oehninge,r S.; Ozgür, K.; Pomeroy, K.O. & Muller C.H. (1999). Assessment of interlaboratory and intralaboratory sperm morphology readings with the use of a Hamilton Thorne Research integrated visual optical system semen analyzer. *Fertil Steril* 71, 80-84.

Colenbrander, B.; Gadella, B. & Stout, T. (2003). The Predictive Value of Semen Analysis in the Evaluation of Stallion Fertility. *Reprod Domest Anim* 38, 305-311.

Cormier, N. & Bailey J.L. (2003). A differential mechanism is involved during heparin-and cryopreservation-induced capacitation. *Biol Reprod* 69, 177-185.

Cormier, N.; Sirard, M.A. & Bailey, J. (1997). Premature capacitation of bovine spermatozoa is initiated by cryopreservation. *J Androl* 18, 457–461.

Correa, J.R. & Zavos, P.M. (1994). The hypoosmotic swelling test: its employment as an assay to evaluate the functional integrity of the frozen-thawed bovine sperm membrane. *Theriogenology* 42, 351-360.

Davis, R.O. & Katz D.F. (1992). Standardization and comparability of CASA instruments. *J Androl* 13, 81-86.

De los Reyes, M.; Palomino J.; de Lange, J.; Anguita, C. & Barros, C. (2009). In vitro sperm penetration through the zona pellucida of immature and in vitro matured oocytes using fresh, chilled and frozen canine semen. *Anim Reprod Sci* 110, 37–45.

Desagher, S. & Martinou, J.C. (2000). Mitochondria as the central control point of apoptosis. *Trends Cell Biol* 10, 369–77.

Didion, B.A.; Dobrinsky, J.R.; Giles, J.R. & Graves, C.N. (1989). Staining procedure to detect viability and the true acrosome reaction in spermatozoa of various species. *Gamete Res* 22, 51-57.

Dobrinski, I.; Smith, T.T.; Suarez, S.S. & Ball, A.B. (1997). Membrane contact with oviductal epithelium modulates the intracellular calcium concentration of equine spermatozoa in vitro. *Biol. Reprod* 56, 861-869.

Donoghue, A.M.; Garner, D.L.; Donoghue, D.J. & Johnson, L.A. (1995). Viability assessment of turkey sperm using fluorescent staining and flow cytometry. *Poult Sci* 74, 1191-1200.

Drummen, G.P.; Van Liebergen, L.C.M.; Op Den Kamp, J.A.F. & Post, J.A. (2002). C11-BODIPY581/591, an oxidation-sensitive fluorescent lipid peroxidation probe: (micro)spectroscopic characterization and validation of methodology. *Free Radical Biol Med* 33, 473-490.

Esteso, M.C.; Fernandez-Santos, M.R.; Soler, A.J. & Garde, J.J. (2003). Head dimensions of cryopreserved red deer spermatozoa are affected by thawing procedure. *Cryo Letters* 24, 261–268.

Esteves, S.C.; Sharma, R.K.; Thomas, A.J. & Agarwal, T.A. (2007). Evaluation of acrosomal status and sperm viability in fresh and cryopreserved specimens by the use of fluorescent Peanut agglutinin lectin in conjunction with hypo-osmotic swelling test. *Int Braz J Urol* 33, 364-376.

Evenson, D.P.; Darzynkiewicz, Z. & Melamed, M.R. (1980). Relation of mammalian sperm chromatin heterogeneity to fertility. *Science* 210, 1131-1133.

Evenson, D.P.; Larson, K.L. & Jost, L.K. (2002). Sperm chromatin structure assay: its clinical use for detecting sperm DNA fragmentation in male infertility and comparisons with other techniques. *J Androl* 23, 25-43.

Evenson, D.P.; Thompson, L. & Jost, L. (1994). Flow cytometric evaluation of boar semen by the sperm chromatin structure assay as related to cryopreservation and fertility. *Theriogenology* 41, 637-651.

Fazeli, A.R.; Zhang, B.R.; Steenweg, W.; Larsson, B.; Bevers, M.M.; van den Broek, J.; Rodriguez-Martinez, H. & Colenbrander, B. (1997). Relationship between sperm-zona pellucida binding assay and the 56-day nonreturn rate of cattle inseminated with frozen-thawed semen. *Theriogenology* 48, 853–863.

Fraser, L.R.; Abeydeera, L.R. & Niwa, K. (1995). Ca^{+2} –regulating mechanisms that modulate bull sperm capacitation and acrosome exocytosis as determined by chlortetracycline analysis. *Mol Reprod Dev* 40, 233-241.

Fraser, L. & Strzeżek J. (2007). Is there a relationship between the chromatin status and DNA fragmentation of boar spermatozoa following freezing–thawing? *Theriogenology* 68, 248–257.

Freeman, M.R.; Archibong, A.E.; James, J.; Mrotek, C.M.; Whitworth, G.A. & Weitzman, G.A. (2001). Hill. Male partner screening before in vitro fertilization: preselecting patients who require intracytoplasmic sperm injection with the sperm penetration. *Fertil Steril* 76, 1113-1118.

Freneau, G.E. ; Chenoweth, P.J. ; Ellis, R. & Rupp, G. (2010). Sperm morphology of beef bulls evaluated by two different methods. *Anim Reprod Sci* 118, 176-181.

Gadella, B.M.; Miller, N.G.; Colenbrander, B.; van Golde, L.M. & Harrison, R.A. (1999). Flow cytometric detection of transbilayer movement of fluorescent phospholipid analogues across the boar sperm plasma membrane: elimination of labeling artifacts. *Mol Reprod Dev* 53, 108-125.

Gadella, B.M. & Harrison, R.A. (2002). Capacitation induces cyclic adenosine 3',5'-monophosphate-dependent, but apoptosis-unrelated, exposure of aminophospholipids at the apical head plasma membrane of boar sperm cells. *Biol Reprod* 67, 340-350.

Garcia-Macias, V.; Martinez-Pastor, F.; Alvarez, M.; Garde, J.J.; Anel, E.; Anel, L. & de Paz, P. (2006). Assessment of chromatin status (SCSA®) in epididymal and ejaculated sperm in Iberian red deer, ram and domestic dog. *Theriogenology* 66, 1921-1930.

Garner, D.L. & Thomas, C.A. (1999). Organelle-specific probe JC-1 identifies membrane potential differences in the mitochondrial function of bovine sperm. *Mol Reprod Dev* 53, 222-229.

Garner, D.L. & Johnson L.A. (1995). Viability assessment of mammalian sperm using SYBR-14 and Propidium Iodide. *Biol Reprod* 53, 276-284.

Garner, D.L.; Thomas, C.A.; Joerg, H.W.; DeJarnette, J.M. & Marshall, C.E. (1997). Fluorometric assessments of mitochondrial function and viability in cryopreserved bovine spermatozoa. *Biol Reprod* 57, 1401-1406.

Glander, H.J. & Schaller, J. (1999). Binding of annexin V to plasma membranes of human spermatozoa: a rapid assay for detection of membrane changes after cryostorage. *Mol Hum Reprod* 5, 109–115.

Gliozzi, T.M.; Zaniboni, L. & Cerolini, S. (2011). DNA fragmentation in chicken spermatozoa during cryopreservation. *Theriogenology* 75, 1613–1622.

Graham, J.K. (2001). Assessment of sperm quality: a flow cytometric approach. *Anim Reprod Sci* 68, 239-247.

Graham, J.K.; Kunze, E. & Hammerstedt, R.H. (1990). Analysis of sperm cell viability, acrosomal integritiy, and mitochondrial function using flow cytometry. *Biol Reprod* 43, 55-64.

Gravance, C.G.; Garner, D.L.; Miller, M.G. & Berger, T. (2001). Fluorescent probes and flow cytometry to assess rat sperm integrity and mitochondrial function. *Reprod Toxicol* 15, 5-10.

Gravance, C.G.; Garner, D.L.; Baumber, J. & Ball, BA. (2000). Assessment of equine sperm mitochondrial function using JC-1. *Theriogenology* 53, 1691-1703.

Gravance, C.G.; Vishwanath, R.; Pitt, C.; Garner, D.L. & Casey, P.J. (1998). Effects of cryopreservation on bull sperm head morphometry. *J Androl* 19, 704–709.

Green, C.E. & Watson, P.F. (2001). Comparison of the capacitation-like state of cooled boar spermatozoa with true capacitation. *Reproduction* 122, 889–898.

Grunewald, S.; Sharma, R.; Paasch, U.; Glander, H.J. & Agarwal, A. (2009). Impact of Caspase Activation in Human Spermatozoa. *Micros Res Techniq* 72, 878–888.

Guérin, P. ; Ferrer, M. ; Fontbonne, A. ; Bénigni, L. ; Jacquet, M. & Ménézo, Y. (1999). In vitro capacitation of dog spermatozoa as assessed by chlortetracycline staining. *Theriogenology* 52, 617-628.

Guthrie, H.D. & Welch, G.R. (2007). Use of fluorescence-activated flow cytometry to determine membrane lipid peroxidation during hypothermic liquid storage and freeze-thawing of viable boar sperm loaded with 4,4-difluoro-5-(4-phenyl-1,3-butadienyl)-4-bora-3a,4a-diaza-s-indacene-3-undecanoic acid. *J Anim Sci* 85, 1402-1411.

Halliwell, B. & Chiroco S. (1993). Lipid peroxidation: its mechanism, measurement and significance. *Am J Clin Nutr* 57, 715–725.

Hammerstedt, R.H.; Graham, J.K. & Nolan, J.P. (1990). Cryopreservation of mammalian sperm: what we ask them to survive. *J Androl* 11, 73-88.

Hay, M.A.; King, W.A.; Gartley, C.J. & Goodrowe, K.L. (1994). Influence of spermatozoa on in vitro nuclear maturation of canine ova. *Biol Reprod* 50, 362.

Hay, M.A.; King, W.A.; Gartley, C.J.; Leibo, S.P. & Goodrowe, K.L. (1997). Effects of cooling, freezing and glycerol on penetration of oocytes by spermatozoa in dogs. *J Reprod Fertil* (Suppl. 51), 99-108.

Hermansson, U.; Ponglowhapan, S.; Linde-Forsberg, C. & Ström-Holst, B. (2006). A short sperm-oocyte incubation time ZBA in the dog. *Theriogenology* 66, 717-725.

Hewitt, D.A. & England, G.C. (1998). An investigation of capacitation and the acrosome reaction in dog spermatozoa using a dual fluorescent staining technique. *Anim Reprod Sci* 51, 321-332.

Hewitt, D.A. & England, G.C.W. (1997). The canine oocyte penetration assay; its use as an indicator of dog spermatozoa performance in vitro. *Anim Reprod Sci* 50, 123-139.

Hidalgo, M.; Rodriguez, I. & Dorado, J.M. (2006). The effect of cryopreservation on sperm head morphometry in Florida male goat related to sperm freezability. *Anim Reprod Sci* 100, 61–72.

Holden, C.A.; Hyne, R.V.; Sathananthan, A.H. & Trounson, A.O. (1990). Assessment of the human sperm acrosome reaction using concanavalin A lectin. *Mol Reprod Dev* 25, 247-257.

Holt, W.V. (2000). Fundamental aspects of sperm cryobiology: the importance of species and individual differences. *Theriogenology* 53, 47-58.

Howarth, B.Jr. (1970). An examination for sperm capacitation in the fowl. *Biol Reprod* 3 338-341.

Iguer-Ouada, M. & Verstegen J. P. (2001). Validation of the Sperm Quality Analyzer (SQA) for dog sperm analysis. *Theriogenology* 55, 1143-1158.

Irvine, D.S.; Twigg, J.P.; Gordon, E.L.; Fulton, N.; Milne, P.A. & Aitken, R.J. (2000). DNA integrity in human spermatozoa: relationships with semen quality. *J Androl* 21, 33-44.

Ivanova, M.; Mollova, M.; Ivanova-Kicheva, M.G.; Petrov, M.; Djarkova, T. & Somlev, B. (1999). Effect of cryopreservation of zona-binding capacity of canine spermatozoa in vitro. *Theriogenology* 52, 163-170.

Jeyendran, R.S.; van der Venn, H.H. & Zaneveld, L.J.D. (1992). The hypoosmotic swelling test: an update. *Arch Androl* 29, 105–116.

Jones, R. & Mann, T. (1977). Damage to ram spermatozoa by peroxidation of endogenous phospholipids. *J Reprod Fert* 50, 261-268.

Kadirvel, G., Kathiravan, P. & Kumar, S. (2011). Protein tyrosine phosphorylation and zona binding ability of in vitro capacitated and cryopreserved buffalo spermatozoa. *Theriogenology* 75, 1630-1639.

Karabinus, D.S.; Evenson, D.P.; Jost, L.K.; Baer, R.K. & Kaproth, M.T. (1990). Comparison of semen quality in young and mature Holstein bulls measured by light microscopy and flow cytometry. *J Dairy Sci* 73, 2364-2371.

Katkov, I.I. & Lulat, A.G-M.I. (2000). Do conventional CASA-parameters reflect recovery of kinematics after freezing?: "CASA paradox" in the analysis of recovery of spermatozoa after cryopreservation. *CryoLetters* 21, 141-148.

Kawakami, E.; Kashiwagi, C.; Hori, T. & Tsutsui, T. (2001). Effects of canine oviduct epithelial cells on movement and capacitation of homologous spermatozoa in vitro. *Anim Reprod Sci* 68, 121-131.

Kawakami, E.; Morita, Y.; Hori, T. & Tsutsui, T. (2002). Lectin-binding characteristics and capacitation of canine epididymal spermatozoa. *J Vet Med Sci* 64, 543-549.

Kawakami, E.; Hori, T. & Tsutsui, T. (1998). Changes in semen quality and in vitro sperm capacitation during various frequencies of semen collection in dogs with both asthenozoospermia and teratozoospermia. *J Vet Med Sci* 60, 607-614.

Kawakami, E.; Vandevoort, C.A.; Mahi-Brown, C.A. & Overstreet, J.W. (1993). Induction of acrosome reactions of canine sperm by homologous zona pellucida. *Biol Reprod* 48, 841-845.

Kim, S-H., Yu D-H. & KimY-J. (2010). Effects of cryopreservation on phosphatidylserine translocation, intracellular hydrogen peroxide, and DNA integrity in canine sperm. *Theriogenology* 73, 282–292.

Kumi-Diaka, J. (1993). Subjecting canine semen to the hype-osmotic teat. *Theriogenology* 39, 1279-1289.

Lemoine, M.; Grasseau, I.; Brillard, J.P. & Blesbois, E. (2008). A reappraisal of the factors involved in in vitro initiation of the acrosome reaction in chicken spermatozoa. *Reproduction* 136, 391–399.

Lewis, S.E. & Aitken, R.J. (2005). DNA damage to spermatozoa has impacts on fertilization and pregnancy. *Cell Tissue Res* 322, 33–41.

Liu, D.Y. & Baker, H.W.G. (1994). A new test for the assessment of sperm-zona pellucida penetration: relationship with results of other sperm tests and fertilization in vitro. *Hum Reprod* 3, 489-496.

Love, C.C. & Kenney, R.M. (1998). The relationship of increased susceptibility of sperm DNA to denaturation and fertility in the stallion. *Theriogenology,* 50, 955-972.

Lucas, X.; Martínez, E.A.; Roca, J.; Vázquez, J.M.; Gil, M.A.; Pastor, L.M. & Alabart, J.L. (2003). Influence of follicle size on the penetrability of immature pig oocytes for homologous in vitro penetration assay. *Theriogenology* 60, 659-567.

Łukaszewicz, E.; Jerysz, A.; Partyka, A. & Siudzińska, A. (2008). Efficacy of evaluation of rooster sperm morphology using different staining methods. *Res Vet Sci* 85, 583-688.

Luvoni, G.C.; Chigioni, S.; Allievi, E. & Macis, D. (2005). Factors involved in vivo and in vitro maturation of canine oocytes. *Theriogenology* 63, 41-59.

Madeddu, M.; Berlinguer, F.; Pasciu, V.; Succu, S.; Satta, V.; Leoni, G.G.; Zinellu, A.; Muzzeddu, M.; Carru, C. & Naitana, S. (2010). Differences in semen freezability and intracellular ATP content between the rooster (Gallus gallus domesticus) and the Barbary partridge (Alectoris barbara). *Theriogenology* 74, 1010–1018.

Maree, L.; du Plessis, S.S.; Menkveld, R. & van der Horst, G. (2010). Morphometric dimensions of the human sperm head depend on the staining method used. *Hum Reprod* 25, 1369 –1382.

Martin, G.; Sabido, O.; Durand, P. & Levy, R. (2004). Cryopreservation induces an apoptosis like mechanism in bull sperm. *Biol Reprod* 71, 28 –37.

Martin, S.J.; Reutelingsperger, C.P.; McGahon, A.J.; Rader, J.A.; van Schie, R.C.; LaFace, D.M. & Green, D.R. (1995). Early redistribution of plasma membrane phosphatidylserine is a general feature of apoptosis regardless of the initiating stimulus: inhibition by overexpression of Bcl-2 and Abl. *J Exp Med* 182, 1545–1556.

Maxwell, W.C.; Catt, S.L. & Evans, G. (1996). Dose of fresh and frozen–thawed spermatozoa for in vitro fertilization of sheep oocytes. *Theriogenology* 45, 261.

Maxwell, W.M. & Johnson, L.A. (1997). Chlortetracycline analysis of boar spermatozoa after incubation, flow cytometric sorting, cooling, or cryopreservation. *Mol Reprod Dev* 46, 408-418.

Mayenco-Aguirre, A.M. & Pérez Cortés, A.B. (1998). Preliminary results of hemizona assay (HZA) as a fertility test for canine spermatozoa. *Theriogenology* 50, 195-204.

Meyers S.A., Yudir A.I., Cherr G.N., VandeVoort C.A., Myles D.G., Primakoff P. & Overstreet J.W., (1997) Hyaluronidase activity of macaque sperm assessed by in vitro cumulus penetration assay. *Mol Reprod Dev* 46, 392-400.

Nagy, S.; Hallap, T.; Johannisson, A. & Rodriguez-Martinez, H. (2004). Changes in plasma membrane and acrosome integrity of frozen-thawed bovine spermatozoa during a 4h incubation as measured by multicolor flow cytometry. *Anim Reprod Sci* 80, 225-235.

Neild, D.; Chaves, G.; Flores, M.; Mora, N.; Beconi, M. & Agüero, A. (1999). Hypoosmotic test in equine spermatozoa. *Theriogenology* 51, 721-727.

Neild, D.M.; Brouwers, J.F.H.M.; Colenbrander, B.; Aguero, A. & Gadella, B.M. (2005). Lipid peroxide formation in relation to membrane stability of fresh and frozen thawed stallion spermatozoa. *Mol Reprod Dev* 72, 230-238.

Niżański, W.; Klimowicz, M.; Partyka, A.; Savić, M. & Dubiel, A. (2009). Effects of the inclusion of Equex STM into Tris-based extender on the motility of dog spermatozoa incubated at 5 degrees C. *Reprod Domest Anim* 44, (Suppl. 2), 363-365.

Ortega Ferrusola, C.; Sotillo Galán, Y.; Varela Fernández, E.; Gallardo Bolaños, J.M.; González Fernández, L.; Tapia, J.A. & Peña, F.J. (2008). Detection of apoptosis like changes during the cryopreservation process in equine sperm. *J Androl* 29, 213-221.

Ortega-Ferrusola, C.; Macías García, B.; Gallardo-Bolaños, J.M.; González-Fernández, L.; Rodríguez-Martinez, H.; Tapia, J.A. & Peña, F.J. (2009). Apoptotic markers can be used to forecast the freezeability of stallion spermatozoa. *Anim Reprod Sci* 114, 393-403.

Parks, J.E. & Graham, J.K. (1992). Effects of cryopreservation procedures on sperm membranes. *Theriogenology* 38, 209-222.

Partyka, A.; Jerysz, A. & Pokorny, P. (2007). Lipid peroxidation in fresh and stored semen of Green-legged Partridge. *E.J.P.A.U.* 10, (2),
http://www.ejpau.media.pl/volume10/issue2/art-08.html

Partyka, A.; Łukaszewicz, E. & Niżański, W. (2011a). Flow cytometric assessment of fresh and frozen-thawed Canada goose (Branta canadensis) semen. *Theriogenology* 76, 843–850.

Partyka, A.; Łukaszewicz, E.; Niżański, W. & Twardoń, J. (2011b). Detection of lipid peroxidation in frozen-thawed avian spermatozoa using C11-BODIPY581/591. *Theriogenology* 75, 1623-1629.

Partyka, A.; Niżański, W. & Łukaszewicz, E. (2010). Evaluation of fresh and frozen-thawed fowl semen by flow cytometry. *Theriogenology* 74, 1019-1027.

Peña Martínez, A.I. (2004). Canine fresh and cryopreserved semen evaluation. *Anim Reprod Sci* 82-83, 209-224.

Peña, A.I.; Johannisson, A. & Linde-Forsberg, C. (2001). Validation of flow cytometry for assessment of viability and acrosomal integrity of dog spermatozoa and for evaluation of different methods of cryopreservation. *J Reprod Fertil* (Suppl. 57), 371-376.

Peña, A.I.; Quintela, L.A. & Herradón, P.G. (1998). Viability assessment of dog spermatozoa using flow cytometry. *Theriogenology* 50, 1211-1220.

Pena, F.J.; Johannisson, A.; Wallgren, M. & Rodriguez-Martinez, H. (2003). Assessment of fresh and frozen-thawed boar semen using an Annexin-V assay: a new method of evaluating sperm membrane integrity. *Theriogenology* 60, 677–689.

Pérez-Llano, B.; Lorenzo, J.L.; Yenes, P.; Trejo, A. & Garcia-Casado, P. (2001). A short hypoosmotic swelling test for the prediction of boar sperm fertility. *Theriogenology* 56, 387–398.

Petrunkina, A.M.; Simon, K.; Günzel-Apel, A.R. & Töpfer-Petersen, E. (2003). Regulation of capacitation of canine spermatozoa during co-culture with heterologous oviductal epithelial cells. *Reprod Domest Anim* 38, 455-463.

Petrunkina, A.M.; Simon, K.; Günzel-Apel, A.R. & Töpfer-Petersen, E. (2004). Kinetics of protein tyrosine phosphorylation in sperm selected by binding to homologous and heterologous oviductal explants: how specific is the regulation by the oviduct? *Theriogenology* 61, 1617-1634.

Ramos, L. & Wetzels, A.M. (2001). Low rates of DNA fragmentation in selected motile human spermatozoa assessed by the TUNEL assay. *Hum Reprod* 16, 703–707.

Rathi, R.; Colenbrander, B.; Bevers, M.M. & Gadella, B.M. (2001). Evaluation of in vitro capacitation of stallion spermatozoa. *Biol Reprod* 65, 462-470.

Rijsselaere, T.; Van Soom, A.; Hoflack, G.; Maes, D. & de Kruif, A. (2004). Automated sperm morphometry and morphology analysis of canine semen by the Hamilton-Thorne analyser. *Theriogenology* 62, 1292–1306.

Rijsselaere, T.; van Soom, A.; Maes, D. & de Kruif, A. (2003). Effect of technical settings on canine semen motility parameters measured by the Hamilton-Thorne analyzer, *Theriogenology* 60, 1553–1568.

Rijsselaere, T.; Van Soom, A.; Tanghe, S.; Coryn, M.; Maes, D. & de Kruif, A. (2005). New techniques for the assessment of canine semen quality: A review. *Theriogenology* 64, 706-719.

Rodríguez-Gil, J.E.; Montserrat, A. & Rigau, T. (1994). Effects of hypoosmotic incubation on acrosome and tail structure on canine spermatozoa. *Theriogenology* 42, 815-829.

Rota, A.; Peña, A.I.; Linde-Forsberg, C. & Rodriguez-Martinez, H. (1999). In vitro capacitation of fresh, chilled and frozen-thawed dog spermatozoa assessed by the chloretetracycline assay and changes in motility patterns. *Anim Reprod Sci* 57, 199-215.

Rubio-Guillén, J.; González, D.; Garde, J.J.; Esteso, M.C.; Fernández-Santos, M.R.; Rodríguez-Gíl, J.E.; Madrid-Bury, N. & Quintero-Moreno, A. (2007). Effects of cryopreservation on bull spermatozoa distribution in morphometrically distinct subpopulations. *Reprod Domest Anim* 42, 354 –357.

Saint-Dizier, M.; Salomon J.F.; Petit C.; Renard J.P. & Chastant-Maillard S. (2001). In vitro maturation of bitch oocytes: effect of sperm penetration. *J Reprod Fertil* (Suppl. 57), 147-150.

Santiago-Moreno, J.; Castano, C.; Coloma, M.A.; Gomez-Brunet, A.; Toledano-Diaz, A.; Lopez-Sebastian, A. & Campo, J.L. (2009). Use of the hypo-osmotic swelling test and aniline blue staining to improve the evaluation of seasonal sperm variation in native Spanish free-range poultry. *Poult Sci* 88, 2661-2669.

Silva, P.F. & Gadella, B.M. (2006). Detection of damage in mammalian sperm cells. *Theriogenology* 65, 958-978.

Sirivaidyapong, S., Cheng F.P., Marks A., Voorhout W.F., Bevers M.M. & Colenbrander B. (2000). Effect of sperm diluents on the acrosome reaction in canine sperm. *Theriogenology* 53, 789-802.

Sprecher, D.J. & Coe, P.H. (1996). Differences in bull spermiograms using eosin-nigrosin stain, feulgen stain, and phase contrast microscopy methods. *Theriogenology* 45, 757-764.

Suarez S.S., Katz D.F. & Overstreet J.W. (1983). Movement characteristics and acrosomal status of rabbit spermatozoa recovered at the site and time of fertilization. *Biol Reprod* 29, 1277-87.

Szász, F.; Sirivaidyapong, S.; Cheng, F.P.; Voorhout, W.F.; Marks, A.; Colenbrander, B.; Solti, A.L. & Gadella, B.M. (2000). Detection of calcium ionophore induced membrane changes in dog sperm as a simple method to predict the cryopreservability of dog semen. *Mol Reprod Dev* 55, 289-298.

Talbot, P. & Chacon, R.S. (1981). A triple-stain technique for evaluating normal acrosome reactions of human sperm. *J Exp Zool* 215, 201-208.

Thuwanut, P.; Axner, E.; Johanisson, A. & Chatdarong K. (2009). Detection of lipid peroxidation reaction in frozen-thawed epididymal cat spermatozoa using BODIPY 581/591 C11. *Reprod Dom Anim* 44, (Suppl. 2), 373-376.

van der Venn, H.H.; Jeyendran, R.S.; Al-Hasani, S.; Pérez-Pelàez, M.; Diedrich, K. & Zaneveld, L.J.D. (1986). Correlation between human sperm swelling in hypoosmotic medium (hypoosmotic swelling test) and in vitro fertilisation. *J Androl* 7, 190–196.

Vazquez, J.M.; Martinez, E.A.; Martinez, P.; Garcia-Artiga, C. & Roca, J. (1997). Hypoosmotic swelling of boar spermatozoa compared to other methods for analysing the sperm membrane. *Theriogenology* 47, 913-922.

Verstegen, J.; Iguer-ouada, M. & Onclin K. (2002). Computer assisted semen analyzers in andrology research and veterinary practice. *Theriogenology* 57, 149–179.

Watson, P.F. (1995). Recent developments and concepts in the cryopreservation of spermatozoa and assessment of their post-thawing function. *Reprod Fertil Dev* 7, 871-891.

World Health Organization. (2010). WHO laboratory manual for the examination and processing of human semen. 5th ed. *WHO Press*.

Yanagimachi, R.; Yanagimachi, H. & Rogers, B.J. (1976). The use of zona-free animal ova as a test system for the assessment of the fertilizing capacity of human spermatozoa. *Biol Reprod* 15, 471-476.

Zambelli, D. & Cunto, M. (2006). Semen collection in cats: techniques and analysis. *Theriogenology* 66, 159-165.

Permissions

The contributors of this book come from diverse backgrounds, making this book a truly international effort. This book will bring forth new frontiers with its revolutionizing research information and detailed analysis of the nascent developments around the world.

We would like to thank Igor I. Katkov, for lending his expertise to make the book truly unique. He has played a crucial role in the development of this book. Without his invaluable contribution this book wouldn't have been possible. He has made vital efforts to compile up to date information on the varied aspects of this subject to make this book a valuable addition to the collection of many professionals and students.

This book was conceptualized with the vision of imparting up-to-date information and advanced data in this field. To ensure the same, a matchless editorial board was set up. Every individual on the board went through rigorous rounds of assessment to prove their worth. After which they invested a large part of their time researching and compiling the most relevant data for our readers. Conferences and sessions were held from time to time between the editorial board and the contributing authors to present the data in the most comprehensible form. The editorial team has worked tirelessly to provide valuable and valid information to help people across the globe.

Every chapter published in this book has been scrutinized by our experts. Their significance has been extensively debated. The topics covered herein carry significant findings which will fuel the growth of the discipline. They may even be implemented as practical applications or may be referred to as a beginning point for another development. Chapters in this book were first published by InTech; hereby published with permission under the Creative Commons Attribution License or equivalent.

The editorial board has been involved in producing this book since its inception. They have spent rigorous hours researching and exploring the diverse topics which have resulted in the successful publishing of this book. They have passed on their knowledge of decades through this book. To expedite this challenging task, the publisher supported the team at every step. A small team of assistant editors was also appointed to further simplify the editing procedure and attain best results for the readers.

Our editorial team has been hand-picked from every corner of the world. Their multi-ethnicity adds dynamic inputs to the discussions which result in innovative outcomes. These outcomes are then further discussed with the researchers and contributors who give their valuable feedback and opinion regarding the same. The feedback is then collaborated with the researches and they are edited in a comprehensive manner to aid the understanding of the subject.

Apart from the editorial board, the designing team has also invested a significant amount of their time in understanding the subject and creating the most relevant covers. They scrutinized every image to scout for the most suitable representation of the subject and create an appropriate cover for the book.

The publishing team has been involved in this book since its early stages. They were actively engaged in every process, be it collecting the data, connecting with the contributors or procuring relevant information. The team has been an ardent support to the editorial, designing and production team. Their endless efforts to recruit the best for this project, has resulted in the accomplishment of this book. They are a veteran in the field of academics and their pool of knowledge is as vast as their experience in printing. Their expertise and guidance has proved useful at every step. Their uncompromising quality standards have made this book an exceptional effort. Their encouragement from time to time has been an inspiration for everyone.

The publisher and the editorial board hope that this book will prove to be a valuable piece of knowledge for researchers, students, practitioners and scholars across the globe.

List of Contributors

Regina Celia Rodrigues da Paz
Departamento de Ciências Básicas e Produção Animal, Faculdade de Agronomia, Medicina Veterinária e Zootecnia, Universidade Federal de Mato Grosso – DCBPA/FAMEV/UFMT, Brazil

Joseph Saragusty
Department of Reproduction Management, Leibniz Institute for Zoo and Wildlife Research, Berlin, Germany

Loredana Zilli and Sebastiano Vilella
Laboratory of Comparative Physiology, Department of Biological and Environmental Sciences and Technologies, University of Salento, Via Provinciale Lecce-Monteroni, Lecce, Italy

Daisuke Kami
National Agricultural Research Center for Hokkaido Region, Japan

Anja Kaczmarczyk, Bryn Funnekotter, Akshay Menon, Pui Ye Phang, Arwa Al-Hanbali and Ricardo L. Mancera
Curtin Health Innovation Research Institute, Western Australian Biomedical Research Institute, Curtin University, Perth, Australia

Anja Kaczmarczyk, Bryn Funnekotter, Akshay Menon, Pui Ye Phang, Arwa Al-Hanbali and Eric Bunn
Botanic Gardens and Parks Authority, Fraser Avenue, West Perth, Australia

Eric Bunn
School of Plant Biology, Faculty of Natural and Agricultural Sciences, University of Western Australia, Australia

K. Nirmal Babu, G. Yamuna, K. Praveen, P.N. Ravindran and K.V. Peter
Indian Institute of Spices Research, Kerala, India

D. Minoo
Providence Women's College, Kerala, India

Marian D. Quain
Council for Scientific and Industrial Research Crops Research Institute, Ghana

Patricia Berjak
School of Biological and Conservation Sciences, University of KwaZulu-Natal, Westville Campus, Durban, South Africa

Elizabeth Acheampong
Tissue Culture Laboratory, Department of Botany, University of Ghana, Legon, Accra, Ghana

Marceline Egnin
Plant Biotechnology and Genomics Research Laboratory, Tuskegee University, USA

Edoardo Lopez, Katiuscia Cipri and Vincenzo Naso
"Sapienza" University of Rome, Rome, Italy

Stephen Butler and David Pegg
Planer plc, University of York, UK

Agnieszka Partyka, Wojciech Niżański and Małgorzata Ochota
Wroclaw University of Environmental and Life Sciences, Poland

Printed in the USA
CPSIA information can be obtained
at www.ICGtesting.com
JSHW011459221024
72173JS00005B/1141

9 781632 395290